High Voltage Engineering and Applications

High Voltage Engineering and Applications

Special Issue Editor

Ayman El-Hag

MDPI • Basel • Beijing • Wuhan • Barcelona • Belgrade

Special Issue Editor
Ayman El-Hag
Department of Electrical and
Computer Engineering,
Univeristy of Waterloo
Canada

Editorial Office
MDPI
St. Alban-Anlage 66
4052 Basel, Switzerland

This is a reprint of articles from the Special Issue published online in the open access journal *Energies* (ISSN 1996-1073) from 2019 to 2020 (available at: https://www.mdpi.com/journal/energies/special_issues/HV_Engineering).

For citation purposes, cite each article independently as indicated on the article page online and as indicated below:

LastName, A.A.; LastName, B.B.; LastName, C.C. Article Title. *Journal Name* **Year**, *Article Number*, Page Range.

ISBN 978-3-03928-716-1 (Pbk)
ISBN 978-3-03928-717-8 (PDF)

Cover image courtesy of Shesha Jayaram.

Contents

About the Special Issue Editor . . . . **vii**

Preface to "High Voltage Engineering and Applications" . . . . **ix**

Alhaytham Alqudsi and Ayman El-Hag
Application of Machine Learning in Transformer Health Index Prediction
Reprinted from: *Energies* **2019**, *12*, 2694, doi:10.3390/en12142694 . . . . **1**

Amir Abbas Soltani and Ayman El-Hag
Denoising of Radio Frequency Partial Discharge Signals Using Artificial Neural Network
Reprinted from: *Energies* **2019**, *12*, 3485, doi:10.3390/en12183485 . . . . **14**

Moein Borghei and Mona Ghassemi
Partial Discharge Analysis under High-Frequency, Fast-Rise Square Wave Voltages in Silicone Gel: A Modeling Approach
Reprinted from: *Energies* **2019**, *12*, 4543, doi:10.3390/en1224543 . . . . **28**

Yiming Zang, Yong Qian, Wei Liu, Yongpeng Xu, Gehao Sheng and Xiuchen Jiang
A Novel Partial Discharge Detection Method Based on the Photoelectric Fusion Pattern in GIL
Reprinted from: *Energies* **2019**, *12*, 4120, doi:10.3390/en12214120 . . . . **41**

Alaa Loubani, Noureddine Harid, Huw Griffiths and Braham Barkat
Simulation of Partial Discharge Induced EM Waves Using FDTD Method—A Parametric Study †
Reprinted from: *Energies* **2019**, *12*, 3364, doi:10.3390/en12173364 . . . . **59**

Michal Krbal, Ludek Pelikan, Jaroslav Stepanek, Jaroslava Orsagova and Iraida Kolcunova
A Physical Calibrator for Partial Discharge Meters
Reprinted from: *Energies* **2019**, *12*, 2057, doi:10.3390/en12112057 . . . . **72**

Jiahong He, Kang He and Bingtuan Gao
Modeling of Dry Band Formation and Arcing Processes on the Polluted Composite Insulator Surface
Reprinted from: *Energies* **2019**, *12*, 3905, doi:10.3390/en12203905 . . . . **82**

Mohamed Ghouse Shaik and Vijayarekha Karuppaiyan
Investigation of Surface Degradation of Aged High Temperature Vulcanized (HTV) Silicone Rubber Insulators
Reprinted from: *Energies* **2019**, *12*, 3769, doi:10.3390/en12193769 . . . . **102**

Muhammad Majid Hussain, Muhammad Akmal Chaudhary and Abdul Razaq
Mechanism of Saline Deposition and Surface Flashover on High-Voltage Insulators near Shoreline: Mathematical Models and Experimental Validations
Reprinted from: *Energies* **2019**, *12*, 3685, doi:10.3390/en12193685 . . . . **118**

Zhijin Zhang, Shenghuan Yang, Xingliang Jiang, Xinhan Qiao, Yingzhu Xiang and Dongdong Zhang
DC Flashover Dynamic Model of Post Insulator under Non-Uniform Pollution between Windward and Leeward Sides
Reprinted from: *Energies* **2019**, *12*, 2345, doi:10.3390/en12122345 . . . . **138**

Mohammed El Amine Slama, Abderrahmane Beroual and Abderrahmane (Manu) Haddad
Surface Discharges and Flashover Modelling of Solid Insulators in Gases
Reprinted from: *Energies* **2020**, *13*, 591, doi:10.3390/en13030591 . . . . **155**

Runhao Zou, Jian Hao and Ruijin Liao
Space/Interface Charge Analysis of the Multi-Layer Oil Gap and Oil Impregnated Pressboard Under the Electrical-Thermal Combined Stress
Reprinted from: *Energies* **2019**, *12*, 1099, doi:10.3390/en12061099 . . . . **169**

Jinsong Li, Hua Yu, Min Jiang, Hong Liu and Guanliang Li
Numerical Modeling of Space–Time Characteristics of Plasma Initialization in a Secondary Arc
Reprinted from: *Energies* **2019**, *12*, 2128, doi:10.3390/en12112128 . . . . **187**

Zhenyu Li and Xuezeng Zhao
Calculation of Ion Flow Field of Monopolar Transmission Line in Corona Cage Including the Effect of Wind
Reprinted from: *Energies* **2019**, *12*, 3924, doi:10.3390/en12203924 . . . . **203**

Jiahong He, Kang He and Longfei Cui
Charge-Simulation-Based Electric Field Analysis and Electrical Tree Propagation Model with Defects in 10 kV XLPE Cable Joint
Reprinted from: *Energies* **2019**, *12*, 4519, doi:10.3390/en12234519 . . . . **216**

WenWei Zhu, YiFeng Zhao, ZhuoZhan Han, XiangBing Wang, YanFeng Wang, Gang Liu, Yue Xie and NingXi Zhu
Thermal Effect of Different Laying Modes on Cross-Linked Polyethylene (XLPE) Insulation and a New Estimation on Cable Ampacity
Reprinted from: *Energies* **2019**, *12*, 2994, doi:10.3390/en12152994 . . . . **238**

Abdul Wali Abdul Ali, Nurul Nadia Ahmad, Normiza Mohamad Nor, Muhd Shahirad Reffin and Syarifah Amanina Syed Abdullah
Investigations on the Performance of a New Grounding Device with Spike Rods under High Magnitude Current Conditions
Reprinted from: *Energies* **2019**, *12*, 1138, doi:10.3390/en12061138 . . . . **260**

Muhd Shahirad Reffin, Abdul Wali Abdul Ali, Normiza Mohamad Nor, Nurul Nadia Ahmad, Syarifah Amanina Syed Abdullah, Azwan Mahmud and Farhan Hanaffi
Seasonal Influences on the Impulse Characteristics of Grounding Systems for Tropical Countries
Reprinted from: *Energies* **2019**, *12*, 1334, doi:10.3390/en12071334 . . . . **278**

About the Special Issue Editor

Ayman El-Hag (University of Waterloo): Dr. El-Hag received his B.S. and M.S. degrees from King Fahd University of Petroleum and Minerals in 1993 and 1998, respectively, and his PhD from the University of Waterloo in 2003. He joined the Saudi Transformer Co. as a Quality Control Engineer from 1993 to 1999. From January to June 2004, Dr. El-Hag worked as a Postdoctoral fellow at the University of Waterloo, after which he joined the University of Toronto as an NSERC Postdoctoral fellow from July 2004 to July 2006. In 2006, Dr. El-Hag joined the electrical engineering department at the American University of Sharjah. He was promoted to associate and then to professor in 2011 and 2016 respectively. Currently, he is a lecturer at the electrical and computer engineering department at University of Waterloo. Dr. El-Hag is an associate editor in the IEEE Transaction of Dielectric and Electrical Insulation and was the Middle-East Regional Editor of the IEEE DEIS Electrical Insulation Magazine from 2016 to 2018. He is a member of the IEEE DEIS Outdoor Insulation Technical Committee and co-founder of the electric power and energy conversion systems (EPECS) conference. Dr. El-Hag's current main areas of interest are condition monitoring and diagnostics of electrical insulation and applications of machine learning in power engineering.

Preface to "High Voltage Engineering and Applications"

High-voltage engineering is a crucial part of our modern society, from the different components in the power grid to the copy machines around us. Recently, different challenges have emerged in high-voltage engineering that need special attention from researchers. In this Special Issue, we received 35 articles where 18 were accepted and 17 were rejected. The accepted articles discussed different topics and issues related to high-voltage engineering. The first article by Haytham addresses the application of machine learning to transformer health index estimation. The authors used different machine learning algorithms to assess the transformer insulation condition using different transformer oil tests. Subsequently, five articles addressed different issues related to partial discharge (PD). Issues like PD denoising, novel techniques to detect PD under AC, and fast transient and modeling of PD propagation are some of the topics related to PD that were discussed. Then, four articles discussed different topics related to outdoor insulators. Field problems in both ceramic and non-ceramic insulators were highlighted, like pollution flashover and aging under field conditions. The presented work is mixed between experimental and simulation studies. Basic studies follow in four different articles that discuss fundamental issues like surface discharge and interfacial polarization. The focus in these articles was on the use of state-of-the-art modeling techniques like finite element method and space charge simulation. Two articles related to underground cables were to follow. The first articles discusses the influence of voids on electric field distribution in underground cables, and the second article addresses several issues related to cable ampacity. Finally, two articles related to different grounding issues are presented. The topical diversity of the articles is apparent, as is the geographical distribution of the authors; we present articles from China, Canada, Malaysia, the Czech Republic, Slovakia, Iran, France, the UK, the USA, India, and the United Arab Emirates.

Ayman El-Hag
Special Issue Editor

energies

MDPI

Article

Application of Machine Learning in Transformer Health Index Prediction

Alhaytham Alqudsi [1] and Ayman El-Hag [2,*]

1 Mechanical Engineering Department, École de technologie supérieure, Montréal, QC H3C 1K3, Canada
2 Electrical and Computer Engineering, University of Waterloo, Waterloo, ON N2L 3G1, Canada
* Correspondence: ahalhaj@uwaterloo.ca; Tel.: +1-519-277-2984

Received: 18 June 2019; Accepted: 12 July 2019; Published: 14 July 2019

Abstract: The presented paper aims to establish a strong basis for utilizing machine learning (ML) towards the prediction of the overall insulation health condition of medium voltage distribution transformers based on their oil test results. To validate the presented approach, the ML algorithms were tested on two databases of more than 1000 medium voltage transformer oil samples of ratings in the order of tens of MVA. The oil test results were acquired from in-service transformers (during oil sampling time) of two different utility companies in the gulf region. The illustrated procedure aimed to mimic a realistic scenario of how the utility would benefit from the use of different ML tools towards understanding the insulation health index of their transformers. This objective was achieved using two procedural steps. In the first step, three different data training and testing scenarios were used with several pattern recognition tools for classifying the transformer health condition based on the full set of input test features. In the second step, the same pattern recognition tools were used along with the three training/testing scenarios for a reduced number of test features. Also, a previously developed reduced model was the basis to reduce the needed number of tests for transformer health index calculations. It was found that reducing the number of tests did not influence the accuracy of the ML prediction models, which is considered as a significant advantage in terms of transformer asset management (TAM) cost reduction.

Keywords: feature selection; insulation health index; machine learning; oil/paper insulation; transformer asset management

1. Introduction and Background

One of the major parameters that define the operation and planning of an electrical utility is the transformer asset health condition. Based on their health condition, electrical utility engineers can predict the transformer useful remnant lifetime. Such an understanding can benefit utility companies to prepare a proper financial plan to estimate the future cost of maintenance and replacement for the transformer units. Significant research has been conducted to help utility companies in cutting their asset maintenance costs. This area of research is commonly referred to as the transformer asset management (TAM) practice [1]. TAM, as explained in [1–3], defines a strategic set of future maintenance and replacement activities for the utility transformer asset based on diagnostic testing methods of the transformer health condition. The ultimate objective of TAM is to ensure the power system reliability within an economic platform. Abu-Elanien in [2] defines the diagnostic testing methods in its two-part forms of condition monitoring (CM) and condition assessment (CA). CM refers to all the electrical, chemical, and physical tests that are used collectively towards CA tools that determine the transformer health condition. Azmi et al. states that TAM practices are at their best when they are comprised of both the CA and financial information [3]. Having knowledge of the transformer history (loading and failure history), associated risk index (based on the load it feeds), current health condition, and all related financial costs (maintenance, operation, and failure) would result in an economic risk management plan with adequate subsequent decisions.

1.1. Insulation Health Index Computation

According to the literature, the transformer health condition is governed by its oil-paper insulation condition [1,4,5]. According to [1], analyzing the transformer oil samples would be more advantageous than testing other transformer components (turns ratio, winding resistance, leakage reactance, etc.) for fault detection and life expectancy. With the health condition of all transformer components being taken into consideration, the overall health condition can be defined using the health index (HI). The health index, as explained by Jahromi et al. [6,7], is a single index factor which combines operating observations, field inspections, and laboratory tests to aid in the TAM cycle. The insulation condition is a vital part of the HI computation that could suffice when limited data is available for the transformer service record and design. Understanding the insulation condition would require thorough transformer oil sample analysis through electrical, chemical, and physical laboratory tests.

All conducted oil insulation laboratory tests fall into one of three major test categories, which are namely the dissolved gases (DGA), oil quality (OQA), and furan (FFA) tests [6,7]. DGA tests are typically conducted for the detection of transformer internal faults (electrical and/or thermal). The dissolved gases include hydrogen (H_2), methane (CH_4), ethane (C_2H_6), ethene (C_2H_4), ethyne (C_2H_2), carbon monoxide (CO), and carbon dioxide (CO_2) [8]. For the second category of tests, OQA, the oil quality is determined by testing the oil breakdown voltage (BDV), acidity, water content, interfacial tension (IFT), dielectric dissipation factor (DDF), and color [9]. Finally, the measurement of FFA determines the extent of paper insulation degradation through determining the furfuraldehyde (or commonly known as furan) content in the transformer oil. Furan is a chemical compound that dissolves in the insulation oil upon the breakdown of the cellulose chain of the paper material [10]. The furan test is a strong indicator of transformer paper insulation ageing. Collectively, laboratory tests on dissolved gases, oil, and paper quality would be used to compute the HI value based on a given formula that is developed by experts in the TAM field. Examples of such different formulas can be found in [6–13], with the method illustrated in [6,7] being solely used for computing the HI for the majority of publications.

1.2. Novel Methods for HI Computation

The approach of using artificial intelligence tools, such as machine learning (ML) technologies and fuzzy logic for the HI computation, was well studied in several publications. In Ref. [14], the assessment of the HI was done using a neuro-fuzzy approach. The aim of the study was to test the performance of the five-layer based neuro-fuzzy model in computing the HI as per the scoring method dictated in [6]. The inputs to the model were the oil test features that have been taken from in-service transformers records. Though limited in the number of available data, the author was able to show a prediction accuracy of more than 50% using the developed neuro-fuzzy model. In other works, such as [15], a general regression neural network (GRNN) was developed for the HI of four-class based condition assessment (namely, very poor, poor, fair, and good). For this particular work, the transformer health was graded based on international oil assessment standards (such as that of [9]). Six key inputs, including the oil total dissolved combustible gas, furan content, dielectric strength, acidity, water content, and dissipation factor, were utilized. An 83% success was reported in predicting the transformer health condition. Zeinoddini-Meymand et al. in [16] used an artificial neural network and neuro-fuzzy models to include the basic oil assessment test features and additional economical parameters that have not been accounted for in many publications. These economical parameters include the transformer percentage economical life-time and ageing acceleration factor. According to [16], the inclusion of such parameters in the HI condition-class problem resulted in an excellent assessment performance, which correlated with that of the field experts.

In the work of [17], probabilistic Markov chain models were used in predicting the future performance of the transformer asset based on their HI computation for a defined span of time. Transition probabilities have been derived (using a non-linear optimization technique) to be the core element of the Markov chain model, which in turn was used to predict the future HI of the transformer asset. The reported results indicate satisfactory prediction performances for a number of tested transformers. An interesting approach was presented by Tee et al. in [18] for determining the transformer health condition using principle component analysis (PCA) and analytical hierarchy process (AHP). Data of the transformer oil quality tests and age have been used in both the aforementioned techniques to rank the transformer asset based on the insulation health condition. The ranking obtained by both techniques showed a comparable performance to that of an expert-based empirical formula assessment for the same transformer asset.

In Ref. [19], a fuzzy-based support vector machine (SVM) was used in HI-condition assessment. The HI condition of a given transformer was determined based on a number of factors, including industry standards and utility expert judgments. The SVM model showed a classification rate of 87.8%. A fuzzy-logic based model was also incorporated in [20], where the HI class condition (three classes) was predicted using the technical oil test features as inputs. In other attempts, such as [21], a general study was conducted using different conventional feature selection methods on oil test features for predicting the HI condition with different ML techniques.

With reference to all the reported works in this paper, a promising future is foreseen for the use of ML in the TAM field. As was indicated earlier, predicting the HI of the transformer asset will substantially impact the financial strategy of the utility company in asset maintenance plans. The objective of this work was to extend our previous research of [22] and establish a platform for using a wide range of ML tools in understanding the transformer health condition.

1.3. Organization of the Presented Work

In the following sections of this paper, the transformer databases used for this study will be introduced. The methodology of computing the HI value and accordingly classifying the health condition for a given sample in the oil databases are illustrated. Thereafter, the different ML tools used in the pattern recognition/classification problem of this study are introduced. The stepwise regression feature selection tool will also be introduced. Accordingly, two major steps will be done in achieving the objectives of the presented work:

- Full-feature modelling: The pattern recognition model will be trained and tested based on the complete number of available test features (which is 10 in this study). Eight different pattern recognition methods will be used with three different training and testing scenarios based on the two different oil databases that were acquired in this study.
- Reduced-feature modelling: Based on the reported work of [22], stepwise regression was used as a feature selection tool for predicting the HI value of a given oil sample. Accordingly, it was concluded which oil test features are of the highest statistical significance in computing the HI value of a given transformer. Only the indicated oil test features from [22] will be used from the two databases in a reduced-feature modelling step. Again, eight different pattern recognition methods will be used along with the same three training and testing scenarios used in the full-feature modelling procedure.

2. Materials and Methods

2.1. Transformer Oil Samples

As mentioned earlier, transformer oil sample databases were acquired from two utility companies in the gulf region. For confidentiality purposes, the two companies will be referred to as *Util1* and *Util2*. In total, 730 transformer oils samples were obtained from *Util1*, while 327 transformer oil samples were obtained from *Util2*. Transformers from both databases are medium voltage distribution transformers.

Util1 transformers are 66/11 kV, 12.5 to 40 MVA, while those of *Util2* are 33/11 kV and 15 MVA. For every transformer in both databases, 10 different oil test results are available, namely: H_2, CH_4, C_2H_6, C_2H_4, C_2H_2, BDV, IFT, water content, acidity, and furan.

2.2. Structuring the HI Database

The transformer databases have a total of 1057 data samples combined. The prime objective of this paper is to be able to estimate the transformer insulation health condition using ML with either the entire oil feature set or a partial part of it. The published work in [6,7] is considered as the base method for computing the HI. In this method, all input test features are assigned a score value based on a predefined scale. The scored test feature is then multiplied or scaled by a predefined weight factor and is quantifiably added to the other scaled test features. With the inclusion of other arithmatic operations, three quantitative factors related to the DGA, OQA, and FFA tests are calculated (denoted by β). The β factors are discrete values that range from 1 to 5 (or 1 to 4 for β_{OQA}), with the ascending order of the values indicating a deteriorating condition of the transformer health. The three factors will then be multiplied by their associated weights (denoted by α) to eventually produce the HI value using Equation (1) as illustrated in [6]:

$$HI = \frac{(\beta_{DGA} \times \alpha_{DGA}) + (\beta_{OQA} \times \alpha_{OQA}) + (\beta_{FFA} \times \alpha_{FFA})}{4 \times (\alpha_{DGA} + \alpha_{OQA} + \alpha_{FFA})} \times 100\%. \quad (1)$$

The HI value ranges from 0% to 100%, with 100% being a transformer in the healthiest possible condition. Once the HI is computed, the data sample is classified into one of three health condition classes, which are: Bad (B), fair (F), and good (G). Table 1 illustrates the HI value range for a given class and the number transformers from each utility company in that class.

To have a better visualization of the 10 input variables and how they are related to the health condition of the transformer data, a 3-D plot of the data using the β factors from Equation (1) is depicted in Figure 1. As can be seen in Figure 1, there is a clear and distinct difference between the three classes of data that is influenced by the three β factors.

Table 1. Health index value range of health condition classes and the corresponding number of samples.

Data Group	Good (G), HI > 85%	Fair (F), 50% < HI ≤ 85%	Bad (B), HI ≤ 50%
Util1	496	206	28
Util2	238	84	5
Total from *Util1* and *Util2*	734	290	33

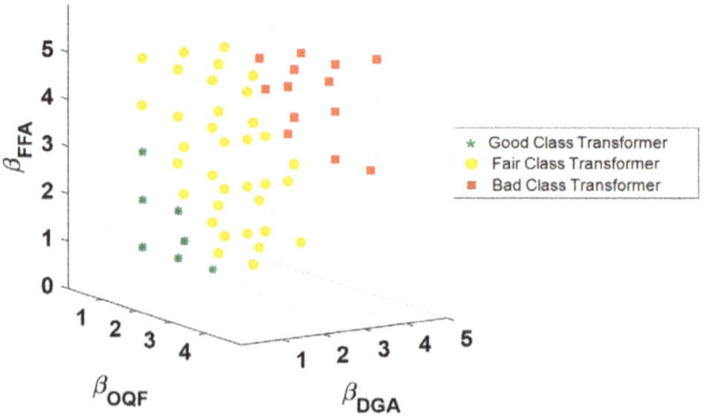

Figure 1. A 3D representation of the input variables through β_{DGA}, β_{OQA}, and β_{FFA}.

2.3. The Machine Learning Methodology

Transformer condition assessment requires the analysis of substantial data, whose instances represent investigated transformers and whose features represent variables measured to predict the transformer HI condition. A pattern recognition or classification model (classifier) can be trained on the dataset so that learning algorithms can operate faster and more effectively; in other words, costs are reduced, and learning accuracy is improved [23]. The methodology presented in this paper is based on feature selection and pattern classification for assessing the HI of power transformers. ML helps in gaining insights into the properties of data dependencies, and the significance of individual attributes in the dataset. Classification is an ML technique used to assign labels (classes) to unlabeled input instances based on discriminant features. Class labels in this study are the three health condition classes. Feature selection techniques determine the important features to include in the classification process of a particular data collection. Ten features are available for this study as was illustrated earlier. The entire process of applying these ML techniques to predict the class of unseen data consists of the typical phases of training and testing.

Consider a binary classification problem with positive (P) and negative (N) classes (i.e., two class problem). As a generalization for such multi-class classification problems, the overall classification accuracy measure is assessed in terms of the quantity of truly and falsely classified samples. Accordingly, a confusion matrix is constructed for recording the frequency of truly positive (TP), truly negative (TN), falsely positive (FP), and falsely negative (FN) samples. Table 2 shows a confusion matrix for binary classification problems, in which the class is either P or N.

Table 2. Confusion matrix of binary classification problems.

		Classified as	
		P	N
Really is	**P**	TP	FN
	N	FP	TN

As illustrated in [24], the overall classification accuracy measure is calculated by:

$$Accuracy\ Rate = \frac{\text{TP} + \text{TN}}{\text{TP} + \text{FP} + \text{TN} + \text{FN}}. \quad (2)$$

To validate the classification model, the k-fold cross-validation technique is used. Wherein, the dataset is randomized then equally subdivided into k subsets, of which k − 1 subsets are used for training and the remaining subset is retained for testing to validate the resulting model. Then, a different subset is used as a test set and the remaining k − 1 are used for training; thus, a second model is built. The process is repeated k times (folds) until k models are built. The final estimation is based on the average of the k results. A 10-fold cross-validation is usually used [25]. This method has the advantage of using all data for both the testing and validation. Moreover, it reduces the standard deviation with random seeds, as compared to methods that split the dataset into two sets, a training set and a testing set.

Feature selection is the process of selecting a subset of relevant, high quality, and non-redundant features for building learning models [26], with improved accuracy [27]. Quite often, datasets contain features with different qualities, which can influence the performance of the entire learning framework. For instance, noisy features can decrease the classifiers' performance. Moreover, some features are redundant and are highly correlated, i.e., they do not give additional information. Considering all available information, provided by all kinds of features, would make it hard for the classifier to discover the real distinguishing characteristics, and would cause it to be overly specified (over fitted) on the examples it is trained with, hence reducing its generalization power drastically. Therefore, it is important to select the appropriate features to base classification on them [28,29]. As explained in the earlier publication of [22] and detailed in [30], stepwise regression deals with studying the statistical significance of a number of test features as they relate to the variable of interest. In other words, stepwise regression will be used to determine which of the 10 oil test features will be adequately sufficient to predict the HI value with the absence of the least significant ones. In the stepwise regression process, a final multilinear regression model is developed for predicting the variable of interest by adding or removing test features in a stepwise manner. The process starts with a single feature present in the regression model. A feature is then added to assess the incremental performance of model in computing the variable of interest (the HI value in this case). In each step, the F-statistic of the added feature in the model is computed. The F-statistic is found by:

$$F_j = \frac{SS_R(\gamma_j|\gamma_0, \gamma_1, \gamma_2 \ldots, \gamma_{j-1})}{MS_E} \tag{3}$$

where γ is the regression coefficient of the associated feature in the multilinear regression model. F_j is the F-statistic value with the inclusion of the jth term in the regression model given the other existing test features in the model. SS_R is the regression sum of squares of the data sets' computed model output as compared to data sets' actual values, and MS_E is the mean square of error of the model with all its existing and currently tested features. During each step, a p-value for the F-statistic of the added feature is determined and tested against the null hypothesis. The null hypothesis rejects the idea that the feature in question is statistically significant to the variable of interest. Thus, when performing the stepwise feature selection in the forward manner and the p-value of the added term to the model is found to be below the pre-defined entrance tolerance, the null hypothesis is rejected and the term is added to the model. Once all the forward stepwise process is done, the stepwise process starts to move in the backward manner. If a given test feature that exists in the model has its p-value for the F-statistic above the exit tolerance, the null hypothesis is confirmed and that feature is removed from the final model.

2.4. Classifiers Used in the Study

In this paper, WEKA version 3.8.2 was used as the platform for the different classifiers. It is a collection of machine learning algorithms for data mining tasks developed at the University of Waikato, New Zealand. A brief description of the different machine learning algorithms used by WEKA in this study is presented below.

- *Random Forest (RForest):* It is a form of the nearest neighbor predictor that starts with a standard ML technique called a decision tree [31,32]. RForest is a meta-estimator that fits a number of decision tree classifiers on various sub-samples of the dataset and uses averaging to improve the predictive accuracy and control over-fitting. RForest is a fast classifier that can process many classification trees [22].
- *Decision Tree (J48):* J48 is the Java implementation of the C4.5 decision tree algorithm in the WEKA data mining software [32]. The algorithm builds decision trees by calculating the information gain of the attributes, and then uses the attribute that has the highest normalized information gain to split the instances into subsets of the same class label (called a leaf node of the tree). The algorithm repeats the same process with all split subsets as children of a node. Finally, the tree is pruned by removing the branches that do not help in classification.
- *Support Vector Machines (SVMs):* SVM classifies instances by constructing a set of hyperplanes to separate the class categories [33]. SVMs belong to the general category of kernel methods [34,35], which depend on the data only through dot products. To ensure that the hyperplane is as wide as possible between classes, the kernel function computes a projection product in some possibly high dimensional feature space. SVMs have the advantages of being less computationally intense than other classification algorithms, a good performance in high-dimensional spaces, and efficiency in handling nonlinear classification using the kernel trick that indirectly transforms the input space into another high dimensional feature space.

- *Artificial Neural Networks (ANNs):* ANNs are bio-inspired methods of data processing that enable computers to learn similarly to human brains [24]. ANNs are typically structured in layers made up of a number of interconnected nodes. Patterns are presented to the network via the input layer, which communicates to one or more hidden layers where the actual processing is done via a system of weighted connections. The hidden layers then link to an output layer where the answer (health index level in our case) is the output. In the learning phase, weights are changed until the best ANN model that fits the input data is built [24].
- *k-Nearest Neighbor (kNN):* The kNN algorithm is a supervised learning technique that has been used in many ML applications. It classifies objects based on the closest training examples in the feature space. The idea behind kNN is to find a predefined number of training samples closest in distance to a given query instance and predict the label of the query instance from them [24]. kNN is similar to a decision tree algorithm in terms of classification, but instead of finding a tree, it finds a path around the graph. It is also faster than decision trees.
- *OneR:* One rule is made for each attribute in the predictor variable in the dataset in which a given class is assigned to the value of that attribute [36]. The rule is created by counting the frequency of target classes that appear for a given attribute in the predictor variable. The most frequent target class is assigned for that attribute and the error for using that one rule for the entire data set is computed. The attribute of the predictor variable with the least error is considered as the final one rule. For the problem in this study, the attributes of each predictor variable are continuous numerical values and thus they must be discretized to create the one rule. The discretization process is thoroughly explained in [36].
- *Multinomial Logistic Regression (MLR):* A linear regression model attempts to fit a linear equation that maps the predictor variables to an estimated response variable [37]. Logistic regression or binary logistic regression, on the other hand, is more of a two-class classification-based model that is considered as a generalized linear model. The algorithm aims to define linear decision boundaries between the different classes [38]. The linear logistic regression model maps the input feature vector into a probability value via the sigmoid logistic function that is set during the training process. Based on the obtained probability, a decision is made of whether the given sample belongs to a particular class or not. For multiple classes, the MLR model is developed, which is a set of binary logistic regression models of a given class against all other classes. The classification of the sample is based on the maximum computed probability amongst the different logistic regression models [39].
- *Naïve Bayes (NB):* The NB classifier is a probabilistic classifier that is based on Bayes theorem [40]. This classifier is based on the assumption that the predictor variables are conditionally independent given the class of the data sample in question. In other words, the posterior probability of the sample being in a particular class given the predictor variables is computed using Bayes theorem [40]. For that, the likelihood of the predictor variable given the class, predictor prior probability, and class prior probability are determined. The samples are classified based on the outcome of the maximum posterior probability computed amongst all the different classes. This type of classifier is easily modelled and is typically suitable for large datasets.

3. Results

As illustrated earlier, two subsequent procedural steps were followed to achieve the main objective of this paper. In the first step, eight different classifiers were modelled for classifying the transformer health condition as being B, F, or G with three different training/testing scenarios involving the two utility databases (*Util1* and *Util2*). The full number of 10 features (as obtained from *Util1* and *Util2*) will be used to model the classifiers, and thus such classifiers will be named as the full-feature classifiers. In the following step, predetermined stepwise regression features from the reported results of [22] were used as the only features in modelling the eight classifiers with the same three training/testing scenarios of the full-feature model. Such classifiers will be named as the reduced-feature classifiers.

An assessment of the performance of both the full-feature and reduced feature classifiers was done by means of the accuracy rate obtained as per Equation (2). The mean accuracy rate (MAR) is shown in the presented results, which is basically the average value obtained for the accuracy rate for 10 trials.

3.1. *Full-Feature Classifier Modelling*

Eight different types of classifiers were used, which are NB, MLR, ANN, SVM, kNN, OneR, J48, and RF. Different training and testing scenarios were designed for the study. For a training/testing scenario, *Tr-Util1, Ts-Util1*, the classifier would be trained on data from *Util1* and tested using the unused data from *Util1*. Similarly is the case with a training/testing scenario, *Tr-Util2, Ts-Util2*.

In order to validate the generalized nature of the classifiers, a training/testing scenario, *Tr-Util1, Ts-Util2*, would have the classifier being trained on data from *Util1* and tested on different data from *Util2*.

To have a better understanding of how the results are obtained, consider the following example. When applying the training/testing scenario *Tr-Util1, Ts-Util1* with the RF classifier, the confusion matrix will indicate the frequency of truly and falsely classified data samples. Table 3 shows the confusion matrix obtained for the *Tr-Util1, Ts-Util1* scenario using the RF classifier. As can be seen in Table 3, 99.2% (492 of 496) for the G class of transformers were correctly classified. Similarly, 94.2% and 68% of the data samples were correctly classified as the F and B class, respectively. The accuracy rate was calculated using Equation (2) as:

$$Accuracy\ Rate = \frac{\text{TG+TF+TB}}{\text{TG+TF+TB+FG+FF+FB}} = \frac{492+194+19}{492+194+19+13+3+9} \times 100\% = 96.58\%, \quad (4)$$

where TG, TF, and TB are the truly classified data samples in the G, F, and B classes, respectively, and FG, FF, and FB are the falsely classified data samples in the G, F, and B classes, respectively. The same simulation was done 10 times and the MAR was noted. Table 4 shows the summary of the obtained MAR results for eight classifier types with the three training/testing scenarios.

Table 3. Confusion matrix for the *Tr-Util1, Ts-Util1* scenario using the Random Forest classifier.

		Classified as		
		G	**F**	**B**
Really is	**G**	492	4	0
	F	9	194	3
	B	0	9	19

Table 4. Summary of full-feature mean accuracy rate results for the eight classifier types with the three training/testing scenarios.

Training/Testing Scenario	NB	MLR	ANN	SVM	kNN	OneR	J48	RF
Tr-Util1, Ts-Util1	92.6%	95.5%	94.9%	92.6%	93.0%	86.7%	95.6%	96.6%
Tr-Util2, Ts-Util2	93.3%	94.8%	92.7%	86.9%	94.5%	85%	98.2%	96.6%
Tr-Util1, Ts-Util2	90.2%	95.4%	95.4%	85.6%	87.8%	85.9%	95.1%	93.6%

3.2. *Reduced-Feature Classifier Modelling*

In the published work of [22], it was concluded that the four test features of furan, IFT, C_2H_6, and C_2H_2 are the concise test features of the highest statistical significance in the regression problem of the HI value. The approach presented in this paper differs than that of [22] such that the ML approach deals with the prediction of the health condition class rather than the HI value (thus a classification problem rather than regression). Thus, the indicated four test features in [22] were used in the reduced-feature classifier modelling. Similar to the approach followed in the full-feature classifier models, eight classifiers with three different training/testing scenarios were used. Table 5 shows the summary of the obtained results for the MAR.

Table 5. Summary of reduced-feature MAR results.

Training/Testing Scenario	NB	MLR	ANN	SVM	kNN	OneR	J48	RF
Tr-Util1, Ts-Util1	94.4%	95.3%	95.1%	92.1%	95.6%	86.7%	95.3%	96.6%
Tr-Util2, Ts-Util2	90.8%	93.3%	91.1%	77.7%	93.6%	86.9%	97.2%	96.9%
Tr-Util1, Ts-Util2	92.4	90.5%	93.3%	86.2%	92.4%	85.9%	92.4%	93.6%

4. Discussions of the Results

Tables 4 and 5 summarize the results obtained for more than 640 simulations combined. This section of the paper will highlight the most important results that would significantly support the objective of using ML for the transformer health index problem.

4.1. Full-Feature Classifier Results

With reference to Table 4, the following points should be noted:

- The overall MAR results obtained for the full-feature classifier models were above 85%. It was observed that the highest classification error was observed for the B class samples. Though many data samples that actually belong to the B class were misclassified, they were always misclassified as being in the F class rather than the G class. This sort of error is of minimal risk in the sense that the classifiers would never give a misleading information of the transformer being in an excellent health condition while truly being in the worst health condition. Table 6 shows examples of the confusion matrices obtained in which a significant number of the B samples were misclassified as being F samples. The misclassification in most cases is attributed to the fact that a limited number of transformer oil samples of the B class are available for training (only 33 samples in total as can be observed in Table 1).
- For the remaining health condition classes, the accuracy rate was high and acceptable for most simulations. Most of the classifiers performed well in distinguishing between an F sample and a G sample. This is mainly attributed to the fact that a significant number of samples are available from both classes for training data.
- One of the extremely important training/testing scenarios would be that of *Tr-Util1, Ts-Util2*. With reference to Table 4, it was observed that most of the classifiers for this particular scenario did not perform as well as in the other training/testing scenarios. This inferior performance is mainly attributed to the fact that the classifier training was done with data from one utility company and testing was done with completely unseen data from another utility company. Still, the MAR results obtained are excellent given the previously unseen testing data. The MLR and ANN classifiers were the best classifiers, which resulted in an MAR of 95.4%. These excellent results support the generalized nature of the proposed approach in such a way that a utility company can use pre-modelled classifiers for their own transformer oil samples.

Table 6. Confusion matrix for (**a**) the *Tr-Util1, Ts-Util1* scenario using the kNN classifier, (**b**) *Tr-Util1 and Util2, Ts-Util1 and Util2* scenario using the ANN.

(a)				
		Classified as		
		G	F	B
Really is	G	483	13	0
	F	22	181	3
	B	0	**13**	**15**

Table 6. *Cont.*

(b)

		Classified as		
		G	F	B
Really is	G	719	15	0
	F	20	262	8
	B	0	17	16

4.2. Feature-Reduced Classifier Results

The selected features are basically the furan (which is an indicator of the paper insulation condition), the IFT (which is an indicator for the oil quality), and C_2H_2 and C_2H_6 that reveal the presence of any faults inside the transformer. Also, the ML can detect any correlation between the tests and hence remove highly correlated tests. With reference to Table 5, the obtained MAR results using most of the feature-reduced classifiers were above 90%. Similar to the full classifier models, the J48 and RF classifiers had the best performance amongst the eight reduced classifier models. These results give a significant conclusion that the feature selection technique of stepwise regression was very successful. The impact of these results would help utility companies reduce the cost of performing TAM practices by reducing the number of test samples required for computing the health index of the transformer asset. However, it is important to note that the frequency of oil sampling and hence overall TAM planning will not be improved by the results of the presented paper.

To illustrate the correlation of the health index with the selected features, Figure 2 shows the plots obtained for the furan, water, IFT, and C_2H_2 content against the health index. Although, furan, IFT, and C_2H_2 were selected by the stepwise regression, the water content was not selected [22]. Water clearly had the least correlation with the health index value, which would explain why it was not selected in either stepwise regression. On the other hand, the other three selected features showed a strong correlation with the health index.

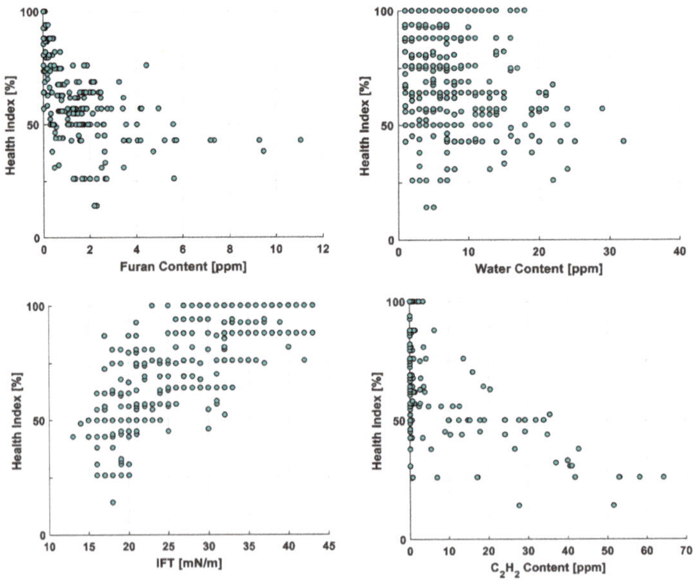

Figure 2. Plots indicating the relationship between a particular predictor variable and the health index value.

An additional study was conducted to observe if the age of the transformer was correlated with the obtained health index value. The age factor was not included in the study due to the fact that the transformer age data was only available for the transformer asset of *Util1*. Figure 3 shows the relationship between the transformer energization date and the health index value. It is apparent that there is no strong correlation between the transformer age and health index value. Merely, the transformer age may not be a good health index indicator without the inclusion of other factors, like the manufacturer, design, loading, and any refurbishing history. This observation is in agreement with that of [18], which basically indicates that a number of factors can influence the transformer conditions based on the service record and fault history, which differs from one transformer to another.

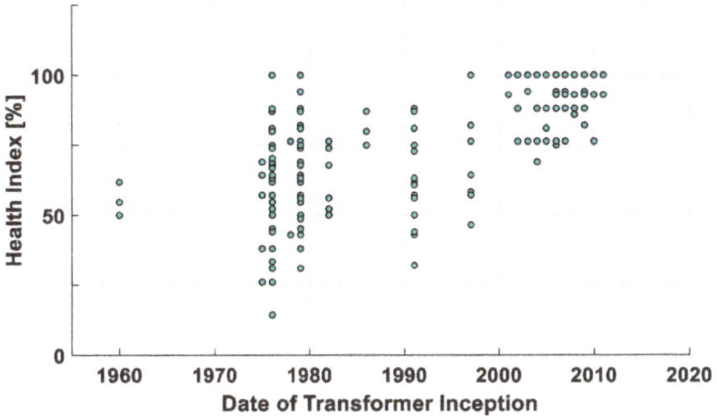

Figure 3. Date of transformer inception versus health index value.

5. Conclusions

The presented work was developed to validate the approach of using ML (through pattern classification tools) for the health index problem. The proposed approach was supported by the significant number of 1000+ transformers that were used for this study from different utility companies. The performance of the classifiers was proven as successful in both the full-feature and reduced-feature classifiers. The average results for all the MAR values were well beyond 85%. The worst results were shown with the OneR classifiers given the fact that these classifiers were trained on one feature. The use of stepwise regression as the feature selection tool for the health index problem was proven successful through the obtained results. The overall conclusion of the reported and discussed work significantly encourages the use of ML-feature selection methodology in the TAM industry for understanding the health condition of the transformer asset.

Author Contributions: Conceptualization, A.A. and A.E.-H.; methodology, A.A.; software, A.E.-H.; validation, A.A.; formal analysis, A.A.; investigation, A.E.-H.; resources A.A.; data curation, A.E.-H.; writing—original draft preparation, A.A.; writing—review and editing, A.E.-H.; visualization, A.A.; supervision, A.E.-H.; project administration, A.A.

Funding: This research received no external funding.

Conflicts of Interest: The authors declare no conflict of interest.

References

1. Zhang, X.; Gockenbach, E. Asset-Management of Transformers Based on Condition Monitoring and Standard Diagnosis. *IEEE Electr. Insul. Mag.* **2018**, *24*, 26–40.
2. Abu-Elanien, A.E.B.; Salama, M.M.A. Asset management techniques for transformers. *Electr. Power Syst. Res.* **2010**, *80*, 456–464. [CrossRef]
3. Azmi, A.; Jasni, J.; Azis, N.; Ab Kadir, M.Z.A. Evolution of transformer health index in the form of mathematical equation. *Renew. Sustain. Energy Rev.* **2017**, *76*, 687–700. [CrossRef]

4. Abu-Siada, A.; Islam, S. A new approach to identify power transformer criticality and asset management decision based on dissolved gas-in-oil analysis. *IEEE Trans. Dielectr. Electr. Insul.* **2012**, *19*, 1007–1012. [CrossRef]
5. Jalbert, J.; Gilbert, R.; Denos, Y.; Gervais, P. Methanol: A Novel Approach to Power Transformer Asset Management. *IEEE Trans. Power Deliv.* **2012**, *27*, 514–520. [CrossRef]
6. Jahromi, A.; Piercy, R.; Cress, S.; Service, J.; Fan, W. An approach to power transformer asset management using health index. *IEEE Electr. Insul. Mag.* **2009**, *25*, 20–34. [CrossRef]
7. Naderian, A.; Cress, S.; Piercy, R.; Wang, F.; Service, J. An Approach to Determine the Health Index of Power Transformers. In Proceedings of the Conference Record of the 2008 IEEE International Symposium on Electrical Insulation, Vancouver, BC, Canada, 9–12 June 2008.
8. Tenbohlen, S.; Coenen, S.; Djamali, M.; Müller, A.; Samimi, M.H.; Siegel, M. Diagnostic Measurements for Power Transformers. *Energies* **2016**, *9*, 347. [CrossRef]
9. Institute of Electrical and Electronics Engineers (IEEE). *IEEE Guide for Acceptance and Maintenance of Insulating Oil in Equipment, IEEE Std. C57.106-2006*; IEEE: Piscataway Township, NJ, USA, 2006.
10. Kachler, A.J.; Hohlein, I. Aging of Cellulose at Transformer Service Temperatures. Part 1: Influence of Type of Oil and Air on the Degree of Polymerization of Pressboard, Dissolved Gases, and Furanic Compounds in Oil. *IEEE Electr. Insul. Mag.* **2005**, *21*, 15–21. [CrossRef]
11. Hernanda, I.G.N.S.; Mulyana, A.C.; Asfani, D.A.; Negara, I.M.Y.; Fahmi, D. Application of health index method for transformer condition assessment. In Proceedings of the TENCON 2014-2014 IEEE Region 10 Conference, Bangkok, Thailand, 22–25 October 2014.
12. Singh, A.; Swanson, A.G. Development of a plant health and risk index for distribution power transformers in South Africa. *SAIEE Afr. Res. J.* **2018**, *109*, 159–170. [CrossRef]
13. Ortiz, F.; Fernandez, I.; Ortiz, A.; Renedo, C.J.; Delgado, F.; Fernandez, C. Health indexes for power transformers: A case study. *IEEE Electr. Insul. Mag.* **2016**, *32*, 7–17. [CrossRef]
14. Kadim, E.J.; Azis, N.; Jasni, J.; Ahmad, S.A.; Talib, M.A. Transformers Health Index Assessment Based on Neural-Fuzzy Network. *Energies* **2018**, *11*, 710. [CrossRef]
15. Islam, M.M.; Lee, G.; Hettiwatte, S.N. Application of a general regression neural network for health index calculation of power transformers. *Int. J. Electr. Power Energy Syst.* **2017**, *93*, 308–315. [CrossRef]
16. Zeinoddini-Meymand, H.; Vahidi, B. Health index calculation for power transformers using technical and economical parameters. *IET Sci. Meas. Technol.* **2016**, *10*, 823–830. [CrossRef]
17. Yahaya, M.S.; Azis, N.; Mohd Selva, A.; Ab Kadir, M.Z.A.; Jasni, J.; Kadim, E.J.; Hairi, M.H.; Yang Ghazali, Y.Z. A Maintenance Cost Study of Transformers Based on Markov Model Utilizing Frequency of Transition Approach. *Energies* **2018**, *11*, 2006. [CrossRef]
18. Tee, S.; Liu, Q.; Wang, Z. Insulation condition ranking of transformers through principal component analysis and analytic hierarchy process. *IET Gener. Transm. Distrib.* **2017**, *11*, 110–117. [CrossRef]
19. Ashkezari, A.D.; Ma, H.; Saha, T.K.; Ekanayake, C. Application of Fuzzy Support Vector Machine for Determining the Health Index of the Insulation System of In-Service Power Transformers. *IEEE Trans. Dielectr. Electr. Insul.* **2013**, *20*, 965–973. [CrossRef]
20. Abu-Elanien, A.E.B.; Salama, M.M.A.; Ibrahim, M. Calculation of a Health Index for Oil-Immersed Transformers Rated Under 69 kV Using Fuzzy Logic. *IEEE Trans. Power Deliv.* **2012**, *27*, 2029–2036. [CrossRef]
21. Benhmed, K.; Mooman, A.; Younes, A.; Shaban, K.; El-Hag, A. Feature Selection for Effective Health Index Diagnoses of Power Transformers. *IEEE Trans. Power Deliv.* **2018**, *33*, 3223–3226. [CrossRef]
22. Alqudsi, A.; El-Hag, A. Assessing the power transformer insulation health condition using a feature-reduced predictor model. *IEEE Trans. Dielectr. Electr. Insul.* **2018**, *25*, 853–862. [CrossRef]
23. Geng, X.; Liu, T.Y.; Qin, T.; Li, H. Feature selection for ranking. In Proceedings of the 30th Annual International ACM SIGIR Conference on Research and Development in Information Retrieval. ACM, New York, NY, USA, 1 January 2007.
24. Witten, I.H.; Frank, E.; Hall, M.A. *Data Mining: Practical Machine Learning Tools and Techniques*, 3th ed.; Elsevier: Amsterdam, The Netherlands, 2016; ISBN 978-0-12-374856-0.
25. Efron, B.; Tibshirani, R. Improvements on Cross-Validation: The 0.632+ Bootstrap Method. *J. Am. Stat. Assoc.* **1997**, *92*, 548–560.
26. Guyon, I.; Elissee, A. An introduction to variable and feature selection. *J. Mach. Learn. Res.* **2003**, *3*, 1157–1182.

27. Wang, H.; Khoshgoftaar, T.M.; Gao, K.; Seliya, N. High-dimensional software engineering data and feature selection. In Proceedings of the 21th IEEE International Conference on Tools with Artificial Intelligence, Newark, NJ, USA, 2–4 November 2009.
28. Mierswa, I.; Wurst, M.; Klinkenberg, R.; Scholz, M.; Euler, T. YALE: Rapid Prototyping for Complex Data Mining Tasks. In Proceedings of the 12th ACM SIGKDD International Conference on Knowledge Discovery and Data Mining (KDD-06), Philadelphia, PA, USA, 20–23 August 2006.
29. Liu, H.; Motoda, H. *Feature Selection for Knowledge Discovery and Data Mining*; Kluwer Academic Publishers: Dordrecht, The Netherland, 2000.
30. Montgomery, D.; Runger, G. *Applied Statistics and Probability for Engineers*; Wiley: Hoboken, NJ, USA, 1994.
31. Rokach, L.; Maimon, O. *Data Mining with Decision Trees: Theory and Applications*; World Scientific Publishing Co., Inc.: River Edge, NJ, USA, 2008.
32. Hall, M.A. Correlation-Based Feature Subset Selection for Machine Learning. Ph.D. Thesis, University of Waikato, Hamilton, New Zealand, 1998.
33. Alpaydin, E. *Introduction to Machine Learning*; MIT Press: Cambridge, MA, USA, 2010; p. 5.
34. Breiman, L. Random Forests. *Mach. Learn.* **2001**, *45*, 5–32. [CrossRef]
35. Quinlan, R. *C4.5: Programs for Machine Learning*; Morgan Kaufmann Publishers: San Mateo, CA, USA, 2014.
36. Holte, R.C. Very Simple Classification Rules Perform Well on Most Commonly Used Datasets. *Mach. Learn.* **1993**, *11*, 63–90. [CrossRef]
37. Department of Statistics & Data Science at Yale University: Online course on Multiple Linear Regression. Available online: http://www.stat.yale.edu/Courses/1997-98/101/linmult.htm (accessed on 20 May 2019).
38. Gudivada, V.N.; Irfan, M.T.; Fathi, E.; Rao, D.L. Cognitive Analytics: Going Beyond Big Data Analytics and Machine Learning. In *Handbook of Statistics*; Gudivada, V.N., Raghavan, V.V., Govindaraju, V., Rao, C.R., Eds.; Elsevier: Amsterdam, The Netherlands, 2016; Volume 35, pp. 169–205.
39. PennState Elberly College of Science-Analysis of Discrete Data: Polytomous (Multinomial) Logistic Regression. Available online: https://newonlinecourses.science.psu.edu/stat504/node/172/ (accessed on 25 May 2019).
40. Mitchell, T.M. Generative and Discriminative Classifiers: Naive Bayes And Logistic Regression. In *Machine Learning*; Mitchell, T.M., Ed.; McGraw Hill: New York, NY, USA, 2015.

energies

MDPI

Article

Denoising of Radio Frequency Partial Discharge Signals Using Artificial Neural Network

Amir Abbas Soltani [1] and Ayman El-Hag [2,*]

[1] Department of Electrical Engineering, Lorestan Branch, Technical and Vocational University, Dorud 1435761137, Iran

[2] Department of Electrical and Computer Engineering, University of Waterloo, Waterloo, ON N2L 3G1, Canada

* Correspondence: ahalhaj@uwaterloo.ca

Received: 21 July 2019; Accepted: 9 September 2019; Published: 10 September 2019

Abstract: One of the most promising techniques for condition monitoring of high voltage equipment insulation is partial discharge (PD) measurement using radio frequency (RF) antenna. Nevertheless, the accuracy of monitoring, classification, localization, or lifetime estimation could be negatively affected due to the interferences and noises measured simultaneously and contaminate the RF signals. Therefore, to achieve high accuracy of PD assessment, exploiting the denoising algorithms is inevitable. Hence, this paper seeks to introduce a new technique to suppress white noise, the most prevalent type of noise, especially for RF signals. In the proposed method, the ability of artificial neural network (ANN) in curve fitting is applied to denoising of different types of measured RF signals emitted from PD sources including 'crack', 'internal void', in the insulator discs and 'sharp points' from external hardware. The processes of denoising for named signals with the proposed method are carried out, and the obtained results are compared with the outputs of a wavelet transform-based method named energy conversation-based thresholding. In all tested signals, the proposed technique showed superior denoising capability.

Keywords: partial discharge; denoising; RF signal; wavelet transform; artificial neural network; curve fitting

---

1. Introduction

High voltage (HV) equipment plays an essential role in power system reliability. Sometimes, there will be irrecoverable damage for industrial or residential customers if a failure in the insulation of HV equipment happens, leading to unexpected outages. Hence, to prevent such problems as well as to exploit the power system in the highest performance, condition monitoring of high voltage equipment insulation systems is considered as a reliable solution [1].

One of the most prevalent, influential, and non-destructive methods for condition monitoring is partial discharge (PD) assessment, which can be used to reveal the weak points of the insulation system at an early stage before complete failure occurrence [1,2]. Nevertheless, the performance of PD monitoring could be reduced due to the negative effect of the various sources of noise or interference. Thus, denoising of PD signals is inevitable during the condition monitoring process [1,3].

The noises added in the process of PD measurement are usually classified into three categories, including pulse shaped, narrowband, and wideband interference. The pulse-shaped intereference is mostly removed from PD signal through a number of pattern recognition approaches such as exploiting artificial neural network (ANN) or support vector machine (SVM) algorithms [3]. For the second type of noise, the narrow band generated by radio waves or telecommunication systems, the most prevalent denoising methods are notch filters and wavelet transform algorithm [3,4].

Wideband noises, sometimes called background noises, have a stochastic nature and depend on the measuring system as more sensitive measuring systems are more prone to wideband noise [3,5]. To remove such noise, digital signal processing algorithms—including mathematical morphology [6,7], empirical mode decomposition [8,9], and wavelet transform [2,10]—can be exploited. Mathematical morphology is a time-domain and effective algorithm with a low computational burden but the determination of the type and length of the structure element has always been a challenge [6]. Empirical mode decomposition is a time-domain algorithm that is recently proposed for white noise suppression. Although this method is able to find the level of signal decomposition through a self-adaptive approach, the computational burden to solve the problem of mode mixing, as well as thresholding process, are the obstacles that are yet not completely solved [9]. Being one of the most prevalent algorithms, wavelet transform (WT) has long been exploited in PD signal denoising. Although significant progress has been made in the WT based methods, this algorithm is still struggling with a number of challenges such as mother wavelet selection, decomposition level determination, and thresholding procedure [2,3,10].

This paper presents a new method utilizing ANN curve fitting and function approximation abilities to remove white noise from RF PD signals. The method is proposed and compared with a WT based algorithm using different types of PD RF signals emitted from three damaged insulator discs, including 'crack', 'internal void', and 'hardware sharp points'.

In the rest of the paper, the laboratory setup for PD signal measurement is provided in Section 2. Then, Section 3 presents both the WT-based and proposed ANN denoising methods. Section 4 is devoted to results and discussion while the conclusion is given in Section 5.

2. Laboratory Setup for PD Signals Measurement

The ceramic insulators used as the samples consist of two different damaged insulator discs where the first disc is intentionally cracked, whereas the second sample has a hole in the disc cap [11]. Moreover, the corona is generated externally using sharp electrodes to mimic defects in overhead line hardware. The overall experimental setup is shown in Figure 1. PD was measured simultaneously using both RF antenna with 1–2 GHz bandwidth and classical PD measurement system. Each damaged disc is added to three intact discs in a string, then using a test transformer, 45 kVrms is applied to the strings; consequently, the wideband horn antenna captures the RF signals. The antenna is connected by a low impedance cable to a 2 GHz oscilloscope. Details of the setup are provided in [11].

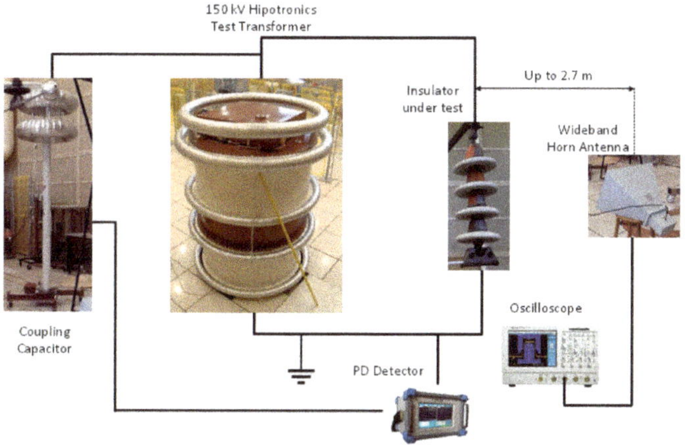

Figure 1. Laboratory setup for simultaneous electrical and RF signals measurement from the source of partial discharge, defected insulator disc.

3. Denoising of Partial Discharge RF Signals

The present section intends to discuss several parameters to quantify the severity of white noise. Also, both the wavelet transform and the proposed method are introduced.

3.1. Peak of Signal to Noise Ratio

In the process of PD signal examination, signal-to-noise ratio (SNR) is normally utilized in order to compare the energies of the original signal and white noise. However, it has been reported that for non-periodic transient signal calculating the peak signal-to-noise ratio (PSNR) is a better measure for the severity of white noise [12]. Therefore, being non-periodic and transient, RF signals better utilize the index of PSNR to assess the severity of white noise.

3.2. Factors for Evaluation of Denoising Algorithms

In terms of exploring and comparing the operation of denoising algorithms, the following parameters are exploited:

1. The Electric Charge Error (QE):

$$QE = \left(\sum_{i=1}^{N}|X(i)| - \sum_{i=1}^{N}|Y(i)|\right) / \left(\sum_{i=1}^{N}|X(i)|\right) \times 100 \quad (1)$$

2. Root Mean Square Error (RMSE):

$$RMSE = \sqrt{\frac{1}{N}\sum_{i=1}^{N}|Y(i) - X(i)|^2} \quad (2)$$

3. Correlation Coefficient (CC):

$$CC = \left(\sum_{i=1}^{N}(X(i) - \overline{X}) \times (Y(i) - \overline{Y})\right) / \sqrt{\sum_{i=1}^{N}(X(i) - \overline{X})^2 \times \sum_{i=1}^{N}(Y(i) - \overline{Y})^2} \quad (3)$$

4. Signal-to-Noise Ratio—Denoised (SNR_D):

$$SNR_D = 10\log_{10}\left(\sum_{i=1}^{N}|X(i)|^2 / \sum_{i=1}^{N}|Y(i) - X(i)|^2\right) \quad (4)$$

where X, Y, and N represent the noise free PD current signal, the denoised RF signal, and the length of the measuring data window respectively.

3.3. Considering Discrete Wavelet Transform for Denoising of RF Signal

3.3.1. Basic Principles

In this section, the process of the discrete wavelet transform (DWT) is highlighted and summarized in the following steps:

Step 1: In the first level of decomposition, the noisy signal $Y(t)$ is applied to the high and low pass filters and coefficients of approximation (A_1) and detail (D_1) sub-bands are obtained. Afterward, in the second level of decomposition, the coefficients of A_1, as a new signal, are applied to the new high and low pass filters. These processes are then continued to reach the desired pre-defined number of decomposition level (NDL), i.e., J.

Step 2: The thresholding procedure (TP) is used to remove the noise in the calculated coefficients of detail sub-bands ($D_{1\text{-}J}$).

Step 3: According to the process shown in Figure 2, the thresholded detailed sub-bands and the last level coefficients of approximation are gathered together by inverse discrete wavelet transform (IDWT) to reconstruct the signal $\hat{Y}(t)$ [13].

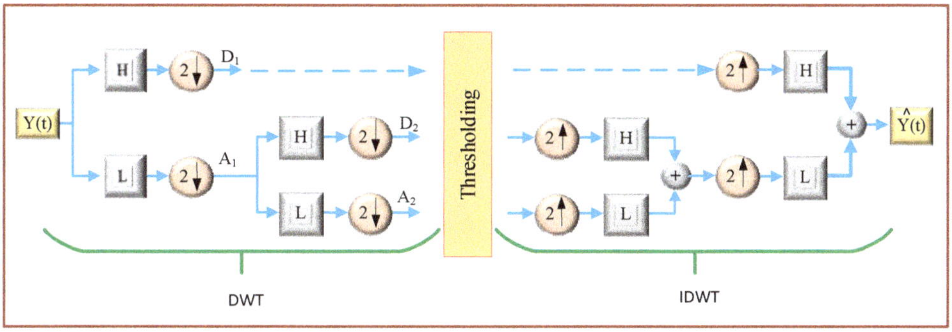

Figure 2. The schematic diagram of the process of denoising by wavelet transform, including decomposition with DWT, thresholding, and inverse DWT.

3.3.2. Mother Wavelet (MW) Selection

In the process of denoising by WT, the target is to keep the coefficients of the signal and noise suppression coefficients in the calculated sub-bands by TP. In the present paper, EBWS method is employed for MW selection, which has shown to give better results in the conducted simulations for denoising of RF signals. In EBWS, the distribution of signal energy in the sub-bands is utilized to select the best MW. In fact, a collection of appropriate MWs is considered; then an index, energy percentage (EP) achieved by (5), is calculated for all MWs. Next, in each level of decomposition, the MW corresponding with the biggest EP is selected as the optimum MW [14]. The selected MWs consist of Daubechies, orders 1–25 (db1–db25), symlets, orders 2–15 (sym2–sym15), and Coiflets, orders 1–5 (coif1–coif5).

$$EP = (\sum_k a_{J,k}{}^2)/(\sum_k a_{J,k}{}^2 + \sum_j \sum_k d_{j,k}{}^2) \tag{5}$$

where a, d, j, and k represent approximation coefficients, detailed coefficients, decomposition level, and length of samples in sub-bands, respectively. It should be noted that in most cases db4, as the MW, has presented better operation for denoising by WT exploited in this work.

3.3.3. Thresholding Procedure (TP)

As is discussed in the previous sub-section, TP must be implemented so that the components of the signal and the noise elements are maintained and eliminated in the detail sub-bands, respectively. Thus, the performance of signal denoising by WT is depended on the way of threshold value calculation. In this paper, the proposed idea by [10], energy conservation-based method (ECBT), considered as one of the promising methods, is implemented as follows:

1. The original noise-free signal is decomposed to reach the predefined level, J, resulting in A_J and $D_{1\text{-}J}$ sub-bands.
2. The mentioned procedure above is repeated for the noisy signal, in order to achieve NA_J and $ND_{1\text{-}J}$ sub-bands as well.

3. The coefficients of NA_J are kept completely, and all components of ND_1 are set on zero. Next, threshold values, λ_J, are estimated for each level of $ND_{2\text{-}J}$, and the coefficients are then thresholded using (6)

$$\lambda_j^{ECBT}(x) = \begin{cases} x, & if|x| \leq \lambda_j \\ \lambda_j & if|x| \geq \lambda_j \end{cases} \quad (6)$$

where, X and λ_j, respectively, present the coefficients and threshold value of level j, obtained by (7).

$$\lambda_j = A_j\left(\left[n_j \times \frac{E_{D_j}}{E_{ND_j}}\right]\right) \quad (7)$$

where A_j, n_j, E_{Dj}, and E_{NDj} are the vector obtained through (8), the length of ND_j, energy of the detail sub-band D_j, and energy of the noisy detail sub-band ND_j, respectively.

$$A_j = Ascend(|ND_j|) \quad (8)$$

Since the original noise-free signal is not available to attain E_{Dj}, a pre-defined lookup table is proposed by authors [10], demonstrating the necessity of a prior knowledge of PD signals for implementation of ECBT.

3.4. Proposed Method

The proposed idea in this section introduces a novel approach to the problem of denoising, employing ANN. In fact, the process of denoising in this algorithm is merely carried out in time-domain instead of currently widespread time–frequency domain algorithms. In other words, the samples of the digital signal are considered as a collection of discrete ones, rather than a continuous signal, intended to be fitted by an appropriate curve and is explained in the present subsections.

3.4.1. Artificial Neural Network Curve Fitting

Artificial neural network is a prevalent algorithm exploited in curve fitting, data classification, and time-series problems [15]. In this work, a multi-layer perceptron network is employed in which the samples of the measured noisy signal are assumed as the inputs and the output is the fitted curve, as depicted in Figure 3. The structure of the employed ANN includes a feed-forward network, in the input and output layers, comprised of sigmoid and linear activation functions, respectively. As shown in Figure 4, a series of samples, which are the ANN input vector, is approximated using an appropriate curve and the output of ANN corresponds to the least square error and obtained by Equation (9). Thus, the acquired curve in the process is considered as the denoised signal.

$$Error = \sum_{i=1}^{n} (e_i)^2 \quad (9)$$

where n and e_i are assumed to be the number of the samples and the error of the ith sample, respectively.

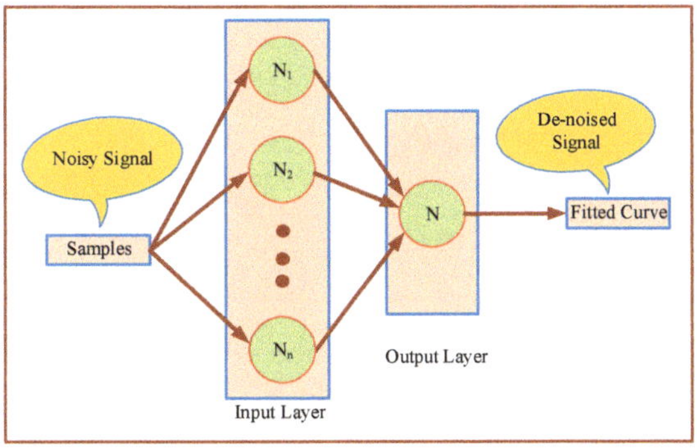

Figure 3. Structure of the ANN in the proposed method.

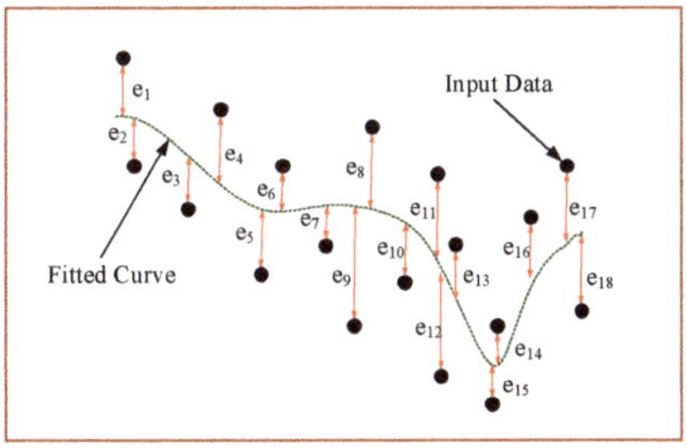

Figure 4. Curve fitting in the procedure of optimization by ANN to achieve the least square error.

3.4.2. Suitable Number of Neurons for the Structure of the ANN

In this paper, a two-layer ANN is considered to avoid the computational burden of higher layers of ANN. Hence, while the output layer is made up by one neroun, the first layer is going to be investigated in this subsection. Therefore, to attain a suitable number of neurons for the first layer, different values will be examined for three types of the RF signals and the one that results in the least RMSE will be selected, shown in Figure 5. It should be noted that in this figure the signals and their frequency spectrum both in noise-free and noisy condition (PSNR = 1) are depicted. It is worth mentioning that the optimization method, exploited in this investigation, is Levenberg–Marquardt. As shown in Figure 6, 100 neurons showed the lowest RMSE for the three different measured RF signals. It should be noted that the length of the data window (DW) is 10,000 samples and it is assumed as a basic data window (BDW) in the paper. Therefore, longer DWs are considered the main data window (MDW) and hence must be divided into the BDWs.

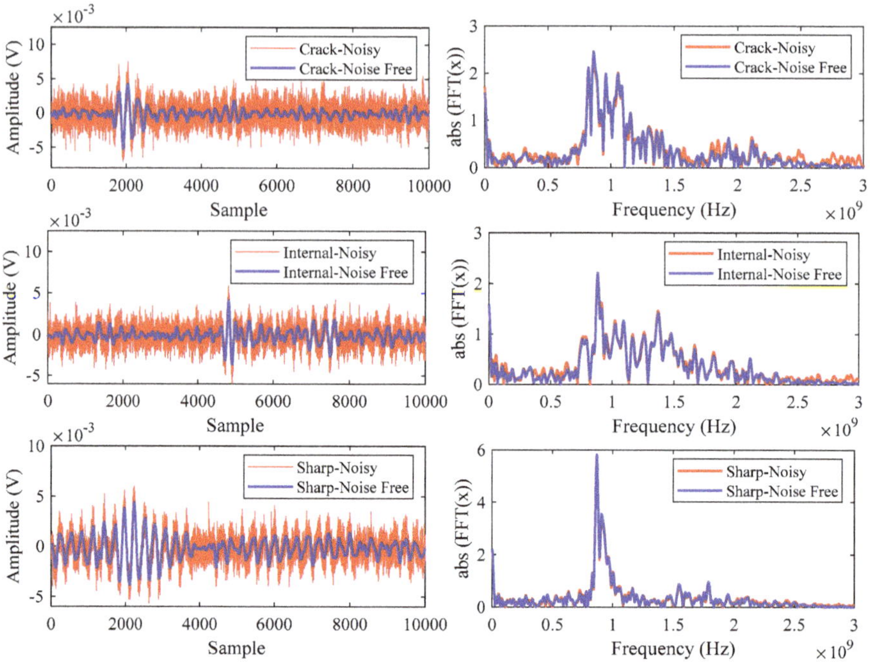

Figure 5. Three RF signals emitted from the different sources of partial discharge in the insulator discs including crack, internal void, and sharp point types and their frequency spectrum both in noisy and noise free conditions.

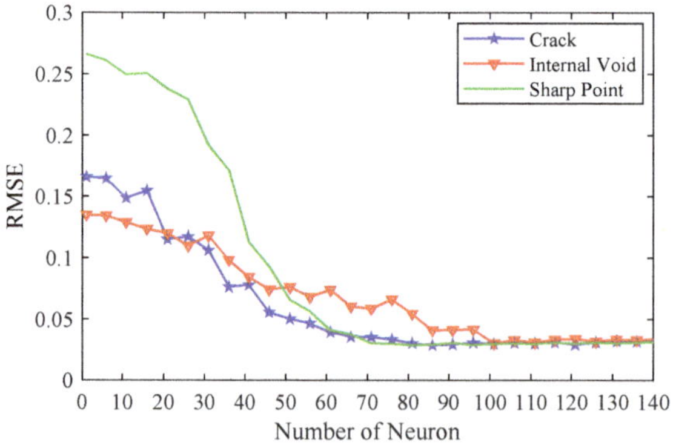

Figure 6. Considering the suitable number of neurons for the three different RF signals.

3.4.3. Optimization Methods

In this subsection, the performance of five optimization methods, including Levenberg–Marquardt, Bayesian regularization, BFGS quasi-Newton, resilient back-propagation, and scaled conjugate gradient for the utilized ANN in the proposed method is investigated. Hence, the operation of these functions for denoising of the RF signal, for instance, the sharp point type, in three noise levels are explored and shown in Table 1. As seen, in all cases, the best performance is obtained by Levenberg–Marquardt; therefore, it is exploited in the structure of ANN in the proposed method as the best optimization method.

Table 1. The results of denoising, the factor RMSE, by five optimization methods and in the cases of three levels of noise (PSNR) including 1, 1.5, and 2.

Optimization Method	PSNR		
	1	1.5	2
Levenberg–Marquardt	0.04102	0.02899	0.02258
Bayesian Regularization	0.04387	0.02966	0.02263
BFGS Quasi-Newton	0.04236	0.03041	0.02261
Resilient Back Propagation	0.04296	0.02918	0.02283
Scaled Conjugate Gradient	0.04165	0.03039	0.02276

3.4.4. Effect of Sampling Rate on the Performance of the Proposed Method

In this subsection, the performance of the proposed method is considered from the sampling rate effect point of view. Hence, the RF signal, for the PD source of a sharp point, in the cases of different sampling rate is exploited in this investigation. Therefore, denoising of the signal in each case is carried out and the factor RMSE is calculated for each, shown in Figure 7. As seen, the more sampling rate for RF signal, the more accuracy in the denoising procedure by the proposed method can be obtained, where the lowest error is observed for the signal by 8 GS/s. However, in this work, the sampling rate of 2 GS/s is used for PD signal measuring, due to the computational burden.

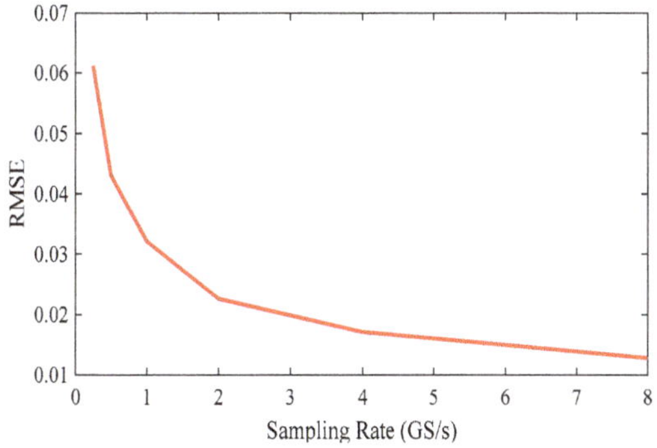

Figure 7. Results of denoising, the factor RMSE, in the various sampling rates of RF signals.

3.4.5. Full Procedure of the Proposed Method

The proposed method can be summarized as follows:

1. Normalizing the RF signal by
$$X_N = X/\max(|X|) \tag{10}$$
where X and X_N are the main RF signal and normalized one, respectively.
2. Dividing MDW into the pre-defined BDW with 10,000 samples lengths each.
3. Separately denoising each BDW, using 100 neurons in the ANN structure.
4. Connecting all BDWs together to attain the complete denoised MDW RF signal.
5. Obtaining the real RF signal by multiplying the maximum value, achieved in Step 1, by the signal denoised in Step 4.

4. Results and Discussions

4.1. Effectiveness of the Proposed ANN-Based Denoisng Technique

In this section, the RF signals are polluted with white noise at different levels, with PSNR ranging from 1 to 2 with steps of 0.25. Moreover, the obtained results from denoising with the proposed method as well as the wavelet-based method are presented. In Figure 8, the result for the first type of RF signal generated from a crack for a severe noise (PSNR = 1) is shown. Additionally, the denoising evaluation factors—including QE, RMSE, CC, and SNR_D—are exploited in order to compare the proposed method with ECBT in the different values of PSNR, as depicted in Figure 9. It is evident that ANN-based denoising technique showed a better performance than ECBT in all calculated parameters. These investigations are repeated for other RF signals, namely internal void, and sharp point types, as well; hence, denoising results for the most severe noise are, respectively, depicted in Figures 10 and 11 for the named damaged types mentioned above. In addition, the proposed method is compared with ECBT in cases of various PSNR, for both RF signals including internal void and sharp point types, as shown in Figures 12 and 13, respectively. As depicted, the proposed method demonstrates its superiority in RF signal denoising.

It is noteworthy that, in ECBT, the proposed method by authors in [10], exploitation of a pre-defined lookup table in the process of denoising is imperative, due to the fact that there is no noise-free original signal for the implementation of the method in the real situation. Whereas, in this work, the noise-free RF signals are exploited instead of using the pre-defined lookup table. That means the proposed method is compared to ECBT with the highest possible performance, being probably too optimistic. Additionally, there is no idea about the optimum number of decomposition levels for DWT, likely to vary for each PD signal. Here, we performed denoising with ECBT in various decomposition levels, including 1 to 8 levels and the optimum results are obtained for the case of 5 levels, utilized in ECBT for denoising of these RF signals.

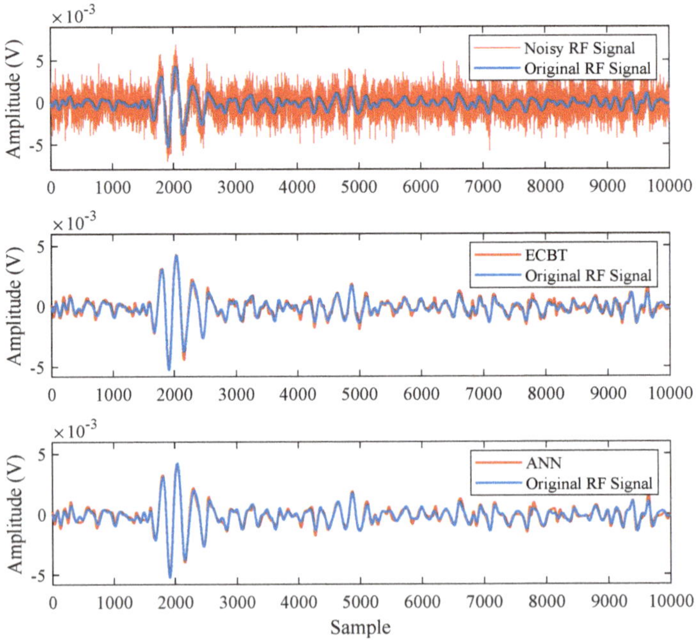

Figure 8. Denoising of RF signal, 'crack' type, in the case of PSNR = 1.

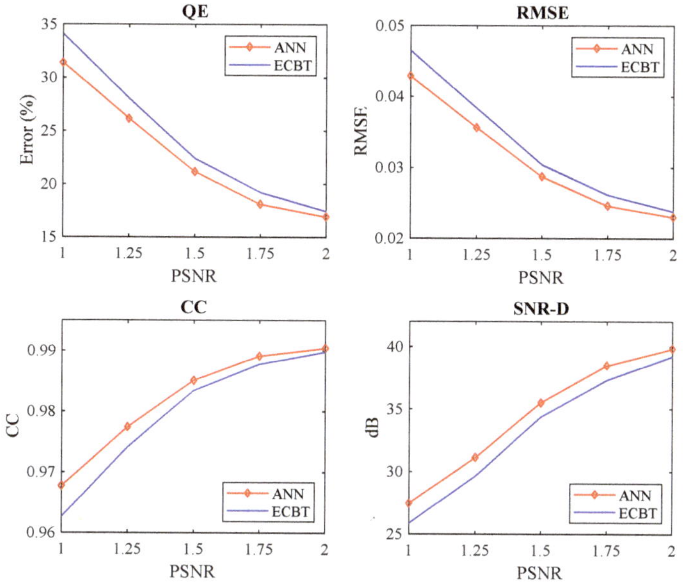

Figure 9. Results of denoising from RF signal in different noise levels where PSNR ranges from 1 to 2 by steps of 0.25, in the case of 'crack' type.

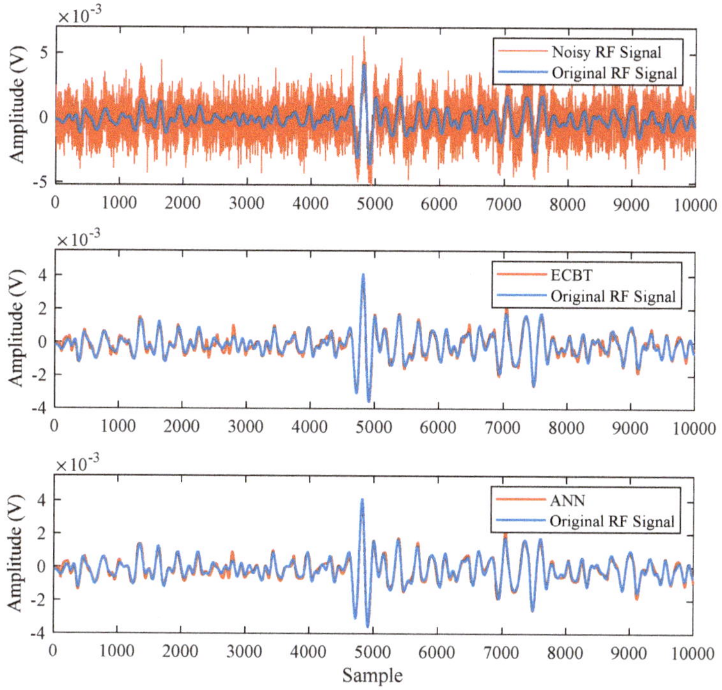

Figure 10. Denoising of RF signal, 'internal void' type, in the case of PSNR = 1.

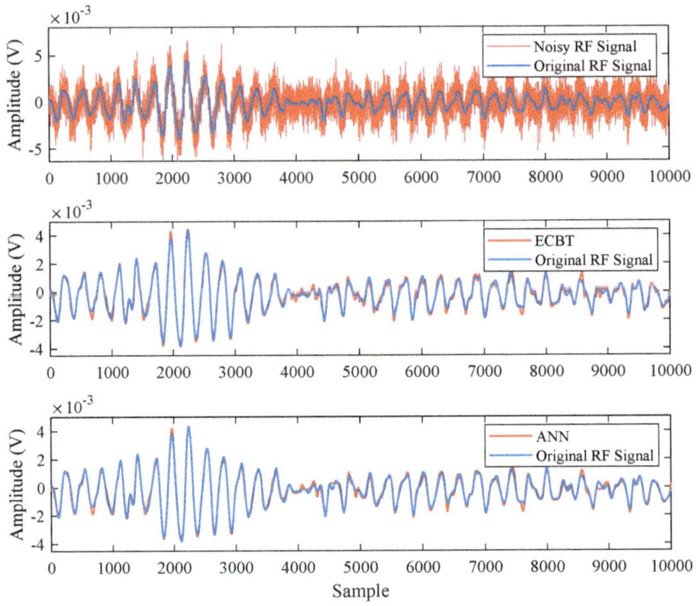

Figure 11. Denoising of RF signal, 'sharp point' type, in the case of PSNR = 1.

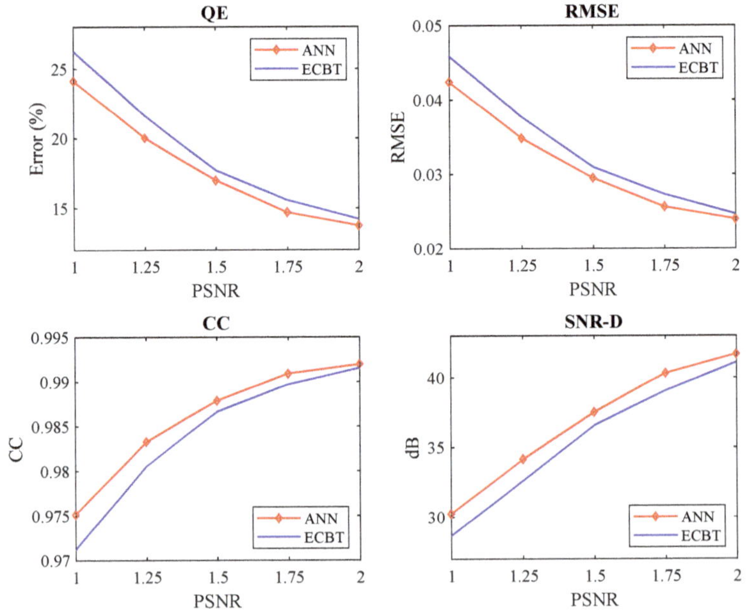

Figure 12. Results of denoising from RF signal in different noise levels where PSNR ranges from 1 to 2 by steps of 0.25, in the case of 'internal void' type.

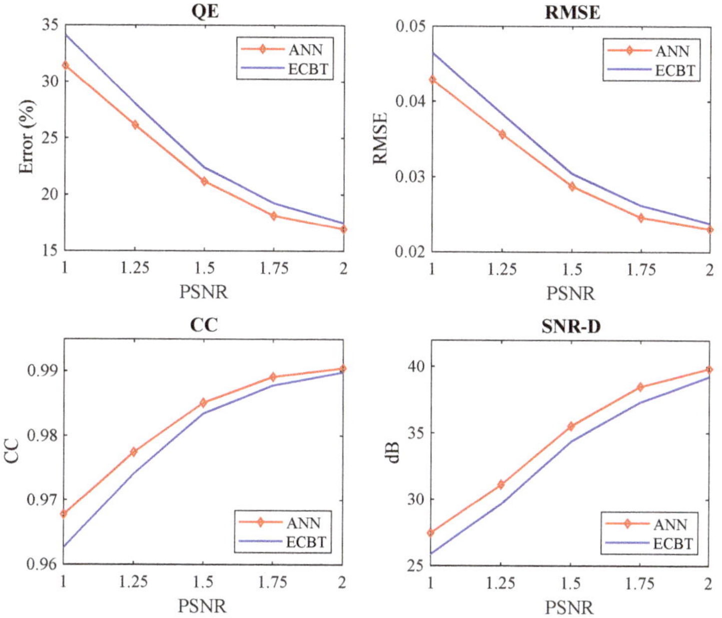

Figure 13. Results of denoising from RF signal in different noise levels where PSNR ranges from 1 to 2 by steps of 0.25, in the case of 'sharp point' type.

4.2. Consideration of Denoising for Combining Two RF Signals

This subsection seeks to consider the performance of the denoising methods in the case of a combination of two RF signals. In fact, it is assumed that in the process of PD signals measuring, two PD sources generate RF signals, simultaneously, and the horn antenna captures both. Thus, the measured signal, for instance, the combination of 'sharp' and 'internal' types, are utilized for comparing the proposed method and ECBT in various noise levels. Hence, the denoising results for severe noises (PSNR = 1) are shown in Figure 14 and the comparison in various noise level is given in Figure 15.

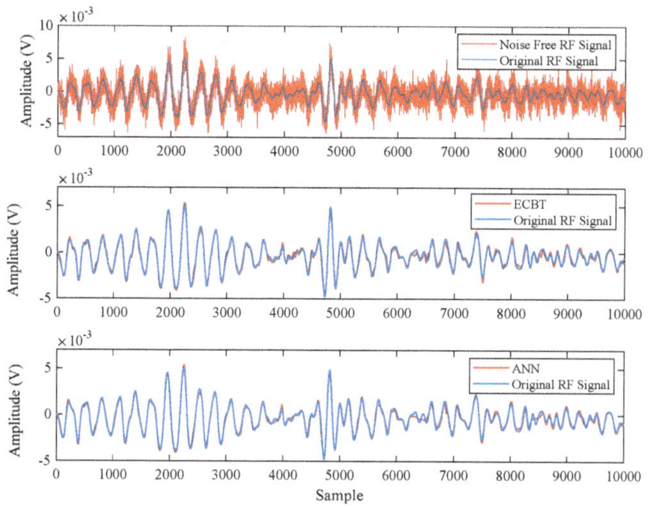

Figure 14. Denoising of RF signal, combination of 'sharp point' and 'internal void' types, PSNR = 1.

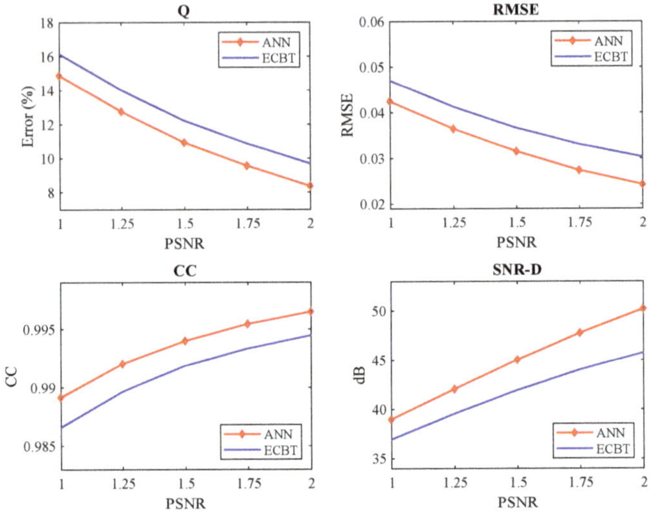

Figure 15. Results of denoising from RF signal in different noise levels where PSNR ranges from 1 to 2 by steps of 0.25, in the case of combination of 'sharp point' and 'internal void' types.

5. Conclusions

In this paper, an ANN-based approach is introduced for white noise suppression. The paper investigates the influence of the main ANN parameters on the performance of the proposed method, namely the appropriate number of neurons and the optimization method for the structure of ANN. Moreover, WT-based algorithm called ECBT is used for comparison, as one of the most popular algorithms for PD signal denoising. The performance of the proposed method is examined by the laboratory-measured RF signals emitted from different PD sources, namely crack, internal void, and hardware sharp points. The RF signals are contaminated with white noise at various levels, with the PSNR ranging from 1 to 2 with the step of 0.25. In all tested cases, the evaluation factors prove a significant superiority of the proposed method for denoising of PD RF signals compared to ECBT. In addition to the ANN superior performance, it is noteworthy that using WT-based algorithms still suffers from other restrictions like mother wavelet selection, determination of the number of decomposition levels, and thresholding procedure. Moreover, to get high performance of WT-based methods, prior knowledge of the signal is needed; whereas, the proposed method is implemented simply without exploiting any prior knowledge.

Author Contributions: Conceptualization, A.A.S. and A.E.-H.; methodology, A.A.S.; software, A.A.S.; validation, A.A.S.; formal analysis, A.A.S.; investigation, A.A.S. and A.E.-H.; resources, A.A.S. and A.E.-H.; data curation, A.E.-H.; writing—original draft preparation, A.A.S.; writing—review and editing, A.E.-H.; visualization, A.A.S. and A.E.-H.; supervision, A.E.-H.; project administration, A.E.-H.

Funding: This research received no external funding.

Conflicts of Interest: The authors declare no conflicts of interest.

References

1. Hussein, R.; Shaban, K.B.; El-Hag, A. Denoising different types of acoustic partial discharge signals using power spectral subtraction. *High Volt.* **2018**, *1*, 44–50. [CrossRef]
2. Ghorat, M.; Gharehpetian, G.B.; Latifi, H.; Hejazi, M.A. A new partial discharge signal denoising algorithm based on adaptive dual-tree complex wavelet transform. *IEEE Trans. Inst. Meas.* **2018**, *67*, 2262–2272. [CrossRef]

3. Carvalho, A.T.; Lima, A.C.S.; Cunha, C.F.F.C.; Petraglia, M. Identification of partial discharges immersed in noise in large hydro-generators based on improved wavelet selection methods. *Measurement* **2015**, *75*, 122–133. [CrossRef]
4. Kopf, U.; Feser, K. Rejection of narrow-band noise and repetitive pulses in on-site PD measurements. *IEEE Trans. Dielectr. Electr. Insul.* **1995**, *6*, 1180–1191. [CrossRef]
5. Sriram, S.; Nitin, S.; Prabhu, K.; Bastiaans, M. Signal denoising techniques for partial discharge measurements. *IEEE Trans. Dielectr. Electr. Insul.* **2005**, *12*, 1182–1191. [CrossRef]
6. Ashtiani, M.B.; Shahrtash, S.M. Feature-oriented denoising of partial discharge signals employing mathematical morphology filters. *IEEE Trans. Dielectr. Electr. Insul.* **2012**, *19*, 2128–2136. [CrossRef]
7. Soltani, A.A.; Shahrtash, S.M. Self-adaptive morphological filter for noise reduction of partial discharge signals. In Proceedings of the 33rd Power System Conference, Tehran, Iran, 22–24 October 2018.
8. Wu, Z.; Huang, N.E. Ensemble empirical mode decomposition: A noise-assisted data analysis method. *Adv. Adapt. Data Anal.* **2009**, *1*, 1–41. [CrossRef]
9. Jin, T.; Li, Q.; Mohamed, A.M. A novel adaptive EEMD method for switchgear partial discharge signal denoising. *IEEE Access* **2019**, *7*, 58139–58147. [CrossRef]
10. Hussein, R.; Shaban, K.B.; El-Hag, A. Energy conservation based thresholding for effective wavelet denoising of partial discharge signals. *IET Sci. Meas. Technol.* **2016**, *10*, 813–822. [CrossRef]
11. Anjum, S.; Jayaram, S.; El-Hag, A.; Jahromi, A.N. Detection and classification of defects in ceramic insulators using RF antenna. *IEEE Trans. Dielectr. Electr. Insul.* **2017**, *24*, 183–190. [CrossRef]
12. Najafipour, A.; Babaee, A.; Shahrtash, S.M. Comparing the trustworthiness of signal-to-noise ratio and peak signal-to-noise ratio in processing noisy partial discharge signals. *IET Sci. Meas. Technol.* **2012**, *7*, 112–118. [CrossRef]
13. Mortazavi, S.; Shahrtash, S. Comparing denoising performance of DWT, WPT, SWT AND DT-CWT for partial discharge signals. In Proceedings of the 43rd International Universities Power Engineering Conference, Padova, Italy, 1–4 September 2008.
14. Li, J.; Jiang, T.; Grzybowski, S.; Cheng, C. Scale dependent wavelet selection for denoising of partial discharge detection. *IEEE Trans. Dielectr. Electr. Insul.* **2010**, *17*, 126–137. [CrossRef]
15. Budiman, F.N.; Khan, Y.; Malik, N.H.; Al-Arainy, A.A.; Beroual, A. Utilization of artificial neural network for the estimation of size and position of metallic particle adhering to spacer in GIS. *IEEE Trans. Dielectr. Electr. Insul.* **2013**, *20*, 2143–2151. [CrossRef]

energies

MDPI

Article

Partial Discharge Analysis under High-Frequency, Fast-Rise Square Wave Voltages in Silicone Gel: A Modeling Approach

Moein Borghei and Mona Ghassemi *

Department of Electrical and Computer Engineering, Virginia Tech, Blacksburg, VA 24061, USA; moeinrb@vt.edu
* Correspondence: monag@vt.edu

Received: 31 October 2019; Accepted: 28 November 2019; Published: 28 November 2019

Abstract: Wide bandgap (WBG) power modules able to tolerate high voltages and currents are the most promising solution to reduce the size and weight of the power management and conversion systems. These systems are envisioned to be widely used in the power grid and the next generation of more (and possibly all) electric aircraft, ships, and vehicles. However, accelerated aging of silicone gel when being exposed to high frequency, fast rise-time voltage pulses that can offset or even be an obstacle for using WBG-based systems. Silicone gel is used to insulate conductor parts in the module and encapsulate the module. It has less electrical insulation strength than the substrate and is susceptible to partial discharges (PDs). PDs often occur in the cavities located close to high electric field regions around the sharp edges of metallization in the gel. The vulnerability of silicone gel to PDs occurred in the cavities under repetitive pulses with a high slew rate investigated in this paper. The objective mentioned above is achieved by developing a Finite-Element Analysis (FEA) PD model for fast, repetitive voltage pulses. This work has been done for the first time to the best of our knowledge. By using the model, the influence of frequency and slew rate on the magnitude and rate of PD events is studied.

Keywords: finite element analysis; partial discharge modeling; high-frequency; fast-rise square wave voltages; silicone gel; wide bandgap power modules

1. Introduction

The growing trend toward electrification has led to a rapid-growing penetration of power electronics into various residential, industrial, and commercial levels. In this regard, WBG power modules, which can tolerate higher voltages and currents than silicon (Si)-based modules, are the most promising solution for reducing the size and weight of power electronics systems. For example, while the highest breakdown voltage capability, known as the blocking voltage, for commercial Si insulated gate bipolar transistors (IGBT) is 6.5 kV, for SiC IGBTs is 15 kV for 80 A, and 24 kV for 30 A [1]. While the blocking voltage of this 15-kV SiC-IGBT is 2.3 times higher than the Si-IGBT, its volume is one-third that for the Si-IGBT [1]. This translates into higher electric stress within the module. The high electric field, as well as its exposure to voltage pulses with high slew and repetition rates, create an extremely poor environment for insulation systems in envisioned high voltage high-density WBG modules where partial discharges (PDs) have the most impact on insulation degradation [2,3]. Two common insulation materials used in power modules are ceramic substrate and silicone gel. As the main insulation system, the ceramic substrate ensures electrical insulation between active components and the baseplate, which is generally grounded. Silicone gel is used for encapsulation to prevent electrical discharges in air and to protect substrates, semiconductors, and connections against humidity, dirt, and vibration. One of the weakest points inside the power electronics module, in terms of resistance against PD, is the region in the silicone gel close to the sharp edges of metallization layers, as shown in Figure 1. These sharp edges lead to high electric field stress in the air-filled voids located in the mentioned regions that may initiate PDs within voids.

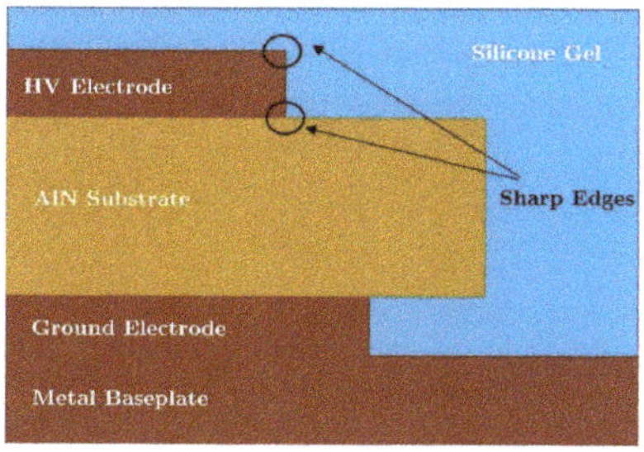

Figure 1. Schematic of an IGBT or a diode substrate.

One of the merits of WBG devices is that their slew rates and switching frequencies are much higher than Si-based devices. However, from the insulation side, frequency and slew rate are two of the most critical factors of a voltage pulse, which influences the level of degradation of the insulation systems that are exposed to such voltage pulses [4]. Solid dielectrics are found to be vulnerable to frequency, while the dominant factor for PD inception in liquid dielectrics is rising [5,6]. In this regard, the study of the insulation degradation of silicone gel, which has the properties of both liquids and solids can be challenging. The modeling of PD gains even more importance when knowing that the detection of PDs becomes very difficult under voltage pulses with short rise-times of tens to hundreds of ns [7].

The existence of voids and air bubbles is an inevitable phenomenon in dielectrics. They can be formed either during the manufacturing process or the streamer propagation itself. Aside from the manufacturing process, the electrical treeing itself causes the formation of air bubbles inside the dielectrics [8]. The experiments have shown that the self-healing characteristic of silicone gel cannot quickly omit the air bubbles. Even the cavity can grow due to the surface discharges occurring in silicone gel [9]. A cavity inside the silicone gel can last up to 10 ms that is 1000 times longer than that of non-viscous liquids [10]. PDs during this time may lead to the gradual degradation of silicone gel.

The air-filled cavities inside a dielectric are weak points due to their lower permittivity than the surrounding material. According to the enhancement factor introduced in Reference [11], the electric field in a spherical void inside the silicone gel can be enhanced by about 30% compared to the void-free case. Therefore, the impact of such voids on the expected lifetime of silicone gel should be taken into account.

Some methods, such as geometric techniques alone or combined with the application of nonlinear field-dependent conductivity layers, have been proposed to address the high electric field issue in envisioned high voltage high-density WBG power modules [12–19]. However, to the best of our knowledge, PD modeling for air-filled cavities in silicone gel, especially under fast, repetitive voltage pulses, still has not been reported. Elucidating and understanding mechanisms and phenomena behind PDs in air-filled cavities in silicone gel helps develop more efficient electric field control methods.

To accurately model the cavity-solid dielectric system, different models have been proposed [20–33]. In this paper, the FEA approach is utilized to dynamically model PDs inside a cavity located in silicone gel and calculate the electric field and potential distribution in different parts of the system. The parameters in the model are chosen and verified to be in agreement with the experimental results reported in the literature.

The model developed in COMSOL Multiphysics (5.4) interfaced with MATLAB (R2019b) provides the opportunity to examine the impact of fast, high-frequency voltage pulses on the initiation of PD from the standpoint of both intensity and number.

2. PD Initiation and Propagation

When a discharge occurs, it takes place by the movement of free electrons and ions. These charged particles are formed due to the collisions between energetic electrons and neutral atoms. The free electrons in the presence of a high electric field gain enough energy to move rapidly and collide with neutral particles to form another electron as well as a positive ion. The electron avalanche process can occur when there is at least one free electron available and the electric field is sufficiently high to energize the electron for upcoming collisions. Later in this section, these two conditions are examined.

2.1. Streamer Inception Criterion

A PD does not completely bridge between the electrodes. PDs are local discharges that occur in defects or air-filled voids inside the dielectric material. From a physical point of view, a discharge occurs when a free electron is available while being energetic enough to ionize further molecules and cause the growth of the streamer. At high electric fields, electrons can gain a sufficient amount of energy for the collision and ionization of further atoms or molecules, as shown in Figure 2.

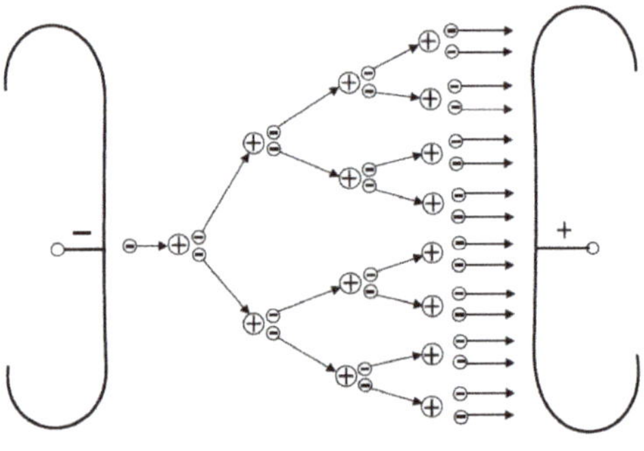

Figure 2. Streamer formation between two electrodes.

Thus, for modeling the inception time of PDs correctly, it is necessary to have an accurate representation of the electric field magnitude at which the impact ionization process can be begun. In this regard, a critical avalanche criterion has been proposed in Reference [21] that probes into the critical number of electrons needed for the head of an avalanche to be self-propagating [11]. Satisfying the mentioned requirement, the streamer inception criterion can be expressed concisely as follows.

$$\frac{E_{inc}}{E_{cr}} = F \quad (1)$$

The ratio of the PD inception field to the critical field at which total breakdown happens is defined by a dimensionless parameter, F, a function of gas inside the cavity, permittivity of the dielectric, and the defect geometry. For a spherical cavity, it can be defined by the equation below.

$$F = 1 + \frac{B}{(pd)^n} \quad (2)$$

where B and n are constants related to gas. For air, they are equal to 8.6 $Pa^{0.5}m^{0.5}$ and 0.5, respectively [11]. The gas pressure p and defect diameter parallel to background field d are the other parameters used in this model.

It should be mentioned that environmental conditions can have a considerable contribution to the ionization process. As found in Reference [27], the existence of humidity can reduce the breakdown voltage of silicone gel.

2.2. Initial Free-Electron

Along with the streamer inception criterion, another condition must also be satisfied to guarantee that the ionization process happens. It evaluates the existence of at least one free electron for the initiation of discharge. Therefore, even after exceeding the inception field, the occurrence of PD might be associated with a statistical time delay related to the second criterion of PD inception.

The initial electron can be provided in various ways. Radiative ionization is one of the processes to generate the initial electrons. Irradiation of energetic photons can ionize molecules, and the rate of electrons per unit of time that can be generated through this mechanism is formulated below [28].

$$\dot{N}_{rad} = K_{rad}\,(\rho/p)_0\, p\, V_{cav}\left(1-\mu^{-0.5}\right) \tag{3}$$

where K_{rad} is a constant characterizing the irradiation process, $(\rho/p)_0$ is the pressure reduced density of the gas, V_{cav} is the volume of the cavity, and μ is the overvoltage factor.

Aside from the photon radiation, the initial electron can be provided through the de-trapping of electrons from the surface cavity. These electrons are remnants of previous PD events that are positioned into the traps at the cavity surface. The number of electrons remaining from a previous PD is obtained as:

$$N_{PD} = \frac{q_{PD}}{e}\, e^{-\frac{t-t_{PD}}{\tau_{dec}}} \tag{4}$$

The number of electrons remained from a PD event is linearly dependent upon the true PD charge magnitude (q_{PD}). It is also affected by time. As the time passes from its occurrence, the number of available electrons is decreased by different decay mechanisms such as moving the electrons into deeper traps from which the electrons cannot be further de-trapped.

The emission of electrons from the surface of the cavity has been found to follow Richardson-Schottky scaling [19]. Thus, the rate of electron emission from the surface of the cavity due to de-trapping of remnant electrons can be formulated by the equation below.

$$\dot{N}_{dtr} = N_{PD}\, \nu_0\, e^{-\frac{\phi - \sqrt{e\frac{|E_{cav}(t)|}{4\pi\varepsilon_0}}}{kT}} \tag{5}$$

In the formulation above, ν_0 is the fundamental phonon frequency, and ϕ is the de-trapping work function. The Boltzmann constant and absolute temperature are denoted by k and T, respectively. Moreover, as shown in Equation (6), the rate of electron emission relies on the magnitude of the electric field inside the cavity. The higher the electric field, the higher the probability of electron detachment.

Then, the total rate of electron generation can be approximately yielded by summation of the electron generation rate due to each of these processes.

$$\dot{N}_e = \dot{N}_{rad} + \dot{N}_{dtr} \tag{6}$$

The final step in the assessment of free electron availability is to address the stochastic nature of the phenomena. At the first step, the probability of the existence of a free electron for PD ignition (P) at each time step ought to be obtained [27].

$$P = 1 - \exp\left(-\int_0^t \dot{N}_e(t)\cdot dt\right) \tag{7}$$

At the second step, a random number R between zero and unity is generated and will be compared to P. If P is greater than R, then the PD happens.

3. Modeling of the PD Process

The actions of a PD are associated with an increase in cavity conductivity, which leads to a voltage drop and a field reduction across the cavity. To accurately model this, once the two conditions for PD ignition are satisfied, the conductivity of the cavity is increased from its initial value (close to zero) to a maximum level ($\sigma_{cav,m}$).

Moreover, one should take into account the charge decay mechanism. The accumulated charges on the surface of the cavity do not have an infinite lifetime. They might decay through conduction along the cavity wall by the movement of electrons from shallow traps into deeper ones [29]. This process is found to be dependent upon the magnitude of charges deposited on the cavity surface. If the magnitude of the physical charge exceeds a certain value, Q_{U}, the surface conductivity is increased to a higher magnitude, $\sigma_{surf,H}$ [30]. Once the charge magnitude drops below the threshold, the surface conductivity comes back to its initial value. Using FEA, the charge magnitude at each instant can be obtained through the integration of field displacement over the cavity surface area.

The entire PD modeling procedure is demonstrated in Figure 3. Starting from the origin of time, the FEA model is run to enable the assessment of the two PD requirements (i.e., streamer criterion as well as the initial electron availability). If any of the two prerequisites are not satisfied, the algorithm goes to the next time step. The procedure of looking for a PD is continued until finding a time instant that meets the requirements or the program runs out of time steps. Once the occurrence of a PD event is guaranteed, the cavity and surface conductivities are updated. Afterward, the FEA is again run to find the extinction time of the PD. Note that a state variable (shown in Figure 3) is employed to distinguish between the state of looking for the initiation of the next PD event (*state* $= 0$) and the state of looking for the extinction time of a current PD (*state* $= 1$). Moreover, it should be noted that the time step does not have a constant value. There are two values for the time step. One is the time step between PD events (Δt_L) and the other is the time step during PD activities (Δt_H). Since the duration of PD events are much smaller than the time interval between PD events, Δt_L is expected to be larger than Δt_H.

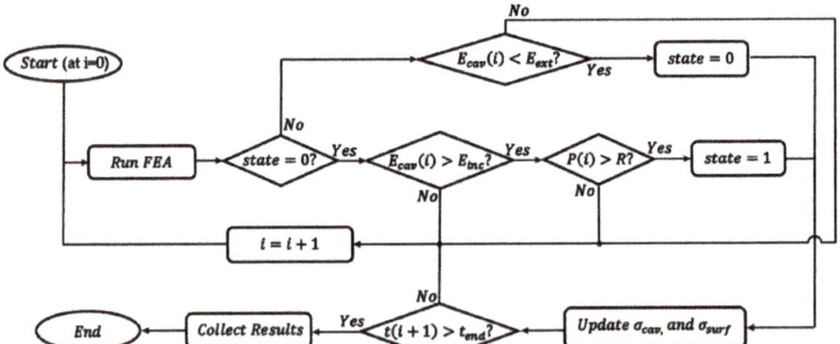

Figure 3. PD modeling flowchart.

4. Results and Key Challenges

4.1. Case Study

The case study considered in this paper consists of a sphere-sphere electrode system that provides a weakly non-uniform electric field. A spherical air-filled cavity is located between the electrodes. The configuration of the system is shown in Figure 4, and the related parameters are listed in Table 1. For FEA, the 2D-axisymmetrical representation of the configuration is used. The parameters related to electron generation rate, time-stepping, and charge decay are listed in Table 2. The details and representation of the mesh are shown in Table 3 and Figure 5, respectively.

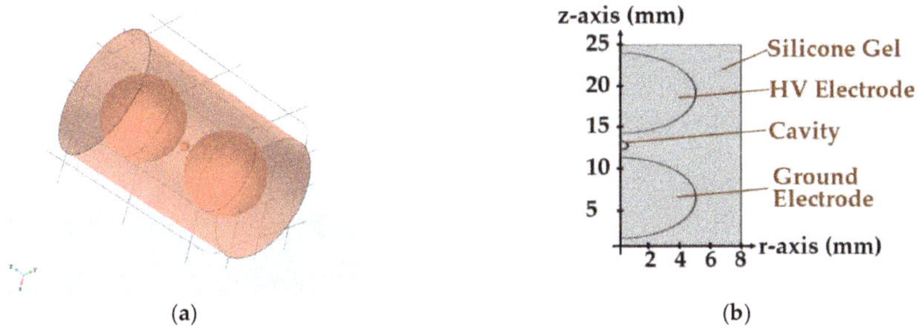

Figure 4. The graphical representation of the case study. (**a**) 3D and (**b**) Axi-symmetrical 2D.

Table 1. Case Study Parameters.

Symbol	Parameter	Value
	Geometrical	
d_{cav}	cavity diameter	1 mm
r_{el}	electrode radius	5 mm
d	distance between electrodes	3 mm
ε_r	Silicone gel relative permittivity	2.8
σ_0	conductivity	10^{-13} S/m
U_{peak}	Applied Voltage voltage peak magnitude	11 kV
f	frequency	20 kHz
r_t	rise-time	100 ns
D	duty cycle	50%

Table 2. Parameters of PD Model.

Symbol	Value
E_{cr}	2.42×10^6 V/m
K_{rad}	2×10^6 kg^{-1} s^{-1}
$(\rho/p)_0$	10^{-5} kg m^{-3} Pa^{-1}
p	1 atm
τ_{dec}	$0.15T$
ϕ	1.3 eV
ν_0	10^{14} s^{-1}
T	300 K
Δt_L	$1/(1000f)$
Δt_H	1 ns
$\sigma_{cav,m}$	0.0004 S/m
$\sigma_{surf,L}$	2×10^{-13} S/m
$\sigma_{surf,H}$	10^{-11} S/m
E_{inc}	3.1 kV/mm
E_{ext}	0.1 kV/mm

Table 3. Mesh Properties.

Quantity	Value
Maximum element size	5×10^{-4} m
Minimum element size	1.88×10^{-6} m
Maximum element growth rate	1.2
Curvature factor	0.25
Resolution of narrow regions	1
Maximum element size	5×10^{-4} m

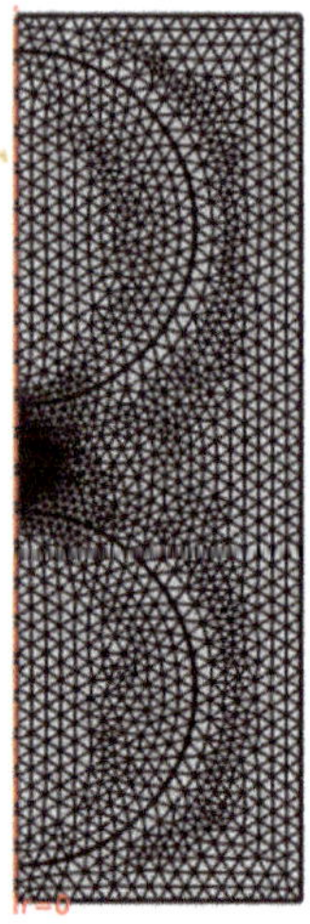

Figure 5. Mesh Pattern of the void-dielectric system.

4.2. Results

4.2.1. Field Distribution before and after PD

The electric field distribution is shown in Figure 6 right before and right after PD occurrence. As expected, the electric field intensity inside the cavity is higher than the surrounding material, and it increases up to a point where the cavity can no longer withstand the electric stress. With the availability of an initial electron, the collision between electrons and molecules is begun, which gives rise to an increase in the cavity conductivity, and loses its insulating property, subsequently.

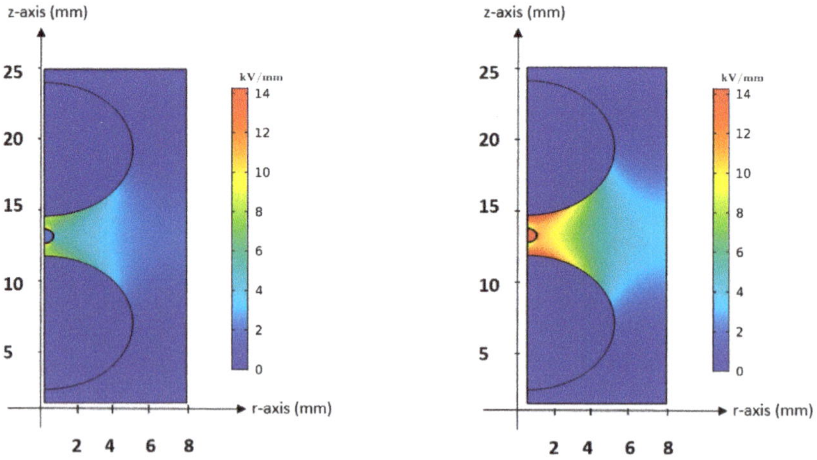

(a) (b)

Figure 6. The electric field distribution (**a**) before, and (**b**) after PD occurrence.

The reduction of cavity electric resistance causes the electric field and voltage across the cavity to drop. It is continued until the electric field has reached the extinction field.

4.2.2. Discharge Pulses over Time for 20 Cycles

A summary of results is reported in Table 4. In this table, the true and apparent charge magnitudes are used to illustrate the intensity of PD activities. The true charge magnitude is the amount of charge accumulated on the upper or lower cavity hemisphere while the apparent charge magnitude is the charge magnitude accumulated over the surface of the ground electrode. In addition, Figure 7 shows the phase-resolved PD (PRPD) pattern that is in good agreement with the experiments reported in Reference [34]. Two PDs occur at each cycle. One occurs at the rising flank of the pulse, and the other one occurs at the falling flank. To date, PD activity was not modeled under fast, high-frequency voltage pulses to the best of our knowledge, and this paper models it for the first time.

Table 4. Results of PD simulation.

Quantity	Value
Number of PDs per cycle	1.9866
Mean duration of PD	503.67 ns
Mean true charge	0.42 μC
Mean apparent charge	14.88 μC
Maximum true charge	2.46 μC
Maximum apparent charge	86.58 μC

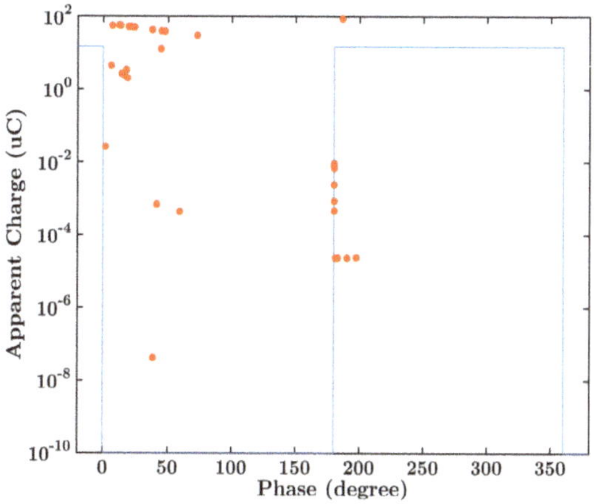

Figure 7. Phase-resolved PD (PRPD) Pattern.

While the magnitudes of PD charges are very similar to each other, Figure 7 conveys the fact that the statistical time lag is much higher in the falling flank. This can be justified through the need for an initial free electron that is harder to obtain at the falling flank due to the charge decay process reducing the emission of electrons from the surface due to the last discharge.

4.2.3. Rise-Time and Frequency Impact

As seen in Figure 8, the destructive impact of frequency and rise-time on PD charge magnitude is clear. However, among these two factors, frequency is more detrimental at its high values. The number of PDs per cycle may decrease by a frequency enhancement, while the severity of PDs may be escalated.

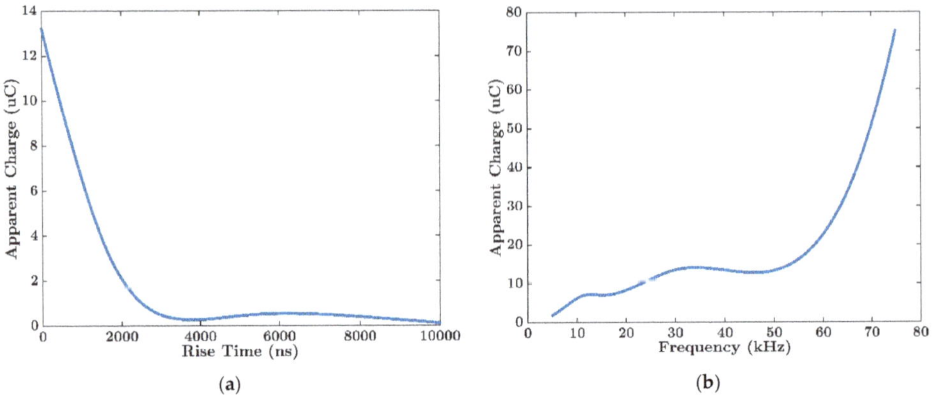

Figure 8. The effect of (**a**) frequency, and (**b**) rise-time on PD intensity.

4.3. Key Challenges

4.3.1. Estimation of Model Parameters

One of the main steps to be taken to reach reliable results for PD simulation is the valid estimation of the physical parameters associated with the model. To accurately determine the value of these parameters, the adjustment of the model with the experimental data can be a feasible criterion.

Each of the physical parameters may contribute to one or more of the PD characteristics (e.g., inception, magnitude, duration, etc.). In this regard, the results reported in Reference [35] were utilized for the purpose of parameter estimation. In the previously mentioned paper, Figure 7 represents the phase-resolved PD (PRPD) pattern for a silicone gel dielectric within two spherical electrodes. The applied voltage is reported to be 11 kV and the PD is found to be activated when the applied voltage reaches 9.5 kV.

The degrees of freedom are determined by the number of indices that are used for the adjustment. Among the parameters, cavity conductivity during PD ($\sigma_{cav,m}$), the cavity inception field (E_{inc}), the cavity extinction field (E_{ext}), and the decay time constant (τ_{dec}) are found to be appropriate for modeling. $\sigma_{cav,m}$ helps control the magnitude and duration of PD events. The higher $\sigma_{cav,m}$ is, the quicker the electric field drops inside the cavity. However, the current density over the cavity wall decreases with the increase in $\sigma_{cav,m}$.

E_{inc} can help determine the inception time. As mentioned, the PD occurs at a voltage of 9.5 kV, and the field at which the voltage reaches this value is chosen as E_{inc}. On the other hand, E_{ext} can contribute to the length of PD, which can also enhance the magnitude of discharges.

Lastly, the time constant can take part in the inception time of PD events. The longer the time constant, the higher the chances of free-electron availability.

Assuming the mean charge magnitude at each half cycle, the proportionality of the number of PDs at half cycles, and the inception voltage, the four degrees of freedom ($\sigma_{cav,m}$, E_{inc}, E_{ext}, and τ_{dec}) are used by fitting the experimental indices.

The values mentioned in Table 4 are determined after adjusting these parameters to the model. Figure 9 demonstrates the PRPD pattern obtained after running the developed model for 50 cycles of the applied voltage, which shows acceptable agreement with the PRPD pattern reported in Reference [35].

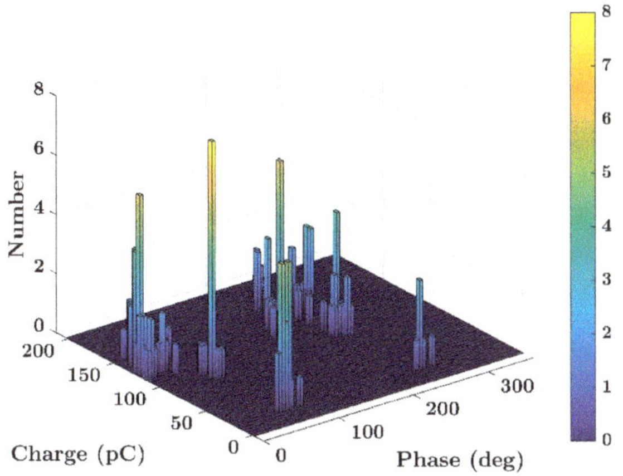

Figure 9. The 3D PRPD pattern for the case study in Reference [12] after 50 cycles.

4.3.2. Convergence Problem

Based on the algorithm described in this paper, the time steps are strictly determined by the algorithm process. While a degree of freedom could help choose those time steps that have higher chances to meet the error tolerance, the strict way of time stepping may enforce convergence problems.

There are multiple ways to overcome these challenges including using a more precise mesh to enhance the convergence chance. In addition, the time-stepping method itself might be capable of providing an opportunity to resolve the convergence problem. In this paper, the time-stepping method employed to solve the time-dependent set of numerous algebraic equations is a method called generalized-α, which is coupled with the first-order and second-order nonlinear differential equations.

This method was first developed by Chung and Hulbert for solving structural dynamics problems in 1993 [36]. This is an implicit method for solving transient problems and attempts to control the damping extent ($0 \leq \xi < 1$) at high-frequencies without a significant influence on the order of accuracy. While this method is more accurate than the Backward Differentiation Formula (BDF) due to its lower damping, it is less stable for the same reason. Thus, alongside the potential of mesh properties, the generalized-α method is used in this paper in order to resolve the convergence problem by increasing the damping frequency.

4.3.3. Burden of Calculation

Since the PD phenomenon has a memory over the previous PD events, an accurate model cannot probe into each event in an isolated manner.

On the other hand, over the course of time, the time instants under investigation grow extensively larger, which makes the FEA more time-consuming. In order to prevent the excessive burden of calculation, one must assess the role of previous PD events on the occurrence and intensity of the upcoming PD event in the first place.

Once the calculations related to a PD event are performed, its contribution to the next discharges can be categorized into two parts. The first one is the role of the PD event in the correct modeling of the changes in voltage distribution and the transition between consecutive PD events. The second contribution is due to its role in the provision of initial free electrons for the upcoming PD, which helps estimate the likelihood of a PD occurrence conducted within the programming section.

In order to accurately model the voltage distribution, the voltage transient associated with each of the discharges should be modeled. However, to model the previous PD events, it is not necessary to have the same number of time instants as in the case of searching for the next PD event. Therefore, a lower number of time instants should be adopted, which is enough to model the transition of the cavity state from non-conducting to conducting and again to a non-conducting state. These changes can be modeled through a lower number of time instants (N_{comp}). In this way, the feasibility of applying the proposed model at a high number of cycles is maintained.

Figure 10 compares the computation time in the case of using the compaction technique $\left(N_{comp} = 20\right)$ to the case of using the exhaustive approach. All the simulations are performed using a computer system with a processor operating at 3.1 MHz with a Random-access memory (RAM) of 24 GBs.

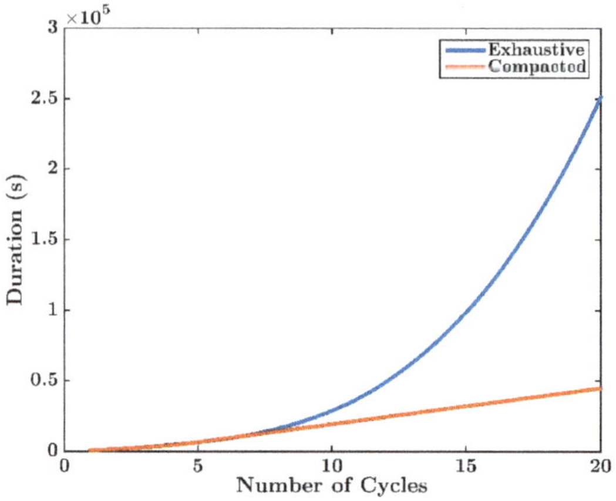

Figure 10. The impact of the compaction technique on computation time.

5. Conclusions

The crucial role of power electronics in the future needs the examination of power modules from different perspectives to ensure the proper operation of the system. As the new generations of modules benefit from WBG semiconductors operating at higher voltages, frequencies, and temperatures, it is necessary to scrutinize the electric stress supported by the insulation system. This paper, using a modeling approach with the aid of simulation tools, probed into a destructive phenomenon, named PD, that can happen inside the power module. In this paper, a PD model was proposed based on the physical processes taking place during the PD phenomena. The model was adjusted to the experimental results reported in the literature. From the obtained results, it is clear that the frequency and rise-time can have a destructive impact on the health of dielectric. The novel modeling approach proposed in this paper, which is done for the first time in the case of fast-rise, high-frequency square voltage, can be the groundwork for any power electronic module design to ensure its reliable operation from the standpoint of insulation performance.

Author Contributions: The authors contributed equally to this work in all parts.

Funding: This research received no external funding.

Conflicts of Interest: The authors declare no conflict of interest.

References

1. Passmore, B.; O'Neal, C. High-Voltage SiC Power Modules for 10–25 kV Applications. *Power Electron. Europe Mag.* **2016**, *1*, 22–24.
2. Ghassemi, M. PD Measurements, Failure Analysis, and Control in High-Power IGBT Modules. *High Volt.* **2018**, *3*, 170–178. [CrossRef]

3. Ghassemi, M. Electrical Insulation Weaknesses in Wide Bandgap Devices. In *Simulation and Modelling of Electrical Insulation Weaknesses in Electrical Equipment*; Albarracin, R., Ed.; IntechOpen: London, UK, 2018; pp. 129–149.
4. Ghassemi, M. Accelerated Insulation Aging Due to Fast, Repetitive Voltages: A Review Identifying Challenges and Future Research Needs. *IEEE Trans. Dielectr. Electr. Insul.* **2019**, *26*, 1558–1568. [CrossRef]
5. Mancinelli, P.; Cavallini, A.; Dodd, S.J.; Chalashkanov, N.M.; Dissado, L.A. Analysis of Electrical Tree Inception in Silicone Gels. *IEEE Trans. Dielectr. Electr. Insul.* **2017**, *24*, 3974–3984. [CrossRef]
6. Madonia, A.; Romano, P.; Hammarström, T.; Gubanski, S.M.; Viola, F.; Imburgia, A. PD characteristics at square shaped voltages applying two different detecting techniques. In Proceedings of the IEEE Conference on Electrical Insulation and Dielectric Phenomena (CEIDP), Toronto, ON, Canada, 16–19 October 2016; pp. 247–250.
7. Hammarström, T.J.Å. Partial Discharge Characteristics at Ultra-Short Voltage Rise Times. *IEEE Trans. Dielectr. Electr. Insul.* **2018**, *25*, 2241–2249. [CrossRef]
8. Dodd, S.J.; Salvatierra, L.; Dissado, L.A.; Mola, E. Electrical trees in silicone gel: A combination of liquid and solid behaviour patterns. In Proceedings of the IEEE Conference on Electrical Insulation and Dielectric Phenomena (CEIDP), Shenzhen, China, 20–23 October 2016; pp. 1018–1021.
9. Sato, M.; Kumada, A.; Hidaka, K.; Yamashiro, K.; Hayase, Y.; Takano, T. Degradation process of silicone-gel by internal surface discharges. In Proceedings of the IEEE International Conference on Dielectric Liquids (ICDL), Bled, Slovenia, 29 June–3 July 2014; pp. 1–4.
10. Wang, N.; Cotton, I.; Robertson, J.; Follmann, S.; Evans, K.; Newcombe, D. Partial Discharge Control in a Power Electronic Module Using High Permittivity Non-Linear Dielectrics. *IEEE Trans. Dielectr. Electr. Insul.* **2010**, *17*, 1319–1326. [CrossRef]
11. Niemeyer, L. A Generalized Approach to Partial Discharge Modeling. *IEEE Trans. Dielectr. Electr. Insul.* **1995**, *2*, 510–528. [CrossRef]
12. Ghassemi, M. Geometrical techniques for electric field control in (ultra) wide bandgap power electronics modules. In Proceedings of the IEEE Electrical Insulation Conference (EIC), San Antonio, TX, USA, 17–20 June 2018; pp. 589–592.
13. Tousi, M.M.; Ghassemi, M. Electric field control by nonlinear field dependent conductivity dielectrics characterization for high voltage power module packaging. In Proceedings of the IEEE International Workshop on Integrated Power Packaging (IWIPP), Toulouse, France, 24–26 April 2019; pp. 54–58.
14. Tousi, M.M.; Ghassemi, M. Nonlinear field dependent conductivity materials for electric field control within next-generation wide bandgap power electronics modules. In Proceedings of the 2019 IEEE Electrical Insulation Conference (EIC), Calgary, AB, Canada, 16–19 June 2019; pp. 63–66.
15. Tousi, M.M.; Ghassemi, M. Electrical insulation packaging for a 20 kV high density wide bandgap power module. In Proceedings of the IEEE Energy Conversion Congress & Exposition (ECCE), Baltimore, MD, USA, 29 September–3 October 2019; pp. 4162–4166.
16. Tousi, M.M.; Ghassemi, M. Nonlinear resistive electric field grading in high-voltage, high-power wide bandgap power module packaging. In Proceedings of the IEEE Energy Conversion Congress & Exposition (ECCE), Baltimore, MD, USA, 29 September–3 October 2019; pp. 7124–7129.
17. Tousi, M.M.; Ghassemi, M. The effect of type of voltage (sinusoidal and square waveform) and the frequency on the performance of nonlinear field-dependent conductivity coatings for electric field control in power electronic modules. In Proceedings of the IEEE Conference on Electrical Insulation and Dielectric Phenomena (CEIDP), Richland, DC, USA, 20–23 October 2019; pp. 601–604.
18. Tousi, M.M.; Ghassemi, M. Combined Geometrical Techniques and Applying Nonlinear Field Dependent Conductivity Layers to Address the High Electric Field Stress Issue in High Voltage High-Density Wide Bandgap Power Modules. *IEEE Trans. Dielectr. Electr. Insul.* **2019**, in press.
19. Tousi, M.M.; Ghassemi, M. Characterization of Nonlinear Field Dependent Conductivity Layer Coupled with Protruding Substrate to Address High Electric Field Issue within High Voltage High-Density Wide Bandgap Power Modules. *IEEE J. Emerg. Sel. Top. Power Electron.* **2019**, in press. [CrossRef]
20. Whitehead, S. *Dielectric Breakdown in Solids;* Clarendon Press: Oxford, UK, 1951.
21. Crichton, G.C.; Karlsson, P.W.; Pedersen, A. Partial Discharges in Ellipsoidal and Spheroidal Voids. *IEEE Trans. Electr. Insul.* **1989**, *24*, 335–342. [CrossRef]

22. Pedersen, A.; Crichton, G.C.; McAllister, I.W. The Theory and Measurement of Partial Discharge Transients. *IEEE Trans. Electr. Insul.* **1991**, *26*, 487–497. [CrossRef]
23. Pedersen, A.; Crichton, G.C.; McAllister, I.W. The Functional Relation Between Partial Discharges and Induced Charge. *IEEE Trans. Dielectr. Electr. Insul.* **1995**, *2*, 535–543. [CrossRef]
24. Gutfleisch, F.; Niemeyer, L. Measurement and Simulation of PD in Epoxy Voids. *IEEE Trans. Dielectr. Electr. Insul.* **1995**, *2*, 729–743. [CrossRef]
25. Achillides, Z.; Danikas, M.G.; Kyriakides, E. Partial Discharge Modeling and Induced Charge Concept: Comments and Criticism of Pedersen's Model and Associated Measured Transients. *IEEE Trans. Dielectr. Electr. Insul.* **2017**, *24*, 1118–1122. [CrossRef]
26. Achillides, Z.; Kyriakides, E.; Georghiou, G.E. Partial Discharge Modeling: An Improved Capacitive Model and Associated Transients along Medium Voltage Distribution Cables. *IEEE Trans. Dielectr. Electr. Insul.* **2013**, *20*, 770–781. [CrossRef]
27. Finis, G.; Claudi, A. On the Dielectric Breakdown Behavior of Silicone Gel Under Various Stress Conditions. *IEEE Trans. Dielectr. Electr. Insul.* **2007**, *14*, 487–494. [CrossRef]
28. Schifani, R.; Candela, R.; Romano, P. On PD Mechanisms at High Temperature in Voids Included in an Epoxy Resin. *IEEE Trans. Dielectr. Electr. Insul.* **2001**, *8*, 589–597. [CrossRef]
29. Illias, H.; Chen, G.; Lewin, P.L. Partial Discharge Behavior Within a Spherical Cavity in a Solid Dielectric Material as a Function of Frequency and Amplitude of the Applied Voltage. *IEEE Trans. Dielectr. Electr. Insul.* **2011**, *18*, 432–443. [CrossRef]
30. Forssen, C. Partial Discharges in Cylindrical Cavities at Variable Frequency of the Applied Voltage. Licentiate's Thesis, KTH Royal Institute of Technology, Stockholm, Sweden, 2005.
31. Borghei, M.; Ghassemi, M. Finite element modeling of partial discharge activity within a spherical cavity in a solid dielectric material under fast, repetitive voltage pulses. In Proceedings of the IEEE Electrical Insulation Conference (EIC), Calgary, AB, Canada, 16–19 June 2019; pp. 34–37.
32. Borghei, M.; Ghassemi, M. Partial discharge finite element analysis under fast, repetitive voltage pulses. In Proceedings of the IEEE Electric Ship Technologies Symposium (ESTS), Arlington, VA, USA, 13–19 August 2019; pp. 324–328.
33. Borghei, M.; Ghassemi, M.; Rodriguez-Serna, J.M.; Albarracin Sanchez, R. Finite Element Analysis and Induced Charge Concept Methods for Internal Partial Discharge Modeling: A Comparison. *IEEE Trans. Dielectr. Electr. Insul.* **2019**, in press.
34. Wang, P.; Cavallini, A.; Montanari, G.C. The Influence of Repetitive Square Wave Voltage Parameters on Enameled Wire Endurance. *IEEE Trans. Dielectr. Electr. Insul.* **2014**, *21*, 1276–1284. [CrossRef]
35. Ebke, T.; Khaddour, A.; Peier, D. Degradation of silicone gel by partial discharges due to different defects. In Proceedings of the International Conference on Dielectric Materials, Measurements and Applications, Edinburgh, UK, 17–21 September 2000; pp. 202–207.
36. Chung, J.; Hulbert, G.M. A Time Integration Algorithm for Structural Dynamics with Improved Numerical Dissipation: The Generalized-α Method. *J. Appl. Mech.* **1993**, *60*, 371–375. [CrossRef]

energies

MDPI

Article

A Novel Partial Discharge Detection Method Based on the Photoelectric Fusion Pattern in GIL

Yiming Zang [1], Yong Qian [1], Wei Liu [2], Yongpeng Xu [1,*], Gehao Sheng [1] and Xiuchen Jiang [1]

1 Department of Electrical Engineering, Shanghai Jiao Tong University, 800 Dongchuan Road, Minhang, Shanghai 200240, China; zangyiming@sjtu.edu.cn (Y.Z.); qian_yong@sjtu.edu.cn (Y.Q.); shenghe@sjtu.edu.cn (G.S.); xcjiang@sjtu.edu.cn (X.J.)

2 Key Laboratory for Sulfur Hexafluoride Gas Analysis and Purification of SGCC, Anhui Electric Power Research Institute of SGCC, Hefei 230022, China; sgccliu@163.com

* Correspondence: xyp3525@sjtu.edu.cn; Tel.: +86-155-0211-2557

Received: 25 September 2019; Accepted: 27 October 2019; Published: 28 October 2019

Abstract: Optical detection and ultrahigh frequency (UHF) detection are two significant methods of partial discharge (PD) detection in the gas-insulated transmission lines (GIL), however, there is a phenomenon of signals loss when using two types of detections to monitor PD signals of different defects, such as needle defect and free particle defect. This makes the optical and UHF signals not correspond strictly to the actual PD signals, and therefore the characteristic information of optical PD patterns and UHF PD patterns is incomplete which reduces the accuracy of the pattern recognition. Therefore, an image fusion algorithm based on improved non-subsampled contourlet transform (NSCT) is proposed in this study. The optical pattern is fused with the UHF pattern to achieve the complementarity of the two detection methods, avoiding the PD signals loss of different defects. By constructing the experimental platform of optical-UHF integrated detection for GIL, phase-resolved partial discharge (PRPD) patterns of three defects were obtained. After that, the image fusion algorithm based on the local entropy and the phase congruency was used to produce the photoelectric fusion PD pattern. Before the pattern recognition, 28 characteristic parameters are extracted from the photoelectric fusion pattern, and then the dimension of the feature space is reduced to eight by the principal component analysis. Finally, three kinds of classifiers, including the linear discriminant analysis (LDA), support vector machine (SVM), and k-nearest neighbor (KNN), are used for the pattern recognition. The results show that the recognition rate of all the photoelectric fusion pattern under different classifiers is higher than that of optical and UHF patterns, up to the maximum of 95%. Moreover, the photoelectric fusion pattern not only greatly improves the recognition rate of the needle defect and the free particle defect, but the recognition accuracy of the floating defect is also slightly improved.

Keywords: partial discharge; optical-UHF integrated detection; photoelectric fusion pattern; GIL; NSCT

1. Introduction

In recent years, gas-insulated transmission lines (GIL) are widely used in the power transmission of hydropower stations and nuclear power plants because of their high efficiency, large transmission capacity, high reliability, and small footprint [1–3].

In the operation of GIL, partial discharge (PD) is a precursor in the deterioration of insulation performance, which is the main cause of a breakdown. In addition, the severity of PD is closely related to the type of discharge defect. Therefore, PD detection and pattern recognition are particularly important in the GIL [4,5]. In order to improve the reliability of PD detection and pattern recognition, some scholars have proposed a method of combining the fluorescent fiber detection and the ultrahigh frequency (UHF) detection in the PD detection of gas-insulated equipment, which has the characteristics of high sensitivity, strong anti-interference ability, and wide application range [6].

Although optical-UHF integrated detection can effectively detect the occurrence of PD, optical detection and UHF detection have certain limitations for the pattern recognition of PD. UHF detection is limited by its detection bandwidth and is susceptible to influence from the external electromagnetic interference, resulting in the loss of UHF signals [7]. For optical detection, the propagation of the optical signal is affected by the structure of the GIL, the location of sensors, the distance of sensors, and the position of the PD source, which will weaken or even shield optical signals reaching the sensor. This can also result in the loss of optical signals [8]. Therefore, both UHF detection and optical detection have the phenomena of signals loss, which will lead to incomplete feature information in the PD pattern. Therefore, if the UHF pattern and the optical pattern are separately used for pattern recognition in the optical-UHF integrated detection, both detections will be affected by the pattern aliasing or the false pattern in the recognition process due to the missing information of the PD characteristic. These adverse effects will reduce the accuracy of PD pattern recognition.

In order to improve the accuracy of PD pattern recognition in the GIL, this study proposes an image fusion algorithm based on improved non-subsampled contourlet transform (NSCT) [9], which can gain the photoelectric fusion phase-resolved partial discharge (PRPD) pattern by fusing the optical PRPD pattern with the UHF PRPD pattern. By applying the photoelectric fusion PRPD pattern to the pattern recognition of PD, the two detection methods are complemented, avoiding the problem of missing PD characteristic information in a single type of the detection pattern.

Compared with the traditional NSCT method [10], the improved NSCT image fusion algorithm with the multiscale, multidirectional, and translation-invariant characteristics can retain more image edge and texture information, which can more completely contain the PD characteristic information of optical and UHF signals. In order to better characterize the PD characteristics of the PD pattern, 28 characteristic parameters, such as moment features and texture features, are extracted from the photoelectric fusion PD pattern and, then, the principal component analysis (PCA) method is used to reduce the feature vector space to eight characteristic parameters. After the process of PCA, linear discriminant analysis (LDA), support vector machine (SVM), and k-nearest neighbor (KNN) were used to verify the pattern recognition accuracy by photoelectric fusion patterns. Therefore, through the method of photoelectric image fusion, it provides a novel idea for the optical-UHF integrated detection of PD.

2. PD Experiment of GIL

2.1. Experimental Platform and Defect Model

The GIL experimental platform used in this study is shown in Figure 1. In order to isolate external electromagnetic interference, the whole experimental platform is located in the metal shielded room. In the platform, the GIL test tank is made of aluminum alloy, and the seal is good without light injection. The photoelectric integrated sensor that is composed of a fluorescent fiber intertwined on the cylindrical UHF detection is installed on the tank wall for PD signals acquisition. The detection frequency band of the UHF sensor is 300–500 MHz. The photomultiplier tube (PMT) adopts the HAMAMATSU-H10722-01 model whose corresponding range of spectrum is 230,920 nm, and the voltage/current conversion coefficient is 1V/μA. The function of PMT is to convert the collected optical signals into electrical signals, which is helpful for signal processing. The digital PD detector (Haffley DDX 9121b) collects standard PD signals as a reference to confirm the loss of optical signals and UHF signals. The oscilloscope uses LeCroy-HDO6000A [11].

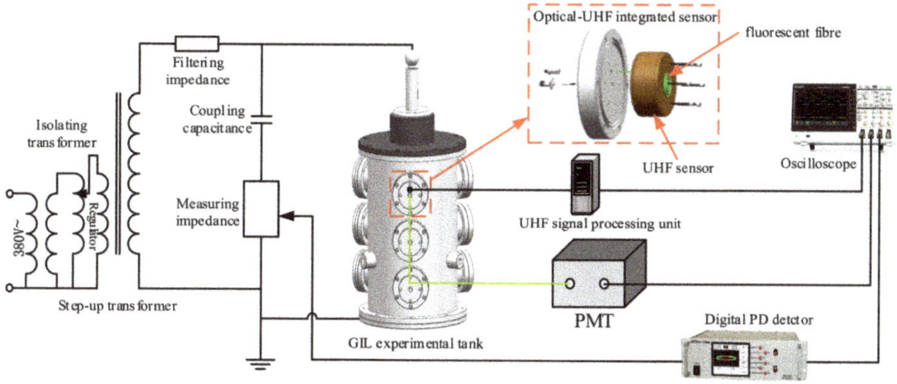

Figure 1. Partial discharge (PD) detection platform of gas-insulated transmission lines (GIL).

In order to simulate the PD defects in GIL, Figure 2 shows three aluminum typical defect models of the needle discharge defect, the floating discharge defect, and the free particle discharge defect designed in this study [12].

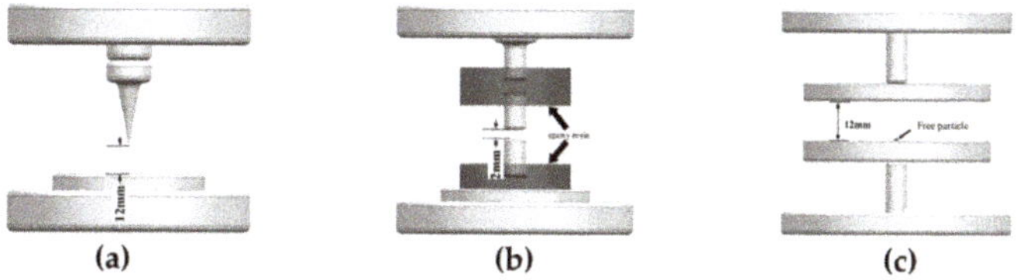

Figure 2. Schematic diagram of three insulation defect models. (**a**) The needle discharge defect, (**b**) the floating discharge defect, and (**c**) the free particle discharge defect.

2.2. Experimental Method

Before the experiment, PD defects and the tank were dedusted. The SF_6 gas was then filled into the GIL experimental tank that was sealed and vacuum until the gas pressure reached 0.2 MPa. According to the above operation, three typical PD defects were placed into the GIL experimental tank for PD experiments. During the experiment, the center of the optical-UHF integrated sensor was facing the defect on the level.

For the data acquisition, PD detection was performed for each defect at several voltage levels. For each defect, 120 detection samples were collected and each sample included 50 PD signals of a power frequency cycle.

2.3. Analysis of the Experimental Results

According to the time domain waveform of Figure 3, comparing the UHF signals and optical signals with the PD detector signals that act as the standard signals confirms that the signal loss is relative to the detection method and discharge defects, rather than external signal interference.

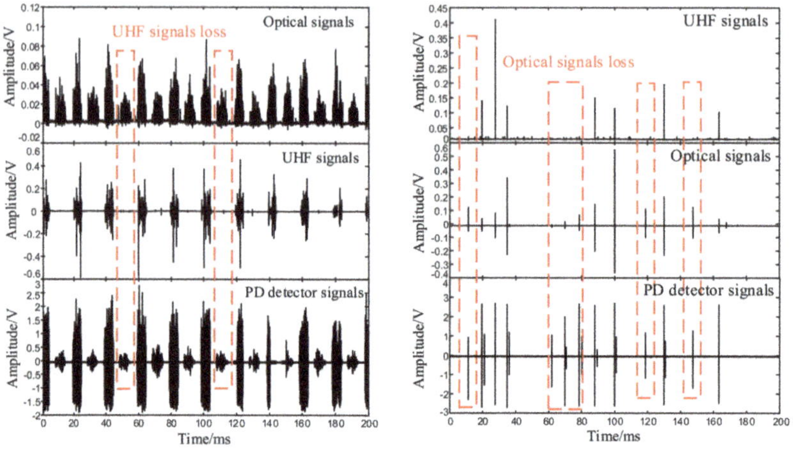

Figure 3. Optical and ultrahigh frequency (UHF) PD signals with different defects. (**a**) Time domain waveform of needle defect and (**b**) time domain waveform of free particle defect.

In the UHF detection, the sensitivity of the UHF detection is low due to the limitation of the detection band and the interference of the environmental noise. Thereby, it is shown that the PD detection of the needle discharge defect has a phenomenon of UHF signal loss in the experiment.

In the optical detection, the uncertainty of the free particle discharge is strong, and the intensity of the free particle discharge is not uniform. In addition, the optical detection is easily affected by the photon propagation path and the reflection condition of GIL inner wall. Therefore, the PD detection of the free particle discharge also has a phenomenon of optical signal loss.

In order to represent the characteristic information of PD more effectively, this study uses the two-dimensional PRPD pattern to describe the phase information of PD. The PRPD pattern is represented by a φ–u two-dimensional pattern, where φ represents the power frequency phase at which PD occurs and u represents the intensity of PD signals [13]. The color of the PRPD pattern represents the discharge density at certain phase and amplitude, which is shown by the color bar. Each PRPD pattern is drawn from a PD signal sample, including 50 PD cycles acquired by the oscilloscope above, which can guarantee that the loss of signal is not an accident.

For the needle defect, the PRPD pattern of the optical-UHF integrated sensor at 20 kV is shown in Figure 4. Optical signals are mainly concentrated near the peaks of positive and negative half cycles, while most UHF signals appear only in the area of the negative half cycle. It can be concluded that the detection sensitivity of the UHF sensor is lower than that of the optical sensor. When the PD intensity of the positive half cycle is small, the UHF sensor cannot detect the PD signals. Therefore, there is a phenomenon of missed UHF signals in the positive half cycle, which can cause the PD characteristic information to be incomplete and interfere with the PD pattern recognition.

For the floating defect, as shown in Figure 5, optical signals and UHF signals are mainly distributed near the peak of the positive and negative half cycles at 20 kV, which has good distribution characteristics of phase concentration. Therefore, the optical and UHF PD signals collected by the optical-UHF integrated sensor has good correspondence.

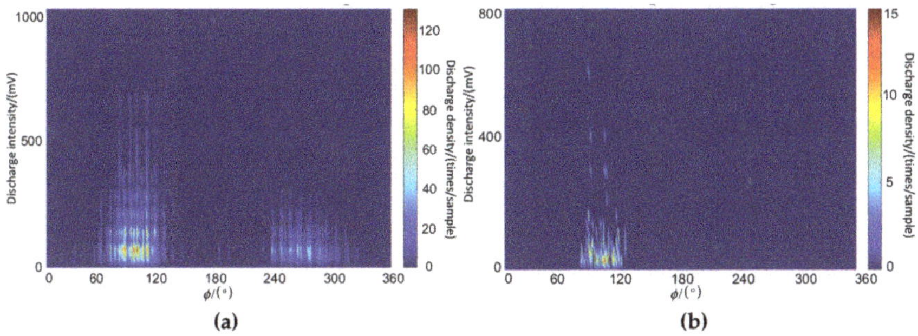

Figure 4. Phase-resolved partial discharge (PRPD) patterns of the needle defect under 20 kV. (**a**) A PRPD pattern of the optical detection and (**b**) a PRPD pattern of the UHF detection.

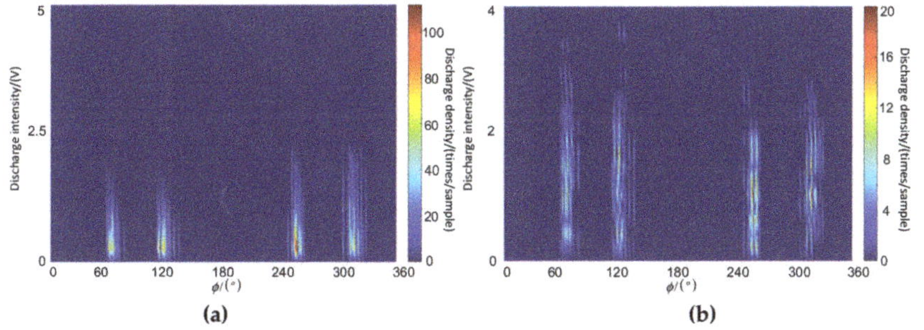

Figure 5. PRPD patterns of the floating defect under 20 kV. (**a**) A PRPD pattern of the optical detection and (**b**) a PRPD pattern of the UHF detection.

For the free particle defect, the PRPD pattern of the optical-UHF integrated sensor at 20 kV is shown in Figure 6. Under the action of the electric field force, the free metal particle undergo random collision movement between the plates and the discharge repetition rate of it is low, which causes the phase distribution of UHF signals to be relatively random in the PRPD pattern. It can be seen from the optical pattern that the randomness of the optical signal distribution is weak. The light spots in the optical pattern are mainly concentrated in the negative half cycle, instead of being distributing randomly. Therefore, the distribution of optical signals has a large difference from the UHF signal pattern. It is indicated that optical signals are attenuated by the occlusion and the reflection phenomenon of the propagation path during the propagation process, which causes optical signals to be missed and incompletely collected [8]. As a result, the loss of PD information leads to a failure to fully represent the PD characteristics of the free particle defect.

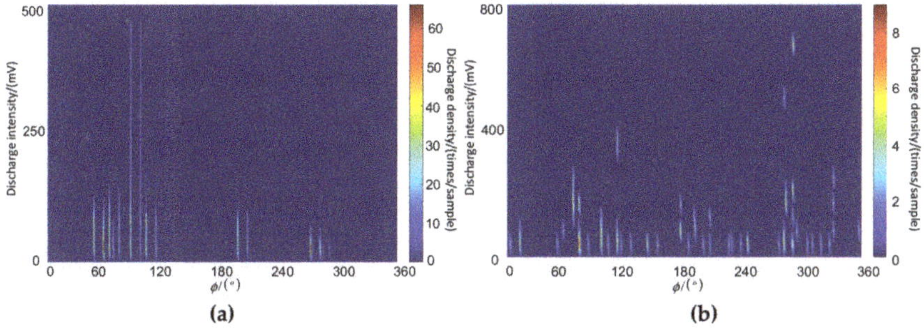

Figure 6. PRPD patterns of the free particle defect under 20 kV. (**a**) A PRPD pattern of the optical detection and (**b**) a PRPD pattern of the UHF detection.

The above experiments prove that the PD characteristic information of UHF and optical PRPD patterns is incomplete. The UHF sensor has the loss of signals on the needle defect, and the optical sensor has the loss of signals on the free particle defect. Therefore, if the two types (UHF patterns and optical patterns) of PRPD patterns are separately used for pattern recognition, the lack of the characteristic information may result in a decrease in recognition accuracy.

3. Image Fusion Algorithm Based on Improved NSCT

3.1. NSCT Structure

For PD patterns, the edge texture plays a key role in identifying the type of PD. Therefore, this study uses an image fusion algorithm based on the improved NSCT method to fuse the optical pattern with the UHF pattern. Its structure can be divided into two parts, non-subsampled pyramid filter banks (NSPFB) and non-subsampled directional filter bank (NSDFB). The structural framework of the NSCT is shown in Figure 7.

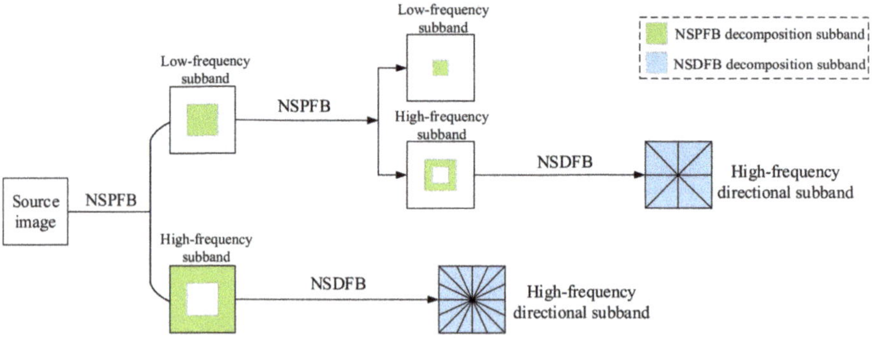

Figure 7. Structure block diagram of non-subsampled contourlet transform (NSCT) decomposition.

NSPFB is used as a filter that is up sampled. In order to achieve multiscale decomposition of the image, NSPFB performs up-sampling processing by the matrix $D = 2I$ iteratively, which can obtain the filter $H(Z^{2I})$. The NSPFB filters the low-frequency subband image of the upper level by the lowpass filter $H_0(Z^{2I})$ and the bandpass filter $H_1(Z^{2I})$, so that each level is decomposed into a low-frequency subband image and a high-frequency subgraph. The definition of the decomposition scale is *j*. In the filtering process, the ideal frequency domain of the lowpass filter at *j* scale is $[-\pi/2j, \pi/2j] \times [-\pi/2j, \pi/2j]$. The corresponding ideal frequency domain of the bandpass filtering is $[-\pi/2j + 1, \pi/2j + 1] \times [-\pi/2j-1, \pi/2j-1]$ [14]. Thus, after the image is decomposed by the *j*-level NSPFB, the image can obtain $j + 1$ subgraph of the same size as the original decomposition image, including one low-frequency subgraph and *j* high-frequency subgraph. Taking the three-level NSPFB decomposition as an example, its structure is shown in Figure 8 [15].

The NSDFB used in the NSCT is based on a fan-shaped filter bank. The two-channel directional filters, $U_0(Z)$ and $U_1(Z)$, with the fan-shaped frequency domain are up-sampled by the sampling matrix *D* to obtain filters $U_0(Z^D)$ and $U_1(Z^D)$. Then $U_0(Z^D)$ and $U_1(Z^D)$ are used to filter the subgraph decomposed in the upper level, which can achieve more accurate directional decomposition of the image in the image of corresponding frequency domain. As shown in Figure 9, taking the two-level directional decomposition as an example, the NSDFB decomposes the two-dimensional frequency domain into several wedge-shaped regions representing directionality [16]. Each wedge-shaped region contains detailed direction features of the image. Therefore, by performing *k*-level directional decomposition on the subgraph at one level of the NSPFB, it is possible to obtain a 2^k directional subgraph with the same size as the source image [17].

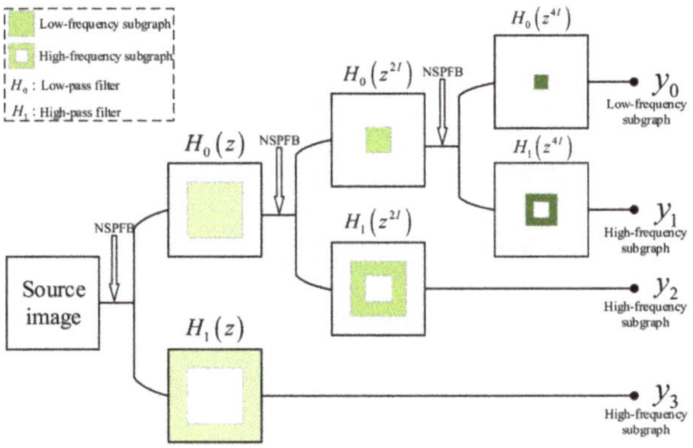

Figure 8. Three-level NSPFB pyramid filter bank.

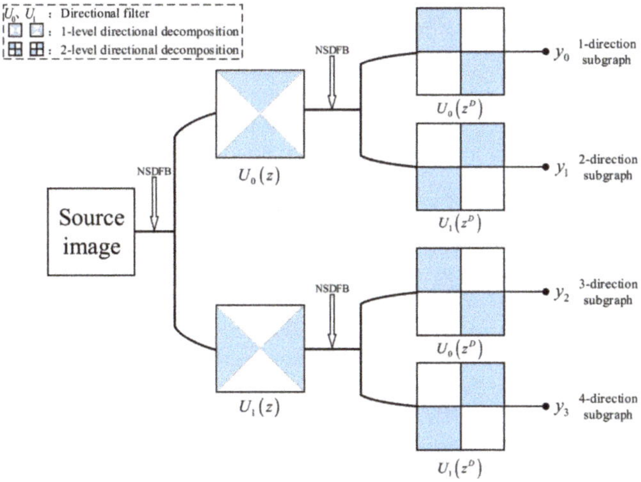

Figure 9. Two-level NSDFB directional filter bank.

3.2. Improved Image Fusion Rules Based on NSCT

According to the NSCT decomposition above, when k_j represents the NSDFB directional decomposition levels of the image on the jth-level of the J-level NSPFB decomposition, the number of subgraph generated by the decomposition can be expressed as $1 + \sum_{j=1}^{j} 2^{k_j}$, including one low-frequency subgraph and $\sum_{j=1}^{j} 2^{k_j}$ high-frequency subgraph. In order to ensure the anisotropy in the NSCT image fusion process, we change the k_j on each level of the NSPFB, which can make high-frequency subgraph on each level of the NSPFB have different directional decomposition. The structural flow of image fusion is shown in Figure 10 [18].

The source images A and B are subjected to grayscale processing before the NSCT conversion. By performing NSCT decomposition on the grayscale images of the source images A and B, the high-frequency subband coefficients, $G_{j,r}^{A}(x,y)$, $G_{j,r}^{B}(x,y)$, and the low-frequency subband coefficients, $L_J^A(x,y)$, $L_J^B(x,y)$, of each source image can be obtained. Among them, $j = (1,2,\ldots,J)$ is the number of decomposition levels of the NSPFB, r is the rth-direction of the NSDFB decomposition on the jth-level ($r = 1,2,\ldots,2^{k_j}$), and the subband coefficient represents the gray value at location (x,y).

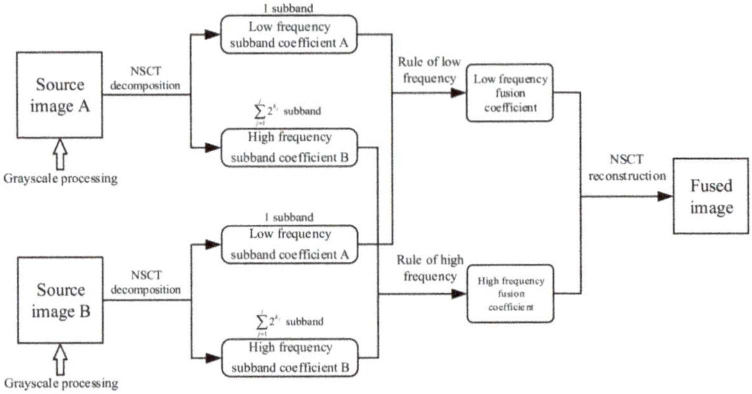

Figure 10. NSCT image fusion structure.

3.2.1. Fusion Rule of Low-Frequency Subgraph

After being decomposed by the NSCT, the contour information of the source image is mainly retained in the low-frequency subgraph. For the PRPD pattern, the contour of PD signals distribution is especially important for pattern recognition. Image fusion of the decomposed low-frequency subgraph is to preserve the contour feature information of the PRPD pattern as completely as possible. Therefore, we proposes a fusion rule of low-frequency subgraph, combining Canny operator and local entropy to better preserve the contour information of the PRPD pattern.

It is known that $L_J^A(x,y)$ and $L_J^B(x,y)$ are two low-frequency subgraphs to be fused, which are the same size. First, the low-frequency subgraph is edge-extracted by the Canny operator to obtain the edge contour binary graphs, $L_{J,Canny}^A(x,y)$ and $L_{J,Canny}^B(x,y)$. Thereby, the contour information of the signal distribution in the low-frequency subgraph is better preserved during the fusion process, which reduces the influence of image texture and sparseness.

In this study, we introduce the concept of image local entropy [19]. The local entropy can reflect the extent of gray dispersion of the image. In the image region with large local entropy, the gray value of this certain region is relatively uniform and contains less feature information. In the image region with small local entropy, the grayscale difference of this certain region is large and contains more feature information. Therefore, the local entropy is larger in the smooth region of the PRPD pattern, while the local entropy is smaller in the boundary contour region of the signal distribution for the PRPD pattern.

$f(x,y)$ is defined as the gray value of location (x,y) in the image, then, an image of size $X \times Y$ whose local entropy $H_{f(x,y)}$ is defined as [20]:

$$H_{f(x,y)} = \sum_{x=1}^{X}\sum_{y=1}^{Y} p_{xy} lg p_{xy} \tag{1}$$

where p_{xy} is the probability of gray distribution at location (x,y), and its expression is as follows:

$$p_{xy} = \frac{f(x,y)}{\sum_{x=1}^{X}\sum_{y=1}^{Y} f(x,y)} \tag{2}$$

Therefore, the fusion method of low-frequency coefficient adopted in this study is summarized as:

1) The local entropy $H_{f(x,y)}^A$ and $H_{f(x,y)}^B$ at location (x,y) of the edge contour binary patterns $L_{J,Canny}^A(x,y)$ and $L_{J,Canny}^B(x,y)$ are calculated by traversing the sampling window of size 3×3.

2) By comparing the magnitude of the local entropy at each location, it is determined how much the sample window contains image contour information. According to this, the fusion weight coefficients, $c_A(x,y)$ and $c_B(x,y)$, of the images, $L_J^A(x,y)$ and $L_J^B(x,y)$, are calculated.

$$c_A(x,y) = \frac{H^A_{f(x,y)}}{H^A_{f(x,y)} + H^B_{f(x,y)}} \tag{3}$$

$$c_B(x,y) = \frac{H^B_{f(x,y)}}{H^A_{f(x,y)} + H^B_{f(x,y)}} \tag{4}$$

3) According to the local entropy of the image and the fusion weight coefficient, the fused low-frequency subgraph $L_J^{fusion}(x,y)$ is calculated. The fusion rules are as follows:

$$L_J^{fusion}(x,y) = c_A \times L_J^A(x,y) + c_B \times L_J^B(x,y) \tag{5}$$

3.2.2. Fusion Rule of High-Frequency Subgraph

After the NSCT decomposition, the detailed texture information of the source image is mainly retained in the high-frequency subgraph, which represents the density of PD signals. Therefore, the key point of the high-frequency subgraph fusion is to enhance the image features, making the high-frequency subgraph more informative.

In this study, phase congruency (PC) is applied to the fusion rule of high-frequency coefficient. PC analyzes the feature points of the grayscale image from the perspective of the frequency domain. The theoretical basis is that the image is subjected to Fourier transform decomposition, and then the points with the most consistent phase of each harmonic component correspond to the feature point of the image [21]. Thus, PC can measure the importance of subgraph features with a dimensionless measurement.

In the fusion of high-frequency subgraphs, the PC value can represent the sharpness of high-frequency subgraphs. Because the subgraph can be regarded as a 2D signal [22], the PC value of the subgraph at location (x,y) can be calculated by Equation (6).

$$PC(x,y) = \frac{\sum_k E_{\theta_k}(x,y)}{\varepsilon + \sum_n \sum_k A_{n,\theta_k}(x,y)} \tag{6}$$

where A_{n,θ_k} is the amplitude of the n-th Fourier component and angle θ_k, θ_k denotes the orientation angle at k, and ε is a positive constant to offset the DC components of subgraph. In this study, the value of ε is set to 0.001 [23]. $E_{\theta_k}(x,y)$ can be calculated by Equation (7).

$$E_{\theta_k}(x,y) = \sqrt{F^2_{\theta_k}(x,y) + H^2_{\theta_k}(x,y)} \tag{7}$$

where $F_{\theta_k}(x,y)$ and $H_{\theta_k}(x,y)$ can be calculated by Equations (8) and (9) respectively.

$$F_{\theta_k}(x,y) = \sum_n e_{n,\theta_k}(x,y) \tag{8}$$

$$H_{\theta_k}(x,y) = \sum_n o_{n,\theta_k}(x,y) \tag{9}$$

where $e_{n,\theta_k}(x,y)$ and o_{n,θ_k} are convolution results of subgraph at location (x,y), which can be calculated by Equation (10).

$$\left[e_{n,\theta_k}(x,y), o_{n,\theta_k}(x,y)\right] = \left[I(x,y) \times M_n^e, I(x,y) \times M_n^o\right] \tag{10}$$

where $I(x,y)$ denotes the pixel value of subgraph at location (x,y). M_n^e and M_n^o represent the even- and odd-symmetric filters of 2D log-Gabor at scale n [23].

PC is a contrast invariant, which cannot reflect local contrast changes [22]. In order to compensate for the lack of PC, a measure of sharpness change (SCM) is introduced as below:

$$SCM(x,y) = \sum_{(x_0,y_0)\in\Omega_0} (I(x,y) - I(x_0,y_0))^2 \tag{11}$$

where Ω_0 denotes a local window with a size of 3×3 that is entered at (x,y). (x_0,y_0) is a pixel point in the local window of Ω_0. In addition, the local SCM (LSCM) is expressed as Equation (12) to determinate the (x,y) neighborhood contrast.

$$LSCM(x,y) = \sum_{a=-M}^{M}\sum_{b=-N}^{N} SCM(x+a,y+b) \tag{12}$$

where $(2M+1)\times(2N+1)$ represents the neighborhood size.

Since PC and LSCM cannot completely reflect the local luminance information, the local energy (LE) is proposed as below.

$$LE(x,y) = \sum_{a=-M}^{M}\sum_{b=-N}^{N} (I(x+a,y+b))^2 \tag{13}$$

Therefore, according to the theory mentioned above, a new activity measure (NAM) is defined using PC, LSCM, and LE to measure various aspects of subgraph information.

$$NAM(x,y) = (PC(x,y))^{\alpha_1}\cdot(LSCM(x,y))^{\beta_1}\cdot(LE(x,y))^{\gamma_1} \tag{14}$$

where α_1, β_1, γ_1 are set to 1, 2, and 2 respectively, which are used to adjust the value of PC, LSCM, and LE in NAM [24].

After the NAM is obtained, the fused high-frequency image can be determined by the Equation (15).

$$H_j^{fusion}(x,y) = \begin{cases} H_A(x,y) & if Lmap_A(x,y) = 1 \\ H_B(x,y) & otherwise \end{cases} \tag{15}$$

where the $H_j^{fusion}(x,y)$ represents the high-frequency fused subgraph of jth-level, $H_A(x,y)$ and $H_B(x,y)$ are high-frequency subgraphs of source image A and B. $Lmap_i(x,y)$ is a decision map for the fusion of high-frequency subgraph, which can be calculated by Equation (16).

$$Lmap_i(x,y) = \begin{cases} 1 & if \lceil S_i(x,y)\rceil > \frac{\widetilde{M}\times\widetilde{N}}{2} \\ 0 & otherwise \end{cases} \tag{16}$$

where the $S_i(x,y)$ is calculated by Equation (17).

$$\begin{aligned} S_i(x,y) = & \{(x_0,y_0)\in\Omega_1 | NAM_i(x_0,y_0) \geq \max(NAM_1(x_0,y_0),\cdots, \\ & NAM_{i-1}(x_0,y_0), NAM_{i+1}(x_0,y_0),\cdots, NAM_K(x_0,y_0))\} \end{aligned} \tag{17}$$

where Ω_1 denotes a sliding window with a size of $\widetilde{M}\times\widetilde{N}$, and (x,y) is the center of it. K is the number of source images.

According to the low-frequency and high-frequency subgraph fusion method mentioned above, the fused low-frequency and high-frequency NSCT coefficients $L_J^{fusion}(x,y)$ and $H_j^{fusion}(x,y)$ can be obtained. Then the fused fusion image F can be reconstructed by inverse NSCT transformation.

4. Photoelectric Image Fusion PD Detection Based on Improved NSCT

4.1. Overall Detection Process

In order to improve the accuracy rate of the pattern recognition, we propose an improved NSCT image fusion algorithm for the problem of missing PD signals in optical and UHF detections, that is, NSCT is used to decompose the grayscale optical and UHF PRPD patterns into low-frequency subgraphs and high-frequency subgraphs accordingly. The above fusion method is used to fuse the optical pattern with the UHF pattern to obtain the photoelectric fusion PD pattern. This pattern is then subjected to a series of processing such as feature extraction, dimensionality reduction, and pattern recognition. The overall process of the test verification is shown in Figure 11.

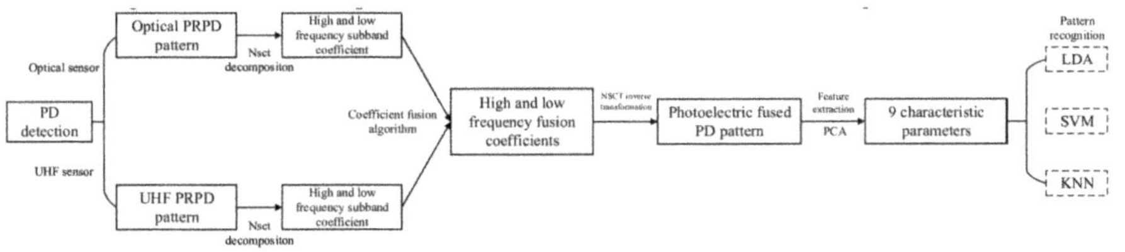

Figure 11. Experiment verification of the overall process.

4.2. Decomposition of PD Patterns Based on Improved NSCT

On the basis of the GIL PD experimental data mentioned above, the PD of the needle defect at 20 kV is taken as an example to perform NSCT decomposition and the fusion of the PD optical pattern and the PD UHF pattern.

In the NSCT decomposition of the PRPD pattern, one low-frequency subgraph and three high-frequency subgraphs can be obtained after the three-level NSPFB decomposition. In order to ensure the anisotropy of the image decomposition and preserve the information of the image in all directions more completely, the high-frequency subgraphs on the NSPFB decomposition of the fjirst-, second- and third-level are decomposed by the NSDFB directional decomposition with 2^1-, 2^2-, and 2^3-direction, respectively. Therefore, each high-frequency subgraph can be decomposed into 2^1-, 2^2-, and 2^3-direction subgraphs.

As a result, $1 + \sum_{k=1}^{3} 2^k$ subgraphs that are equal in size to the original image can be obtained after the optical pattern and the UHF pattern are decomposed by the NSCT, respectively, as shown in Figures 12 and 13. It can be seen that the PD pattern is decomposed into multiscale, multidirectional high-frequency subgraphs and low-frequency subgraphs, which can retain the contour and detailed information of the source image well.

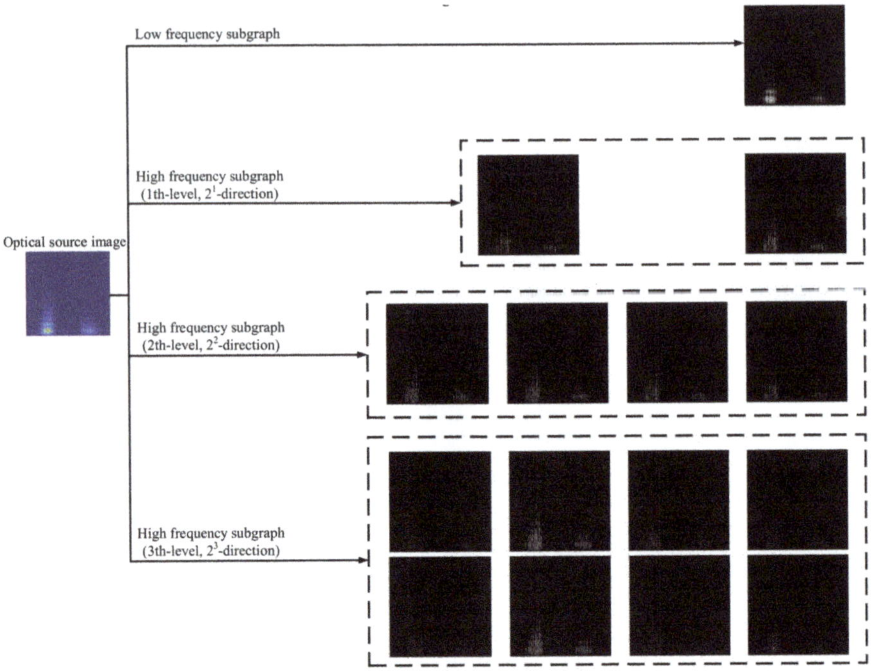

Figure 12. NSCT decomposition of the optical PD pattern.

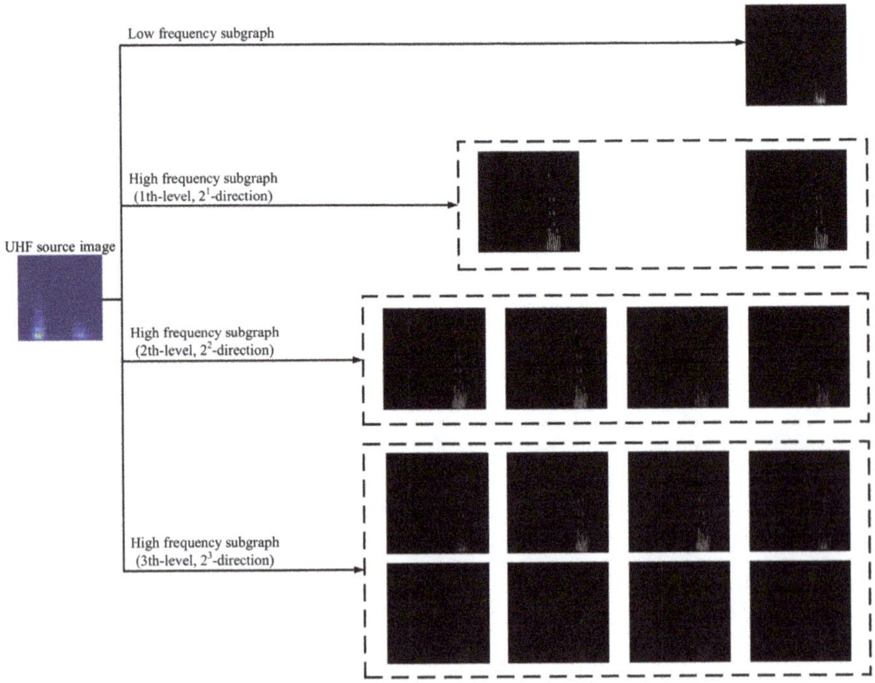

Figure 13. NSCT decomposition of the UHF PD pattern.

4.3. Fusion of the Photoelectric PD Pattern

According to the image fusion rule in Section 3.2., the corresponding fusion rule is performed on each of the low-frequency subgraph and the high-frequency subgraph, respectively, which can obtain one low-frequency photoelectric fusion subgraph and $\sum_{k=1}^{3} 2^k$ high-frequency photoelectric fusion subgraphs, as shown in Figure 14. The inverse NSCT transform is performed on the fused photoelectric subgraph to obtain a photoelectric fusion PD pattern, as shown in Figure 15.

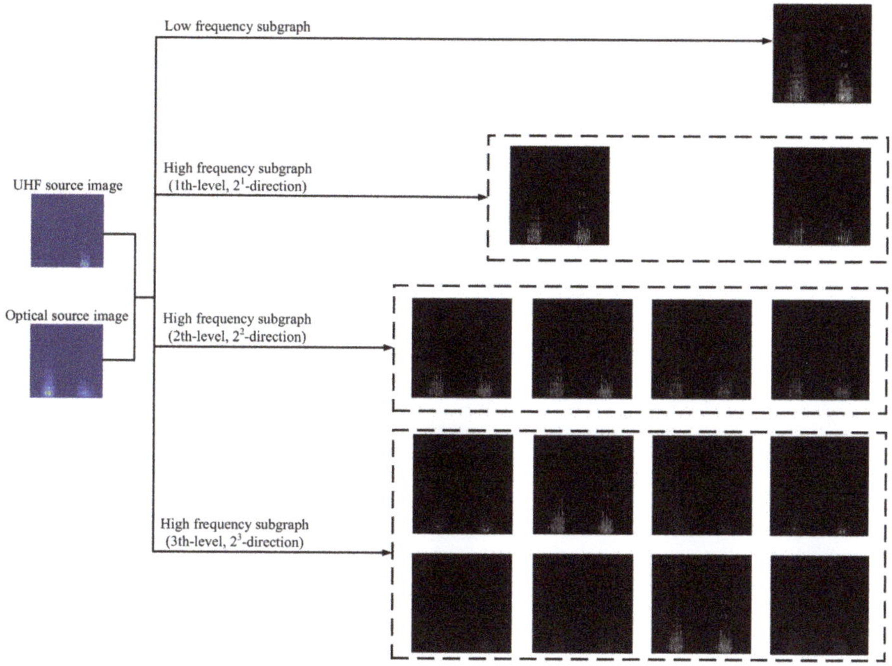

Figure 14. Fusion framework of optical and UHF patterns.

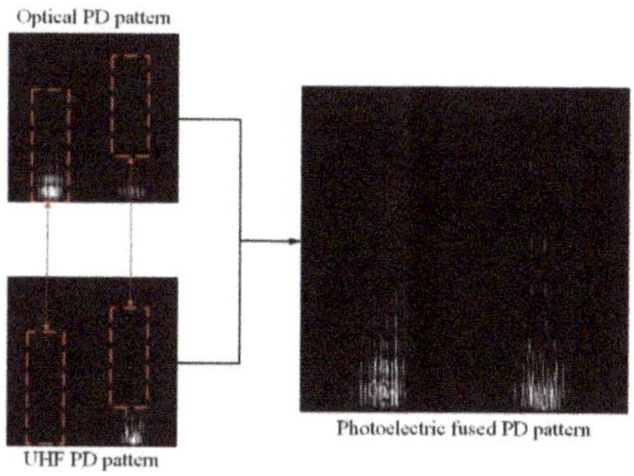

Figure 15. Photoelectric fusion PD pattern.

4.4. Feature Extraction and Dimension Reduction

In order to effectively reflect the image features, the feature parameters of the photoelectric fusion PD pattern are extracted. In this study, the eigenvector space is formed by Tamura texture features, gray-gradient symbiotic matrix, Hu invariant moment, and color moment of the image, a total of 28 characteristic parameters.

Among them, Tamura texture features theoretically include six components. But according to the Tamura feature extraction results of photoelectric fusion images, the characteristic parameters of Tamura texture features that are insensitive to photoelectric fusion images are ignored, and only the roughness, contrast, and directionality are used as features of pattern recognition. Then, 15 gray-gradient symbiotic matrix features, seven texture features, and three color moment features are, respectively, extracted to form the eigenvector space of the photoelectric fusion PD pattern [25–28].

However, there may be overlapping information between different features, resulting in multiple collinearity between the feature parameters. Meanwhile, too many dimensions of the feature vector can easily stress the training of the model, resulting in a lower recognition rate.

Therefore, in order to more fully characterize the characteristic information of the PD pattern and reduce the burden of the model, we use the principal component analysis (PCA) method to reduce the dimension of the eigenvector space. First, the factor correlation analysis of eigenvector space is carried out, gaining the value of KMO: 0.8356. The value of Bartlett spherical test is 132.96. It can be seen that there is strong partial correlation between feature vectors, which is suitable for dimensionality reduction by PCA [29]. In this study, according to the contribution rate of the feature factors, the first eight principal component factors with the cumulative contribution rate of 98% are selected as the input parameters of the recognition model. The PCA results are shown in Table 1.

Table 1. Contribution rate of principal component factors.

Factor Number	1	2	3	4	5	6	7	8	9	10	...	28
Contribution rate/%	65.00	11.46	7.78	6.01	3.02	2.21	1.58	0.86	0.64	0.58	...	0.01
Cumulated contribution rate/%	65.00	76.46	84.24	90.25	93.27	95.48	97.06	97.92	98.56	99.14	...	100

4.5. Pattern Recognition Results of Different PD Patterns

On the basis of the principal component factors described above, different classifiers are used to identify the PD experimental data. The classifiers used, in this study, were linear discriminant analysis (LDA) [30], k-nearest neighbor (KNN) [31], and support vector machine (SVM) algorithm [32]. LDA is a dimensionality reduction technology for supervised learning, which projects the sample onto a sorting line determining the category of the new sample based on the position of the projected point. KNN is a classification and regression method, which determines the classification by calculating which category of k-nearest samples in the feature space of a sample most belongs. The SVM algorithm maps points in low-dimensional space to high-dimensional spaces making them linearly separable, and then classifying them by the principle of linear partitioning.

In order to verify the applicability of the photoelectric fusion PD pattern proposed in this study, the above three classifiers were used to test the photoelectric fusion PD pattern samples, comparing with the recognition results of optical patterns and UHF patterns. At the same time, in order to verify the influence of different training sample numbers, four pattern recognition tests were carried out, respectively. In these four tests, the total number of training samples was 300, 240, 180, and 120, respectively, including all three types of defects. Therefore, the number of testing samples in these four tests were 60, 120, 180, and 240 correspondingly, including all three types of defects as well. It can be seen from the recognition results that different types of PD patterns have a significant influence on the recognition rate, as shown in Figure 16. The definition of recognition rate is:

$$Recognition\ rate = \frac{sample\ number\ of\ correct\ recognition}{number\ of\ testing\ sample}$$

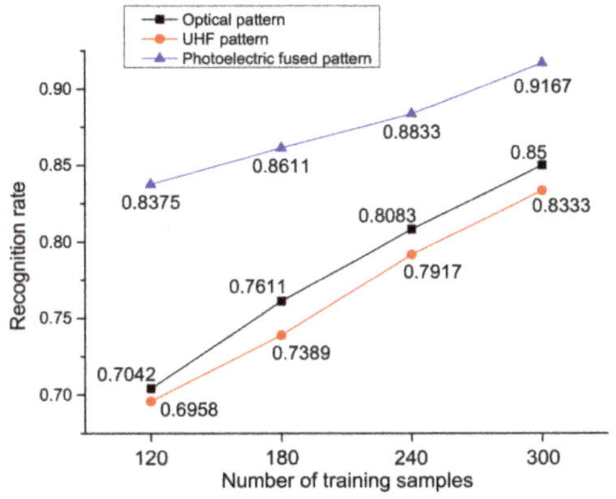

(a)

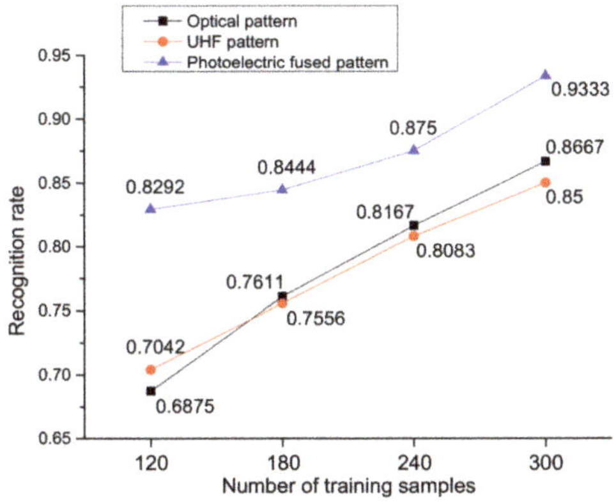

(b)

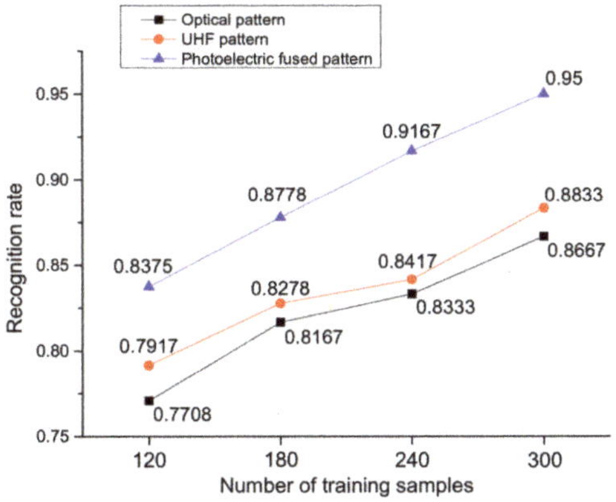

(c)

Figure 16. Recognition results of different classifiers. (**a**) linear discriminant analysis (LDA) classifier, (**b**) k-nearest neighbor (KNN) classifier, and (**c**) support vector machine (SVM) classifier.

According to Figure 16, the recognition rate of the photoelectric fusion pattern is the highest using any classifier. For all three types of PD patterns, the recognition rate increases with the increase of training samples. When using the SVM classifier with 300 training samples, the recognition rate of the photoelectric fusion pattern can reach up to 0.95. Moreover, the recognition rate of the photoelectric fusion pattern can still reach about 0.83 when the number of samples is only 120. Therefore, it can be concluded that the photoelectric fusion pattern can significantly improve the accuracy of PD pattern recognition.

In addition, when the number of training samples is the same, the accuracy of the recognition of the photoelectric fusion pattern by the three classifiers is higher than that of the optical pattern and the UHF pattern. Because photoelectric fusion pattern proposed in this study contains more abundant PD characteristic information, it can effectively improve the accuracy of the recognition. When the number of training samples is only 120, the average recognition rate of the three classifiers in each case is calculated in Figure 17, where three types of PD defects are, respectively, identified by different patterns. It can be seen that the recognition rate of the needle defect by the UHF pattern is especially low because of the loss of UHF signals. Moreover, the recognition rate of the free particle defect by the optical pattern is lower than the others because of the loss of optical signals. However, pattern recognition using photoelectric fusion pattern not only greatly increases the recognition rate of the needle defect and the free particle defect, but also slightly improves the recognition rate of the floating defect. Therefore, the proposed photoelectric fusion pattern has practicality.

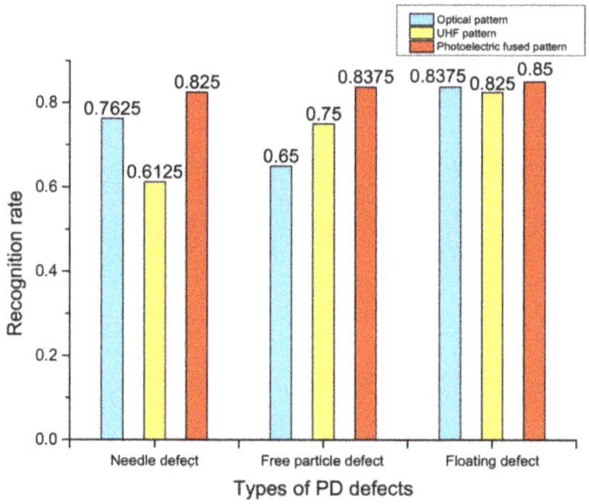

Figure 17. Average recognition rate of each defect under different kinds of patterns (with 120 training samples).

5. Conclusions

In this study, the optical-UHF integrated PD detection of GIL experimental platform is built, collecting PD patterns of three typical PD defects. Aiming at the PD pattern recognition problem of the GIL, an image fusion algorithm based on NSCT is proposed, which can fuse the optical PD pattern with the UHF PD pattern. Finally, three different classifiers are used to identify the photoelectric fusion PD pattern, which can verify the feasibility of this method. The conclusions are summarized as follows:

1) Due to the limitation of the PD detection principle and the influence of the GIL structure, the UHF pattern of the needle PD defect and the optical pattern of the free-particle PD defect have the loss of PD signals in the GIL. This phenomenon results in the reduction of the effective characteristic information in the PD pattern, which can reduce the pattern recognition accuracy of the PD.
2) The photoelectric fusion pattern can effectively avoid signals loss of UHF detection and optical detection in some situations, which can reduce the negative influence of false mode and pattern

aliasing on recognition. Through the photoelectric fusion pattern, the characteristic information of optical patterns and UHF patterns can complement each other, improving the accuracy and reliability of PD pattern recognition.

3) Compared with the optical pattern and the UHF pattern, the photoelectric fusion pattern can significantly improve the recognition rate of PD pattern recognition under the three kinds of classifiers, which can reach up to 0.95. In addition, when the number of training samples is small, the recognition rate can still reach about 0.83. Furthermore, the photoelectric fusion pattern not only greatly improves the recognition rate of the needle defect and the free particle defect, but the recognition accuracy of the floating defect can also be slightly improved. Therefore, the photoelectric fusion pattern has a good application effect.

Author Contributions: Data curation, Y.Z.; formal analysis, Y.Z.; funding acquisition, W.L.; methodology, Y.Z.; supervision, Y.Q., Y.X., G.S., and X.J.; writing—original draft, Y.Z.

Funding: This work was supported by the National Key R & D Program of China (2017YFB0902500) and the State Grid Science & Technology Project (the key technology of environment-friendly gas-insulated transmission line).

Conflicts of Interest: The authors declare no conflict of interest.

References

1. Volcker, O.; Koch, H. Insulation co-ordination for gas-insulated transmission lines (GIL). *IEEE Trans. Power Deliv.* **2001**, *16*, 122–130. [CrossRef]
2. Magier, T.; Tenzer, M.; Koch, H. Direct Current Gas-Insulated Transmission Lines. *IEEE Trans. Power Deliv.* **2018**, *33*, 440–446. [CrossRef]
3. Miyazaki, A.; Takinami, N.; Kobayashi, S.; Yamauchi, T.; Hama, H.; Araki, T.; Nishima, H.; Hata, H.; Yamaguchi, H. Line constant measurements and loading current test in long-distance 275 kV GIL. *IEEE Trans. Power Deliv.* **2001**, *16*, 165–170. [CrossRef]
4. Schoffner, G. In A directional coupler system for the direction sensitive measurement of UHF-PD signals in GIS and GIL. In Proceedings of the 2000 Annual Report Conference on Electrical Insulation and Dielectric Phenomena (Cat. No.00CH37132), Victoria, BC, Canada, 15–18 October 2000; pp. 634–638.
5. Okubo, H.; Yoshida, M.; Takahashi, T.; Hoshino, T.; Hikita, M.; Miyazaki, A. Partial discharge measurement in a long distance SF/sub 6/gas insulated transmission line (GIL). *IEEE Trans. Power Deliv.* **1998**, *13*, 683–690. [CrossRef]
6. Li, J.; Han, X.; Liu, Z.; Yao, X. A Novel GIS Partial Discharge Detection Sensor With Integrated Optical and UHF Methods. *IEEE Trans. Power Deliv.* **2018**, *33*, 2047–2049. [CrossRef]
7. Han, X.; Li, J.; Zhang, L.; Pang, P.; Shen, S. A Novel PD Detection Technique for Use in GIS Based on a Combination of UHF and Optical Sensors. *IEEE Trans. Instrum. Meas.* **2019**, *68*, 2890–2897. [CrossRef]
8. Yongpeng, X.; Yong, Q.; Gehao, S.; Xiuchen, J.; Xiaoli, Z.; Zijie, W. Simulation analysis on the propagation of the optical partial discharge signal in I-shaped and L-shaped GILs. *IEEE Trans. Dielectr. Electr. Insul.* **2018**, *25*, 1421–1428. [CrossRef]
9. Cunha, A.L.D.; Zhou, J.; Do, M.N. The Nonsubsampled Contourlet Transform: Theory, Design, and Applications. *IEEE Trans. Image Process.* **2006**, *15*, 3089–3101. [CrossRef]
10. Wang, X.; Song, R.; Song, C.; Tao, J. The NSCT-HMT Model of Remote Sensing Image Based on Gaussian-Cauchy Mixture Distribution. *IEEE Access* **2018**, *6*, 66007–66019. [CrossRef]
11. Hanai, M.; Kojima, H.; Hayakawa, N.; Mizuno, R.; Okubo, H. Technique for discriminating the type of PD in SF6 gas using the UHF method and the PD current with a metallic particle. *IEEE Trans. Dielectr. Electr. Insul.* **2014**, *21*, 88–95. [CrossRef]
12. Ren, M.D.M.; Zhang, C.; Zhou, J. Partial Discharge Measurement under an Oscillating Switching Impulse: A Potential Supplement to the Conventional Insulation Examination in the Field. *Energies* **2016**, *9*, 623. [CrossRef]
13. Firuzi, K.; Vakilian, M.; Phung, B.T.; Blackburn, T.R. Partial Discharges Pattern Recognition of Transformer Defect Model by LBP & HOG Features. *IEEE Trans. Power Deliv.* **2019**, *34*, 542–550.

14. Mahyari, A.G.; Yazdi, M. Panchromatic and Multispectral Image Fusion Based on Maximization of Both Spectral and Spatial Similarities. *IEEE Trans. Geosci. Remote. Sens.* **2011**, *49*, 1976–1985. [CrossRef]
15. Yang, Y.; Tong, S.; Huang, S.; Lin, P. Multifocus Image Fusion Based on NSCT and Focused Area Detection. *IEEE Sens. J.* **2015**, *15*, 2824–2838.
16. Lei, T.; Feng, Z.; Zong-Gui, Z. In The nonsubsampled contourlet transform for image fusion. In Proceedings of the 2007 International Conference on Wavelet Analysis and Pattern Recognition, Beijing, China, 2–4 November 2007; pp. 305–310.
17. Bhatnagar, G.; Wu, Q.M.J.; Liu, Z. Directive Contrast Based Multimodal Medical Image Fusion in NSCT Domain. *IEEE Trans. Multimed.* **2013**, *15*, 1014–1024. [CrossRef]
18. Bhateja, V.; Patel, H.; Krishn, A.; Sahu, A.; Lay-Ekuakille, A. Multimodal Medical Image Sensor Fusion Framework Using Cascade of Wavelet and Contourlet Transform Domains. *IEEE Sens. J.* **2015**, *15*, 6783–6790. [CrossRef]
19. Shao, L.; Kirenko, I. Coding Artifact Reduction Based on Local Entropy Analysis. *IEEE Tran. Consum. Electron.* **2007**, *53*, 691–696. [CrossRef]
20. Jia-Shu, Z.; Cun-Jian, C. In Local variance projection log energy entropy features for illumination robust face recognition. In Proceedings of the 2008 International Symposium on Biometrics and Security Technologies, Islamabad, Pakistan, 23–24 April 2008; pp. 1–5.
21. Zhang, L.; Zhang, L.; Mou, X.; Zhang, D. FSIM: A Feature Similarity Index for Image Quality Assessment. *IEEE Trans. Image Process.* **2011**, *20*, 2378–2386. [CrossRef]
22. Li, H.; Qiu, H.; Yu, Z.; Zhang, Y. Infrared and visible image fusion scheme based on NSCT and low-level visual features. *Infrared Phys. Technol.* **2016**, *76*, 174–184. [CrossRef]
23. Zhu, Z.; Zheng, M.; Qi, G.; Wang, D.; Xiang, Y. A Phase Congruency and Local Laplacian Energy Based Multi-Modality Medical Image Fusion Method in NSCT Domain. *IEEE Access* **2019**, *7*, 20811–20824. [CrossRef]
24. Qu, X.-B.; Yan, J.-W.; Xiao, H.-Z.; Zhu, Z.-Q. Image Fusion Algorithm Based on Spatial Frequency-Motivated Pulse Coupled Neural Networks in Nonsubsampled Contourlet Transform Domain. *Acta Autom. Sin.* **2008**, *34*, 1508–1514. [CrossRef]
25. Amadasun, M.; King, R. Textural features corresponding to textural properties. *IEEE Trans. Syst. Man Cybern.* **1989**, *19*, 1264–1274. [CrossRef]
26. Tamura, H.; Mori, S.; Yamawaki, T. Textural Features Corresponding to Visual Perception. *IEEE Trans. Syst. Man Cybern.* **1978**, *8*, 460–473.
27. Weng, T.; Yuan, Y.; Shen, L.; Zhao, Y. In Clothing image retrieval using color moment. In Proceedings of the 2013 3rd International Conference on Computer Science and Network Technology, Dalian, China, 12–13 October 2013; pp. 1016–1020.
28. Soh, L.; Tsatsoulis, C. Texture analysis of SAR sea ice imagery using gray level co-occurrence matrices. *IEEE Trans. Geosci. Remote Sens.* **1999**, *37*, 780–795. [CrossRef]
29. Gonzalez-Audicana, M.; Saleta, J.L.; Catalan, R.G.; Garcia, R. Fusion of multispectral and panchromatic images using improved IHS and PCA mergers based on wavelet decomposition. *IEEE Trans. Geosci. Remote Sens.* **2004**, *4*, 1291–1299. [CrossRef]
30. Wei-Shi, Z.; Jian-Huang, L.; Yuen, P.C. GA-fisher: A new LDA-based face recognition algorithm with selection of principal components. *IEEE Trans. Syst. Man Cybern. Part B* **2005**, *35*, 1065–1078.
31. Xiong, J.; Zhang, Q.; Sun, G.; Zhu, X.; Liu, M.; Li, Z. An Information Fusion Fault Diagnosis Method Based on Dimensionless Indicators With Static Discounting Factor and KNN. *IEEE Sens. J.* **2016**, *16*, 2060–2069. [CrossRef]
32. Laufer, S.; Rubinsky, B. Tissue Characterization With an Electrical Spectroscopy SVM Classifier. *IEEE Trans. Biomed. Eng.* **2009**, *5*, 525–528. [CrossRef]

energies

MDPI

Article

Simulation of Partial Discharge Induced EM Waves Using FDTD Method—A Parametric Study †

Alaa Loubani, Noureddine Harid *, Huw Griffiths and Braham Barkat

APEC Centre, Khalifa University, Abu Dhabi 127788, UAE

* Correspondence: noureddine.harid@ku.ac.ae

† This paper is an extended version of our paper published in 2018 IEEE International Conference on High-Voltage Engineering and Applications (ICHVE 2018), Athens, Greece, 10–13 September 2018, doi:10.1109/ICHVE.2018.8642074.

Received: 14 July 2019; Accepted: 2 August 2019; Published: 1 September 2019

Abstract: This paper reports the results of a parametric study on the characteristics of electromagnetic (EM) waves propagated due to surface- and cavity-type partial discharges (PD) in materials using the finite-difference time domain (FDTD) method. First, the EM waves emitted by such discharges in material samples were measured using a broadband aperture antenna. The measurements showed that the frequency range of the measured signals lay within the ultra-high frequency (UHF) range, suggesting that by carefully choosing the UHF antenna characteristics and its location it might be possible to apply this method to characterize the PD-emitted waves; and hence, to potentially use it to detect and monitor PD defects. In this context, the FDTD simulations were used here to simulate the experimental set-up and examine the propagation characteristics of EM waves emitted by such discharges under uniform and non-uniform test electrode configurations. Using an approximation of the exciting PD current pulses, the electromagnetic field components and the voltage signals captured on a simulated monopole sensor were computed in the time domain at various locations. To explore the limits of the application of the UHF method for detecting these PD types, a parametric study was carried out to clarify how the captured signals are influenced by the PD intensity, the frequency content of the exciting PD pulse, the type of insulation material, the dimensions and the position of the UHF antenna. One of the challenges that needs further investigation is the accurate simulation of the actual PD current pulse produced by such discharges, and hence its frequency content, as there is limited or no measured data available. The results showed that while the amplitude of the captured EM signals increase with the PD intensity, no appreciable signal is detected when the PD pulse width is higher than about 4ns, which may not occur often in unbounded air insulated systems. Equally important is the location and orientation of the UHF sensor—the results showed improved sensitivity when the sensor is vertically polarized and placed in close proximity in the lateral direction with reference to the discharge path.

Keywords: partial discharge; surface discharge; UHF sensor; FDTD simulation; cavity discharge

1. Introduction

Partial discharge (PD) is known to be one of the key factors affecting the operation of electric power equipment. Its detection has been, and continues to be a priority for utilities in asset management strategies for ensuring plant life longevity. PD occurs as a result of long-term operating electrical and environmental factors stressing equipment insulation and is characterized by small, high-frequency currents that are difficult to detect by standard substation instrumentation. The topic of PD detection and measurement has long attracted the interest of researchers, and current international standards [1] give recommendations for the measurement of different types of PD and help with the diagnosis of defects associated with them. One of the electrical methods that has been the focus of research interest is the UHF method, which is based on the measurement of electromagnetic waves generated by PD and propagating in the surrounding medium. A substantial amount of work has been published on this method, mainly with respect to PD detection in gas-insulation systems (GIS) [2–9]. Researchers have used a variety of sensors, in the form of discs [6] or broadband antennas to measure the PD-emitted signals [7–9]. However, there is limited research on using UHF sensors for the measurement of PD in systems other than GIS, for example, PD occurring inside solid insulation cavities, surface discharge in cable terminations, cable joints and outdoor insulation. Equally, theoretical simulation of PD using numerical techniques has been at the center of researchers' interest for many years. It has often been used to model the propagation characteristics of partial discharge in GIS [10–15], but some researchers have also used it to study PD phenomena in power transformers [16] and surface discharge in air [17]. It is also a useful tool for designing and calibrating UHF sensors for PD measurement [18,19].

In this paper, the characteristics of the EM signals emitted by surface discharge in air and a cavity discharge in different insulation samples were investigated using FDTD simulations. This work builds on an initial study carried out by the authors on surface discharge [17], and extends to include the analysis of EM waves emitted by cavity discharges occurring in insulation samples of different permittivities. A summary of the measured data obtained on real samples tested in laboratory test cells to simulate both types of discharge is presented. The detection of EM signals generated by PD in air using UHF sensors is particularly challenging considering the low-magnitudes of emitted signals that are embedded in a noisy environment and the relatively long pulse risetimes associated with them. These can occur, for example, on the surface of outdoor insulators and bushings, in cable joints and cable terminations and their detection depends on several parameters. In the present study, several parameters were varied: (i) those related to the discharge including the PD intensity, its pulse width and path size, and (ii) those related to the UHF sensor characteristics including its geometry, frequency band, sensitivity, and location with respect to the discharge source. A parametric study was performed where most of the above parameters were varied within practical ranges and their effect on the signals captured on a UHF sensor were analyzed. Square-shaped cross-linked polyethylene (XLPE), Teflon and Acrylic samples with different thicknesses placed between uniform and non-uniform field electrode configurations were considered. The results highlight, on one hand, some of the limits imposed by the discharge itself, and on the other hand, the required UHF sensor characteristics and location that give measurable signals.

2. PD Measurement Using the UHF Method

Partial discharge measurements were carried out to study the frequency characteristics of EM signals emitted by the discharge. For this purpose, two types of PD were generated: surface discharge on XLPE, Teflon and acrylic samples, and cavity discharge on XLPE samples of different thicknesses. The former was created under uniform and non-uniform field electrode configurations. The PD measurements were carried out using the IEC 60270 method, the 40MHz HFCT and the UHF method simultaneously. Figure 1 shows the experimental set-up, with the test cells used for the uniform and non-uniform electrode gaps showing material samples in the insert. The UHF signals were captured using a Schwarzbeck 1/4λ double ridge aperture antenna with a frequency band between 0.8GHz and 5 GHz, and a 120 MHz–900 MHz monopole antenna to capture the lower UHF range signals. This section summarises the frequency characteristics of the signals measured with the aperture antenna. The details of the test results for the other methods and their analyses will be reported separately. Figure 2a shows examples of measured time domain UHF signals generated by a surface discharge when a 9.2 kV AC voltage was applied across the non-uniform gap with a 6-mm thick Teflon sample in between. The UF sensor was vertically polarized (i.e., such that the E-field is parallel to the double ridge of the antenna) and located at two different horizontal distances and one vertical distance from the test object.

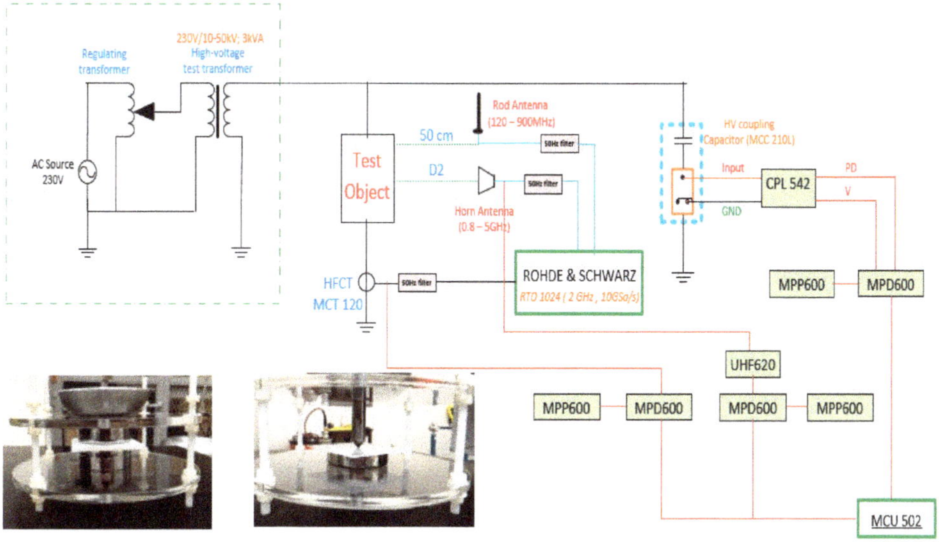

Figure 1. Experimental set-up for partial discharge (PD) measurement.

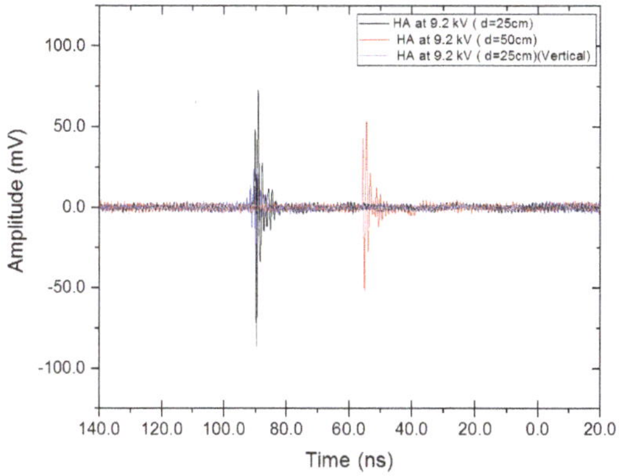

(a)

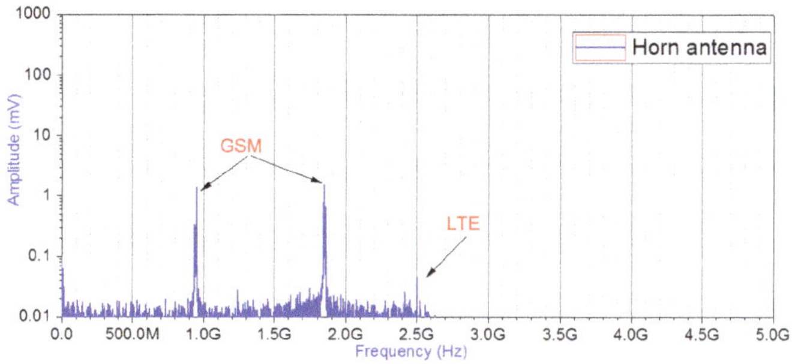

(b)

Figure 2. *Cont.*

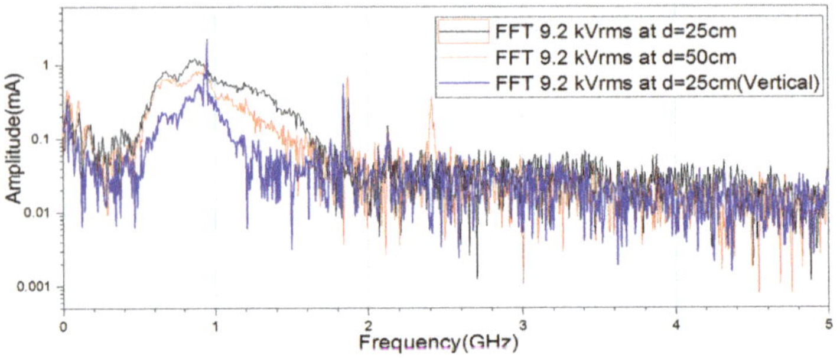

(c)

Figure 2. Examples of surface discharge signals measured using the ultra-high frequency (UHF) sensor: (**a**) time domain signals (**b**) noise level and (**c**) frequency spectra.

The frequency spectrum of the noise level (Figure 2b) and that of the measured signals shown in Figure 2c confirm the frequency range of the EM signals emitted by the surface discharge. Peaks detected at 0.9 GHz, 1.85 GHz and 2.5 GHz are superimposed GSM and long-term evolution (LTE) signals. The results show that the UHF detection can be improved by placing the sensor closer to the discharge source in a horizontal direction. Signals produced by a cavity discharge were measured by using a specially made XLPE block samples with a needle-plane electrode system and were also found to be within the UHF band.

3. Surface Discharge Simulation Using the FDTD Method

3.1. FDTD Principle

The FDTD method is a powerful electromagnetic simulation tool that has been extensively used for studying radio wave propagation in multiple media. Maxwell's equations in the time domain are solved using finite-difference time approximations. Since the electric and magnetic fields are related in time and space, both space segmentation and time stepping are required. Space segmentation takes the form of box-shaped cells whose size are constrained by the wavelength of the EM excitation signals, and its maximum must be less than 1/10 of the smallest wavelength for greater accuracy, giving $L_{max} = 0.1(c/f)$, where c is the velocity of light and f the frequency. For propagation in insulation materials, c is reduced and sol the cell size will be reduced. The electric field (E) and magnetic field (H) components are staggered, with the E-field components centered on the edges of the box and the H-field components centered on the faces to form what is known as the Yee cell [19], as shown in Figure 3. The modelled volume space, including any material constituting the model is built by interconnection of these cells, forming the FDTD mesh.

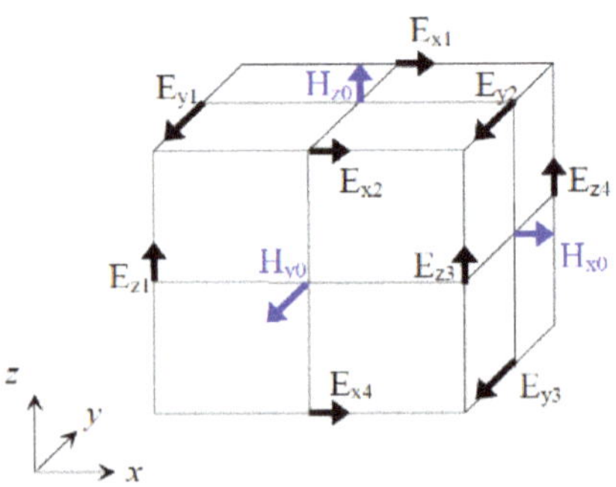

Figure 3. The Yee cell with the labelled field components.

Time stepping is achieved by quantizing time into small steps equal to the time required for the field to travel from one cell to the next. Like the E-field and H-field are offset in space, their values are also offset in time. These field values are updated using a centered-difference approximation leapfrog scheme where the electric fields and then the magnetic fields are computed at each step in time. In this work, the XFdtd® simulation package [20] was used to simulate the PD excitation source, the electrode assembly, the insulation sample with the location and extent of the discharge, and the UHF sensor. The excitation is introduced by applying a sampled waveform at one location. At each step in time, the value of the waveform is used to compute the field value. The surrounding fields propagate throughout the FDTD mesh depending on the characteristics of each cell and the material properties. The field computations continue until a state of convergence is reached. In this study, this is set by either specifying the threshold below which the computed fields decay (around −30 dB in this application) or setting a maximum number of iterations. The extent of outer boundaries and the associated boundary conditions, the cell size and simulation time are set according to the modelled problem.

3.2. PD Current Source and Point Sensor

The PD current can be represented by a pulse whose magnitude and risetime depend on various factors such as the PD type, its intensity and speed of propagation, and the surrounding medium. For example, typical risetimes of a protrusion PD in oil range between 0.7 ns and 2.0 ns, whereas a floating particle produces PD pulses having risetimes in the range 2.5–2.7 ns, and bad contact defects produce pulses up to 17 ns [1]. These PD pulses have spectra in the UHF frequency range of 300–3000 MHz [7].

In this work, a current filament having a Gaussian shape defined by Equation (1) was used to simulate the PD source [21].

$$i(t) = I_{max} e^{-t^2/2\sigma^2} \quad (1)$$

where I_{max} = magnitude of the peak current and σ = pulse width measured at half of the maximum value. The PD current is expressed in the frequency domain as:

$$I(\omega) = I_{max}\sigma\sqrt{2\pi}e^{-(\frac{\omega^2\sigma^2}{2})} \quad (2)$$

The pulse rise time (T) of the Gaussian pulse is the time required for signal magnitude to change from 10% to 90% and is calculated as

$$T = t_{90\%} - t_{10\%} = \sqrt{2}\sigma\ (\sqrt{ln10} - \sqrt{\ln\left(\frac{10}{9}\right)}) \quad (3)$$

Figure 4 shows examples of Gaussian pulses of different risetimes and their respective frequency spectra.

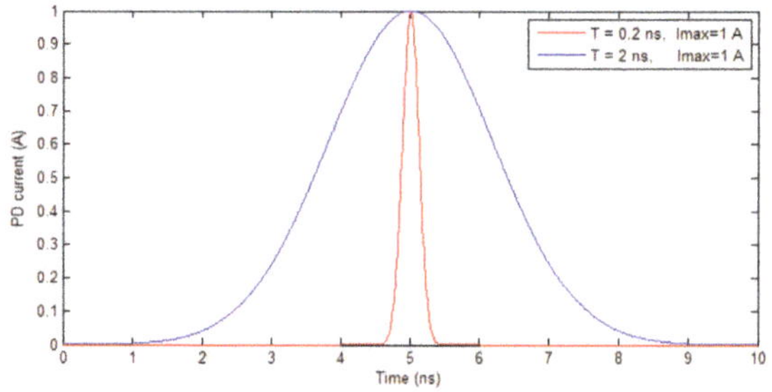

(a)

Figure 4. *Cont.*

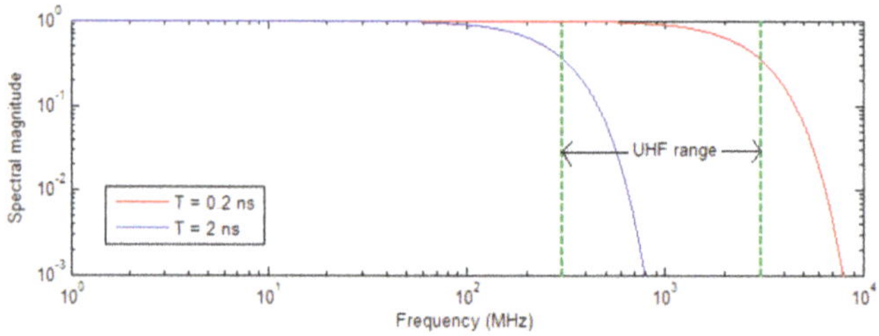

(b)

Figure 4. Examples of Gaussian pulses (**a**) and their respective frequency spectra (**b**).

The PD excitation is represented by a current filament which extends over several FDTD segments that determine the PD path (Figure 5). The filament is oriented horizontally in the positive y-axis direction 0.5 mm above the insulating surface. The source of this filament is assumed to be a current source component having 50 Ω output impedance. A Gaussian type waveform with a given pulse width and amplitude is associated with this source to excite the frequencies up to the upper limit of the UHF range. A "point sensor", which is a data storage facility, is used to store the computed field and current density values at any desired location within the FDTD mesh. The field values returned by this point sensor can be at the edges of the cells if the point sensor location is snapped to the FDTD grid, or extrapolated if it is between two edges.

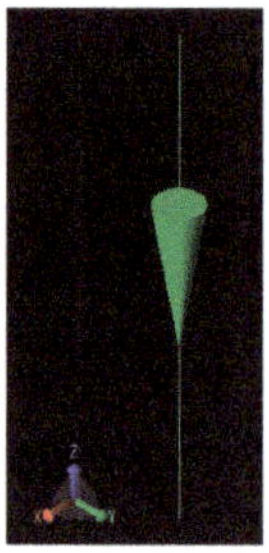

Figure 5. Simulated current filament.

3.3. UHF Sensor and Probe Coupler

The UHF sensor is modelled by a monopole antenna with a 50 Ω coupler located vertically to the ground plane, in the positive y-direction with respect to the discharge location. The 50-Ω probe coupler is matched to the current source to avoid wave reflections. The sensor length should be within the scope of the entire computing spectrum and should be chosen depending on the frequencies excited by the PD pulse.

3.4. Surface Discharge Model

Figure 6 shows the non-uniform point-plane configuration model used for simulating the surface discharge. A similar configuration consisting of two circular electrode plates is used to create a uniform field condition. The high voltage and ground electrodes are made of stainless steel. The insulator sample, a 90 mm × 90 mm square having variable thickness, is placed symmetrically between them, i.e., with its axis of symmetry coincident with the electrode's vertical axis. The vertical lines on the positive y-axis direction show the UHF sensor placed at different distances, with variable length and height. The reference simulation values are taken as: (i) pulse width, 0.4 ns, (ii) pulse amplitude, 1 A, (iii) PD source length, 10 mm and (iv) sensor length, 80 mm.

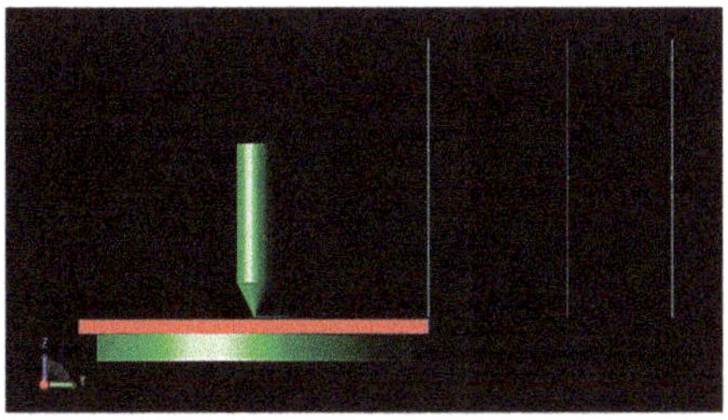

Figure 6. Point-plane configuration.

4. Results of Surface Discharge Simulations

In this section, the PD pulse width (PW), its amplitude (PA), and its path length are parameterized at each computation sequence for a given sensor location and length. The sensor location and length are then varied in sequence to complete the parametric computation. An example of the time-domain voltage signal across the antenna sensor placed at a distance 900 mm from the origin of the discharge is shown in Figure 7 for different pulse widths for the XLPE material sample. For fast-rising pulses, the captured signals are large but attenuate very quickly, whereas for slower pulses the signals are much smaller. The time delays observed between the waveforms are due to the different times of arrival of EM waves at the sensor. The maximum peak-to-peak value of the waveform is taken in what follows as the amplitude of the captured sensor signal, and represents the PD intensity. It was found that the amplitudes of the signals computed with the three different sample materials having the same dimensions did not reveal any appreciable differences, as shown in the example of Figure 8, except in a few cases which will be discussed later. This was the case over the entire range of parameter values considered in this study. For a given sequence, one parameter was varied and all other parameters were kept constant at their reference values. The results are grouped in Figures 9 and 10 for the non-uniform and uniform electrode configuration, respectively, and for three different sensor locations, represented by distance (d). The first conclusion from these results is that signals in the non-uniform point-plane configuration (Figure 9) have higher amplitudes than those measured in the uniform plane-plane configuration (Figure 10). Furthermore, it was found that locating the sensor along the vertical axis (i.e., at a distance above or below the PD source) produces smaller signal amplitudes compared with the horizontal y-axis location—only the results related to the latter are shown here.

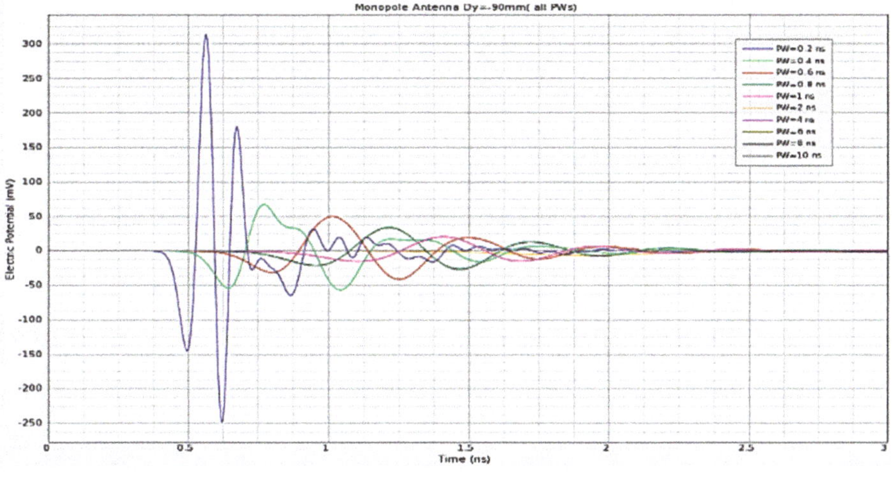

Figure 7. Sensor voltage for different pulse widths, sensor location d = 900 mm.

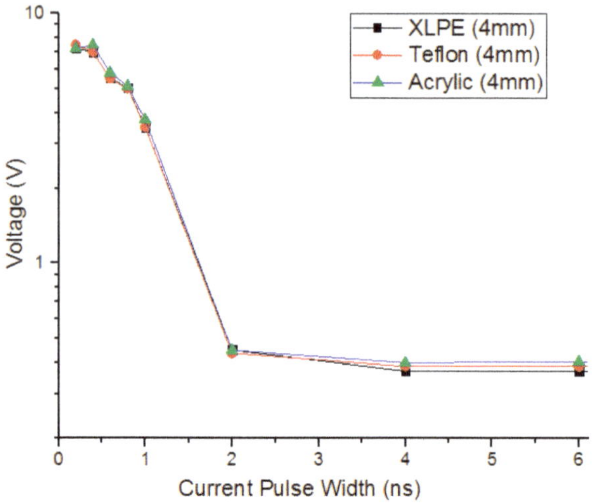

Figure 8. Effect of current pulse width on the sensor signal for three different materials (d = 450 mm).

4.1. Effect of PD Pulse Width

The top row of three graphs in Figures 9 and 10 show the effect of varying the PD pulse width on the sensor signal amplitude voltage while keeping the pulse amplitude, the length of the PD source and the length of the UHF sensor at their base values. As can be seen, the amplitude of the emitted signal is strongly affected by the pulse width. For a given PD pulse width of 2 ns or less, and with the uniform electrode configuration (Figure 10), the signals emitted by Acrylic samples are smaller compared to Teflon and XLPE samples, whereas for the non-uniform field configuration, the signals are approximately equal for all material samples. When the pulse width becomes larger than about 4 ns, their frequency spectrum narrows, and hence it becomes more difficult to detect their radiated signals within the UHF frequency spectrum. Further investigation is needed to clarify the apparently non-linear decrease in signals amplitudes with pulse width. This presents a challenge in unbounded systems such as surface discharge in outdoor insulation or corona discharge in air. As expected, the signal amplitudes for all samples decrease as the sensor distance increases from the PD source.

4.2. Effect PD Intensity

The results presented in the second row of Figures 8 and 9 show that pulse amplitudes below 10 mA emit hardly detectable signals. A steady increase in pulse amplitude with PD intensity can be seen within the range between 10 mA and 100 mA. However, this increase becomes much steeper when the current increases between 100 mA to 1 A. These signals of course depend in practice on the discharge intensity, which is related to the amount of charge deposited, and the speed of its progression along the surface path. Discharge paths are usually not uniform and their progress is affected by other interfacial and surface condition factors. Multiple discharges of different intensities can also originate from different locations on the insulation surface and may be simultaneous or time delayed, with their associated EM waves appearing as complex superimposed signals on the UHF sensor.

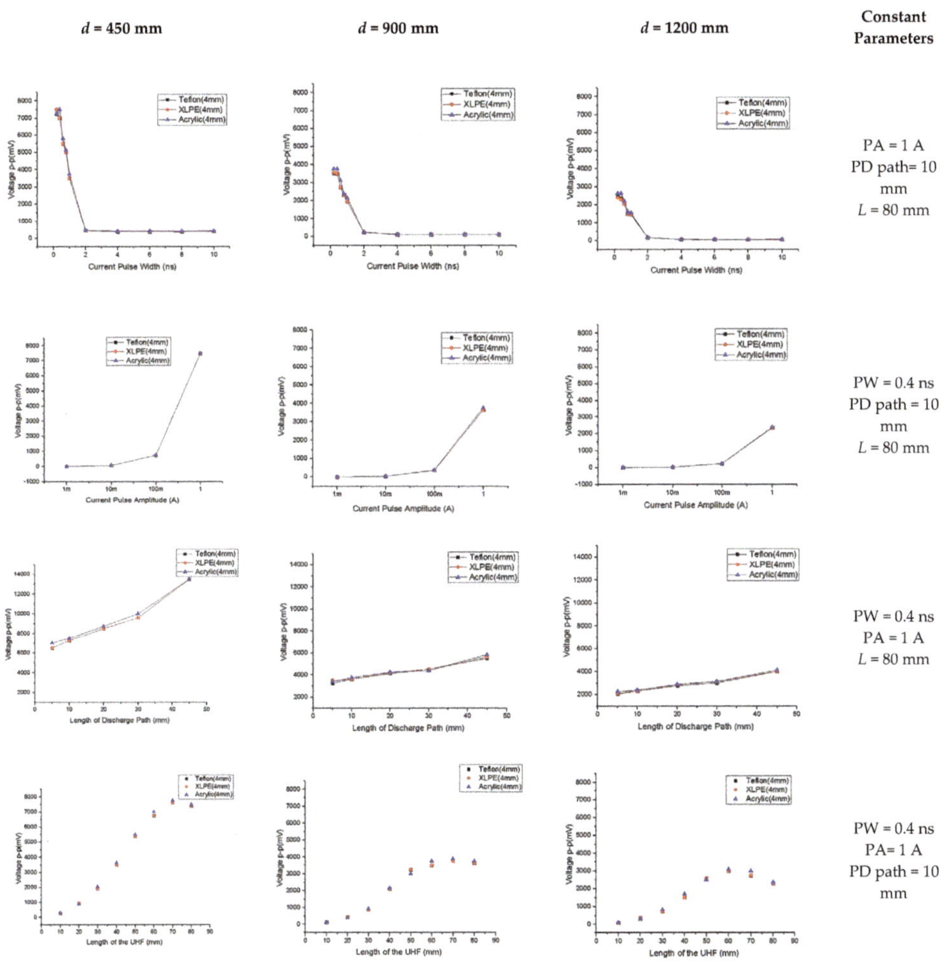

Figure 9. Results of the parametric study, surface discharge, non-uniform field electrode configuration.

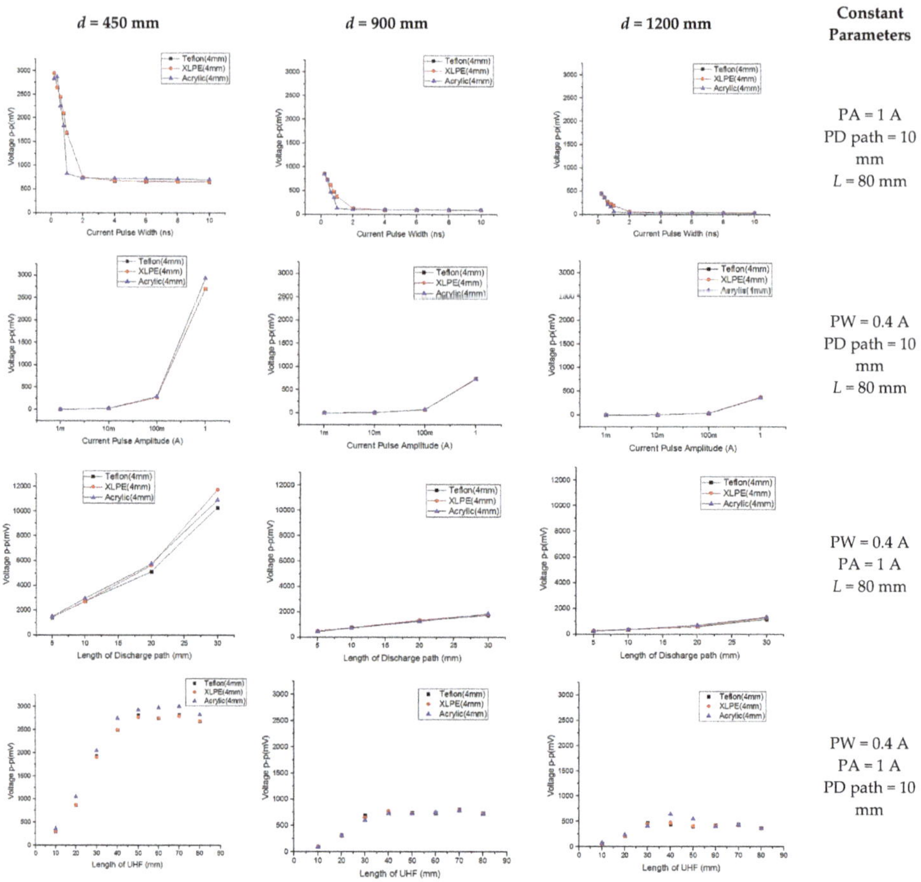

Figure 10. Results of parametric study, surface discharge, uniform field electrode configuration.

4.3. *Effect of Length of the Discharge Path*

The length of the current filament determines the discharge path over which the charge is deposited. The results presented in the third row of Figures 9 and 10 show a proportional increase in UHF signal magnitude with the PD path for a given pulse width and intensity. With the shortest UF sensor distance, XLPE showed slightly lower signal magnitudes compared with Teflon and Acrylic for the non-uniform field configuration (Figure 9). For the uniform field configuration, the Teflon samples showed the smallest magnitudes, with the difference between materials increasing for higher intensity discharges.

4.4. *Effect of the Length of the UHF Sensor*

The length of the sensor was varied from 10 to 80 mm. The graphs in the bottom rows of Figures 9 and 10 show an increase in signal amplitude with UHF sensor length up to a certain value. The sensor length should be smaller than this, and in any case, it should be smaller than the minimum wavelength (λ_{min}) of the EM waves excited by the PD pulse. Since this is related to the PD pulse width, which is 0.4 ns in this case, the sensor length should be much less than λ_{min} = 120 mm. Clearly, satisfactory measurements can be achieved for lengths of 30 mm or less for all three positions. Beyond this length, there is no benefit in obtaining reflection-free signals.

5. Cavity Discharge Simulation Results

The cavity discharge was simulated by introducing a current filament flowing in the vertical direction in the middle of the XLPE sample. The PD path was varied between the minimum value of 1.5 mm, constrained by the FDTD cell size within the material, and the maximum value of 2.5 mm was constrained by the sample thickness. The non-uniform field, point-plate electrode configuration was used for this analysis. Figure 11 shows examples of typical time-domain signals generated by a 1-A pulse, with a width of 1 ns and a path length of 2 mm, demonstrating the need to place the UHF sensor closer to the discharge for increased sensitivity.

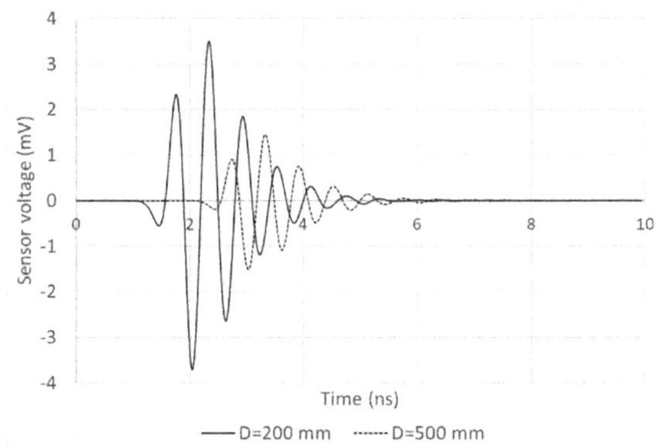

Figure 11. Typical time-domain signals captured by the UHF sensor, cavity discharge, pulse width = 1 ns, pulse path = 2 mm, Pulse amplitude = 1 A.

The main parameters that were varied in this case are the pulse width, the PD pulse path and the UHF sensor location. The PD pulse amplitude was kept constant at 1 A. The results for two different UHF sensor positions are shown in Figure 12. First, it should be noted that the amplitudes of the radiated signals from the cavity discharge are much smaller than those obtained with surface discharge. This is because the EM waves propagate through the XLPE insulation and the surrounding air medium, which causes further attenuation. In practice, additional attenuation due to metallic enclosures in cable joints and cable metallic sheaths makes the detection of PD from internal cavities even more challenging because they shield the EM waves from the outside medium. The results show signal amplitudes increasing with PD path, with the highest increase being associated with fast PD pulses. A combination of measurements and FDTD simulations would enable a better understanding of the cavity discharge characteristics, such as deducing the PD pulse shape that would best fit the measurement results.

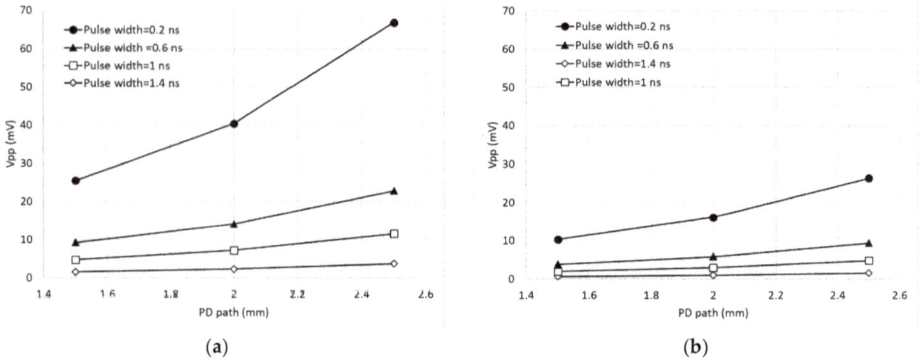

Figure 12. Computed signal amplitudes for the cavity discharge, partial discharge (PD) pulse amplitude = 1 A, UHF sensor positions (**a**) d = 200 mm (**b**) 500 mm.

6. Conclusions

The characteristics of EM waves emitted by surface discharge and cavity discharge were studied using FDTD simulations. Laboratory measurements using test cells designed to simulate these two types of discharge showed that the frequencies over which these discharges excite EM waves lie within the UHF range. The FDTD simulations were then used as computational tool to investigate the use of a UHF sensor for measuring EM waves emitted by such discharges. This allowed various parameters related to both the discharge characteristics and the sensor characteristics to be varied and their influence on the measured signals clarified. The results showed the limits of using a monopole UHF sensor for this type of measurement, highlighting the importance of characterizing the PD pulse on one hand, and the choice of the sensor size and location on the other. These type of studies are of great help in understanding the propagation characteristics of EM waves for PD, and for designing UHF sensors destined for these types of measurements.

Author Contributions: Conceptualization, N.H., H.G. and B.B.; methodology, N.H. and B.B.; software, A.L.; validation, N.H., H.G. and B.B. formal analysis, N.H. and H.G.; investigation, A.L.; data curation, A.L.; writing—A.L. and N.H.; writing—review and editing, N.H., H.G.; supervision, N.H. and H.G. and B.B.; project administration, N.H.

Funding: This research received no external funding.

Conflicts of Interest: The authors declare no conflict of interest.

References

1. International Electrotechnical and Commission. *IEC International Standard 60270: High Voltage Test Techniques—Partial Discharge Measurements*, 3rd ed.; IEC: Geneva, Switzerland, 2000.
2. Judd, M.D.; Hampton, B.F.; Farish, O. Modelling partial discharge excitation of UHF signals in waveguide structures using Green's functions. *IEEE Proc. Sci. Meas. Technol.* **1996**, *143*, 63–70. [CrossRef]
3. Zhan, H.M.; Liu, Y.Z.; Zheng, S.S. Study on the Disc Sensor Based on the Cavity Mold Theory. *IEEE Sens. J.* **2016**, *16*, 5277–5282.
4. Shibuya, Y.; Matsumoto, S.; Tanaka, M.; Muto, H.; Kaneda, Y. Electromagnetic waves from partial discharges and their detection using patch antenna. *IEEE Trans. Dielectr. Electr. Insul.* **2010**, *17*, 862–871. [CrossRef]
5. Tang, J.; Xu, Z.R.; Zhang, X.X.; Sun, C.X. GIS partial discharge quantitative measurements using UHF microstrip antenna sensors. In Proceedings of the 2007 Annual Report—Conference on Electrical Insulation and Dielectric Phenomena, Vancouver, BC, Canada, 14–17 October 2007. [CrossRef]
6. Khotimah, K.; Khayam, U. Design of Dipole Antenna Model for Partial Discharge Detection in GIS. In Proceedings of the 2015 International Conference on Electrical Engineering and Informatics (ICEEI), Denpasar, Indonesia, 10–11 August 2015; pp. 186–191.

7. Reid, A.J.; Judd, M.D.; Fouracre, R.A.; Stewart, B.G.; Hepburn, D.M. Simultaneous measurement of partial discharges using IEC60270 and radio-frequency techniques. *IEEE Trans. Dielectr. Electr. Insul.* **2011**, *18*, 444–455. [CrossRef]
8. Tenbohlen, S.; Denissov, D.; Hoek, S.; Markalous, S.M. Partial discharge measurement in the ultra-high frequency (UHF) range. *IEEE Trans. Dielectr. Electr. Insul.* **2008**, *15*, 1544–1552. [CrossRef]
9. Judd, M.D.; Yang, L.; Hunter, I.B.B. Partial discharge monitoring of power transformers using UHF sensors. Part I: Sensors and signal interpretation. *IEEE Electr. Insul. Mag.* **2005**, *21*, 5–14. [CrossRef]
10. Judd, M.D. Using finite difference time domain techniques to model electrical discharge phenomena. In Proceedings of the 2000 Annual Report Conference on Electrical Insulation and Dielectric Phenomena, Victoria, BC, Canada, 15–18 October 2000; pp. 518–521.
11. Rong, M.Z.; Li, T.H.; Wang, X.H.; Liu, D.X.; Zhang, A.X. Investigation on Propagation Characteristics of PD-induced Electromagnetic Wave in T-Shaped GIS Based on FDTD Method. *IEICE Trans. Electron.* **2014**, *97*, 880–887. [CrossRef]
12. Ohki, H.; Nagaoka, N.; Takeuchi, M.; Tanahashi, D.; Okada, N.; Baba, Y.; Tanaka, H. Finite-difference time-domain simulation of partial discharges in a gas insulated switchgear. *High Volt.* **2016**, *1*, 52–56.
13. Liu, J.H.; Wang, J.; Qian, Y.; Duan, D.P.; Huang, C.J.; Zeng, Y.; Jiang, X.C. Simulation analysis on the propagation characteristics of electromagnetic wave in GIS. *Gaodianya Jishu/High Volt. Eng.* **2007**, *33*, 139–142.
14. Prasetia, H.; Khayam, U.; Itose, A.; Kozako, M.; Hikita, M. Simulation analysis of surface current as TEV signal caused by partial discharge on post insulator in bus duct. In Proceedings of the 2016 IEEE International Conference on Dielectrics (ICD), Montpellier, France, 3–7 July 2016.
15. Yao, R.; Zhang, Y.B.; Si, G.Q.; Wu, C.J.; Yang, N.N.; Wang, Y.X. Simulation analysis on the propagation characteristics of electromagnetic wave in T-branch GIS based on FDTD. In Proceedings of the IEEE 15th Mediterranean Microwave Symposium (MMS), Lecce, Italy, 30 November–2 December 2015; pp. 1–4.
16. Xu, B.; Li, Y.M.; Li, J.H.; Zhao, J.H.; Wang, X.J. Simulation of the propagation characteristic of Ultra High-Frequency signals excited by partial discharge in power transformers. In Proceedings of the 2015 Chinese Automation Congress (CAC), Wuhan, China, 27–29 November 2015; pp. 1209–1210.
17. Loubani, A.; Harid, N.; Griffths, H. Analysis of UHF sensor response to EM waves excited by surface discharge in air using FDTD simulation. In Proceedings of the 2018 IEEE International Conference on High Voltage Engineering and Application (ICHVE), Athens, Greece, 10–13 September 2018. [CrossRef]
18. Ishak, A.M.; Ishak, M.T.; Jusoh, M.T.; Syed Dardin, S.F.; Judd, M.D. Design and Optimization of UHF Partial Discharge Sensors Using FDTD Modeling. *IEEE Sens. J.* **2017**, *17*, 127–133. [CrossRef]
19. Inan, U.S.; Robert, M.R. *Numerical Electromagnetics: The FDTD Method*; Cambridge University Press: Cambridge, UK, 2011.
20. XFdtd. *Ful-wave, 3D Electromagnetic Analysis Software, Reference Manual 7.5.1*; Remcom: State College, PA, USA, 2016.
21. Kreuger, F.H. *Industrial High DC Voltage 1. Fields 2. Breakdowns 3. Tests*; Delft University Press: Delft, The Netherlands, 1995.

energies

Article

A Physical Calibrator for Partial Discharge Meters

Michal Krbal [1,*], Ludek Pelikan [1], Jaroslav Stepanek [1], Jaroslava Orsagova [1] and Iraida Kolcunova [2]

1 Department of Electrical Power Engineering, Brno University of Technology, 601 90 Brno, Czech Republic; xpelik15@stud.feec.vutbr.cz (L.P.); xstepa41@stud.feec.vutbr.cz (J.S.); orsagova@feec.vutbr.cz (J.O.)
2 Department of Electric Power Engineering, Technical University of Kosice, 042 00 Kosice, Slovakia; iraida.kolcunova@tuke.sk
* Correspondence: krbal@feec.vutbr.cz; Tel.: +420-541-146-243

Received: 21 March 2019; Accepted: 21 May 2019; Published: 29 May 2019

Abstract: This article offers an alternative method of calibrating partial discharge meters for research and teaching purposes. Most current modern calibrators are implemented as precise voltage pulse sources with a coupling capacitor. However, our calibrator is based on the physical principles of dielectric materials distributed in a plane or space. Calibrator design is unique and there is an attempt to get closer to the behavior of the measured real objects. The calibration impulses are created by energy from a high voltage power supply at the specific or nominal value of the applied voltage. At the same time, it is possible to simulate the value and quantity of the discharges and their position in the object relative to the input electrodes. The calibrator creates conditions as a real measured object with adjustable parameters. This paper describes a design of this type of calibrator, its implementation, numerical simulation of discharge values and laboratory measurements with functional verification using the Tettex 9520 calibrator and galvanic measured system DDX 7000/8003 and DDX 9121b. All measurements are carried out using the CVVOZEPowerLab Research Infrastructure equipment.

Keywords: partial discharges (PD); partial discharge; calibrator; Tettex 9520; DDX 8003; DDX 9121b

1. Introduction

Measurement of partial discharges (PD) is nowadays conventionally used as the diagnostic method for medium and high voltage electrical devices [1]. The method is primarily intended for the measurement of insulation quality and localization of isolation defects in MV and HV cables, instrument transformers and rotating machines. However, research in this field is still ongoing and the full use of PD diagnostic methods is a matter of the future [2]. The most commonly used method of partial discharge measurement is the galvanic method according IEC 60270, which is very accurate and sensitive in comparison with other options [1]. However, the accuracy of this method depends heavily on the level of interferences from the surroundings, the quality of the pulse discrimination system (PDS), input filter settings and last but not least on well performed calibration [3,4]. Most modern systems for PD measurement are equipped with a PDS. The suppression of interference from the surroundings is allowed by the PDSs, which are based on eliminating signals not coming from the tested objects. An example may be our Haefely DDX 8003 PD system (Haefely Hipotronics, Brewster, NY, USA), which can use two measuring channels and the antenna for the elimination of external influences. The piezo ultrasonic microphone may be an alternative to external corona elimination for other systems. Internal partial discharges occur in all tested objects from the initial voltage value and they are the most often created at sites of mechanical isolation faults, in the areas of material impurities, mechanical cracks and gas bubbles [4]. PD calibrators are used for the purpose of the calibration process, which is described in Section 2. In addition, physical calibrators can also be used to replace a test object with a defined failure. The defined fault in the test object causes a PD with a certain charge value, amplitude symmetry and phase spectrum depending on the applied voltage value. Knowledge of the PD amplitude and phase spectrum of predefined faults is the basis for its diagnostics in real objects. The aim of this research is to create a database of MV and HV cable faults simulated by physical calibrators and their verification. The verification of simulated faults was performed using the circuit calculation. The circuit is simulated in the Electronic Workbench software and the static state calculations before and during the fault are performed in Matlab Simulink software.

2. Calibration Process

The calibration process must be performed before each PD measurement. Each change of the measured circuit (change of the measured object, interconnecting conductors, input filter settings) leads to a change in the conditions of RF current pulses propagation that are generated by the failure in the test object [4]. The calibration process involves injection of a known amount of charge into the test object electrodes with the use of standard external calibrators. The value of charge is an order of magnitude consistent with the expected value of the measured object. One or more pulses are generated by these calibrators within one period of the power grid signal and their amplitude is related to the size of the set charge level of partial discharge and capacity of coupling capacitor of PD measuring system. Calibration impulses can be also generated synchronously with the power line frequency. This feature is provided by advanced calibrators, such as the Tettex 9520 (Haefely Test AG, Basel, Switzerland). The calibration is always based on the creation of pulses predefined by the calibrator, which is parallel connected to the tested object. The principle of the calibration process is the assignment of the absolute value of the PD charge Q_P or Q_{IES} by the measuring system to the set charge value generated by the calibrator.

Calibrators can be realized as external devices or may be formed by impulse generators and the HV injection capacitors. For the PD system, these impulses are identical to the impulses formed inside of the tested object. Unfortunately, the creation of impulses by test objects is significantly more complex. PD within the test objects causes charge flows from the space charge of the dielectric to the point of failure. The part of the charge is also compensated from the power supply, or from the PD system coupling capacity. Moreover, only this part of the charge is expressed as the high frequency current impulses, which are measured by the system [1].

However, most of external calibrators are able to calibrate only the charge level, but without consideration of fault location inside the test object. Flows of charges inside the test object are influenced by the location of the fault. The level of asymmetry of charge value within the positive and negative polarity is generally caused due to unbalanced faults with respect to the electrodes (ground electrode and HV out electrode). The measured PD charge value of real objects is changed depending on the position of the discharge location relative to the input electrodes. An example may also be medium or high voltage cables according to IEC 62067:2001. The measured value of the charge is also dependent on the position of the calibrator at the beginning or the end of the cable. The value can vary by approximately 20% at 20 m MV cable. Overall, the accuracy and sensitivity of the measurement is influenced by the asymmetry and complexity of the measured object structure. The asymmetric localization of a partial discharge in the measured object affects the symmetry of the phase spectrum in the positive and negative polarity parts. Simulation of these phenomena is impossible by standard low voltage calibrators. That is also why our calibrator was developed, which allows setting the PD charge value, producing more different pulses within one period and operating at the nominal voltage level of the test objects. Asymmetrical failures can be simulated as well. In the paper, a design and implementation of extended external galvanic calibration systems for the PD measurement is described. The proposed calibrators are based on numerical models of partial discharge inside the solid dielectric materials, but they are also designed as the physical models represented by the capacity networks and mechanically modified solid dielectric materials. Tested MV cable samples were mechanically damaged to form faults. The position of faults in damaged cables coincided with the fault position simulated by the calibrator. Also, more than 10 used cables with failures were tested and analyzed.

These disorders originated naturally—by local mechanical and thermal degradation and aging of the insulation material. The test results at different voltage levels have proved the ability of the calibrated measuring system to locate insulation defects.

2.1. Currently Used PD Calibrators

All currently used calibrators are based on a simple low voltage generator of impulse or saw-tooth voltage waveform with a defined polarity and the voltage level. Examples are the Tettex 9520 calibrator and the Robinson Miniature Discharge 753US-1PD simulator (Haefely Test AG, Basel, Switzerland). Most of these calibrators are supplied from the internal accumulators. Therefore, they are implemented with galvanic separation. The principle of these modern calibrators is described in [5]. Generating asymmetrical or partially symmetrical pulses and the value of the charge in the PD depending on the supplied voltage is not possible by none of the commercially offered calibrators. Moreover, normal laboratory calibrators cannot work as a reference, with exact dependencies of the PD value on certain HV voltage level. PD calibrators developed in our laboratory meet the following parameters. These calibrators can also be connected in parallel to the test object at nominal voltage or they can be used separately for collecting information into a failure knowledge database.

2.2. Physical Calibrator Requirements

The aim of the physical calibrator is the simulation of charge caused by the failure on real test objects. The calibrator must be designed for the same test voltage as the test object (nominal voltage of power distribution networks components, e.g., 10 kV or 22 kV). The range of the produced charge should be from units of pC for insulators and HV cables, up to tens or hundreds of nC for power transformers and mechanically more complex objects. Requirements for the calibrator are summarized in the following points:

- The operating voltage difference potential from 0 V to approximately 30 kV AC.
- The option to choose the location of the fault (symmetrical or asymmetrical).
- Setting the charge value of PD in the range of 0.1 to 1000 pC depending on the applied voltage.
- Low corona design and easy reconfigurability of charge value parameters or fault position.

3. Schematic and Numerical Model of Calibrator

The new calibrators are based on the capacitor network. Electrical parameters (dissipation factor, permittivity) of commonly used ceramic or PP high voltage capacitors are approximately identical with the properties of solid PE, XLPE dielectrics. We suppose that the model dielectrics can be implemented as a flat (2D) or three-dimensional (3D) arrangement of electronic elements. The 3D model corresponds more to the reality of the measured objects parameters, however, the possibilities of its implementation and parameters settings are more complicated for users. The flat model is a sufficient solution for PD simulation and testing of measurement systems. In addition, rotationally symmetric objects can be satisfactorily replaced by this 2D model. The capacitor network is realized with a certain step that corresponds to the physical dimensions of the real dielectric in the order of several tenths or units of mm.

The network of our Calibrator "Model A" is implemented as flat model with 5 × 12 partition steps. PD, breakdowns or short circuits between the junctions can be implemented with an external circuit, which can be connected between any junctions of the circuit. Schematic of this network for simulating is shown in Figure 1. The numerical calculation of individual model nodes is performed in Matlab software, Matlab Simulink is used to calculate steady state.

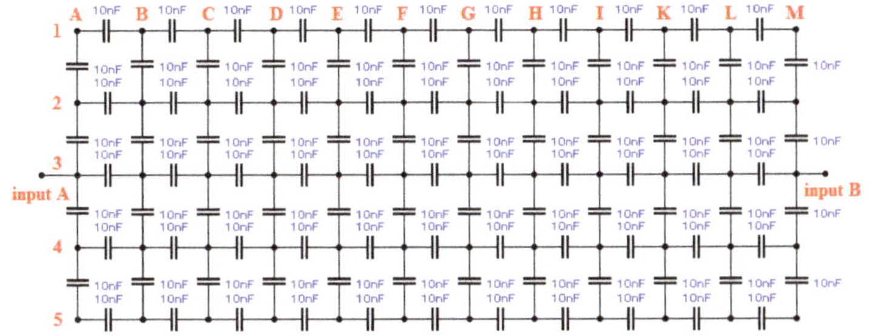

Figure 1. Schematic of the capacitor network "Model A" used for simulation in Workbench software.

The second calibrator "Model B" is designed and implemented as a three-dimensional device. This model is made of transparent solid PMMC dielectric material. Part of the circuit between input electrodes is removed and can be replaced by other solid, liquid or gaseous dielectrics. The last calibrator "Model C" is the latest and most sophisticated. This calibrator is a flat model of MV cable and its scheme is shown in Figure 2. The design of "Model A" and "Model C" calibrators was based on knowledge of [6–8] and the physical dimensions of the MV cable faults.

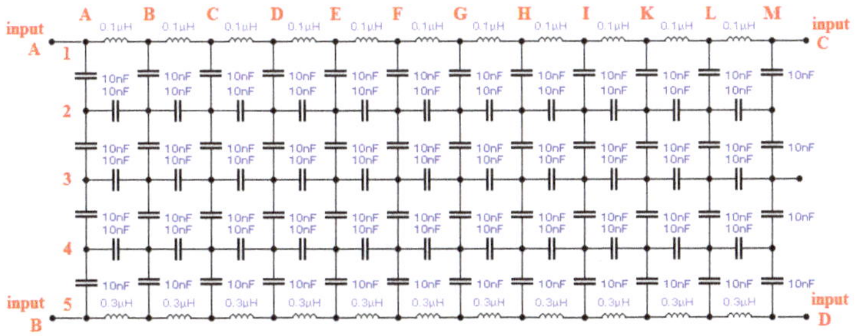

Figure 2. Schematic of the capacitor network "Model C" used for simulation in Workbench software.

Capacitance and inductance values are selected based on the experience, physical properties of PE and XLPE used dielectrics and applicable simulation values. The one step capacity corresponds to the order of about 0.3 mm of a XLPE material. Values of partial inductances are determined by experimental measurement of core and shielding inductance of 20 kV 185 mm^2 MV cables.

The simulation of PDs is performed by a fast change of part capacity between two junctions. The behavior of the circuit is based on the Gemant and Philippoff model [9]—three capacitance circuit schematically illustrated in Figure 3. More precise models can be based on Böning [3,10]—five capacitance model or Eberhard Lemke dipole model based on Paderson assumptions [2].

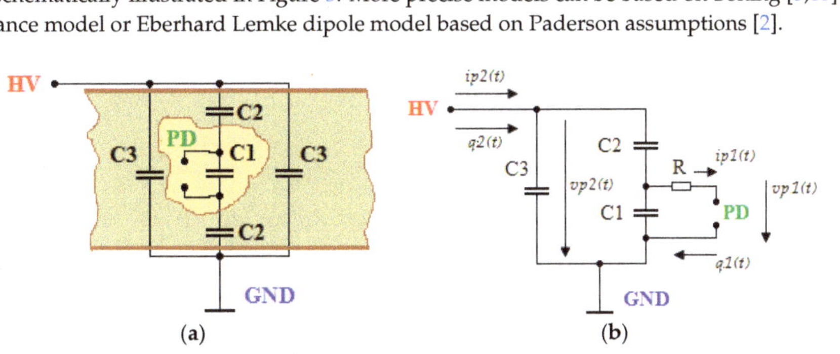

Figure 3. Gemant and Phillippoff Model: (**a**) Schematic of Insulation; (**b**) Equivalent Electrical Circuit.

The part of the charge is then quickly transferred in the calibrator components. Thus the fast current impulse is flowing from the supply circuit to compensate the change of capacitance charge. These impulses can be created several times during one period. This type of PD can be simulated numerically by quickly short connecting two junctions in the capacitor model for a short time period. This connection can be synchronized with the phase of the applied voltage period. The practical implementation of a fast, lossless switch is complicated. In a physical model, using fast, voltage controlled semiconductor elements or gas filled overvoltage protections is recommended.

At present, "Models A" and "C" were designed, simulated, implemented and laboratory tested. The first "Model A" is formed by a network of capacitors and allows simulating the partial discharge in one or more places between junctions at the same time or within one period. The model allows creation of a partial discharge $q_2(t)$ with a set value from 0.1 to 1000 pC. This PD can be created in a variety of locations nearby the input electrodes, in the central part of the model or on its sides. The calculation of overall charge $q_2(t)$ change is possible by the simulation software (e.g., Electronic Workbench software). The external circuit can be turned on at the set value of voltage $vp_1(t)$. The value of the voltage is proportional to the parameters of the impurities in the material, to the physical size of the inclusions and the mechanical failures and their chemical composition. Thus the real behavior of dielectric material is simulated in this way.

Most of the faults in real objects are not symmetrical. Typically, they are not located in the vicinity of the axis of the object or its central part. They can be also located at its edges or near the input electrodes. The voltage drop on the fault or the current taken from the power supply is not symmetrical. An example may be the connection of external circuit between 2B and 2C junctions. The set peak value of PD $q_2(t)$ is 10 pC. The voltage drop on the external circuit $vp_1(t)$ and the impulse current $ip_2(t)$ taken from the power supply is shown graphically in Figure 4. Red curve—voltage drop waveform on fault and current pulses, blue curve—waveform of applied voltage.

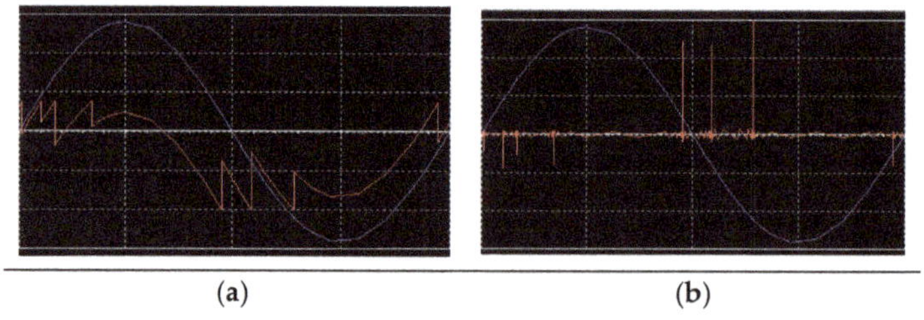

Figure 4. 10 pC PD Asymmetrical Simulation: (**a**) Voltage drop; (**b**) Current from power supply.

The rotationally symmetrical 3D "Model B" is implemented purely from a dielectric material. The electrical diagram of this model in the flat section is shown in Figure 5. Also, it is possible to use the ANSYS software simulator to determine the approximate values of transferred PD charge. However, this model is mainly experimental for verification and comparison of the different dielectric and composite material properties. The model is partially based on the experience of [11].

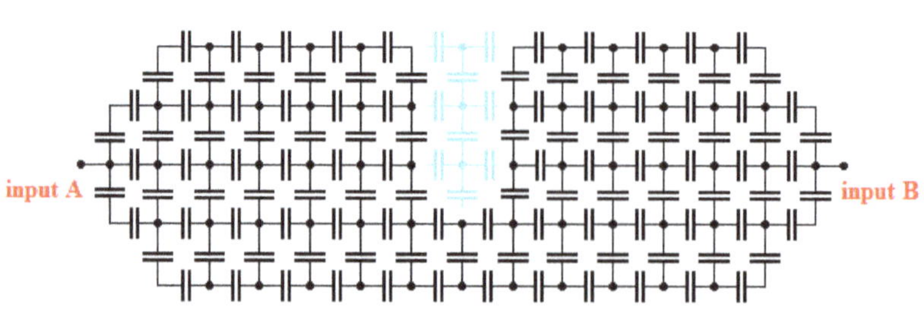

Figure 5. Electrical Diagram of "Model B".

4. Implementation of a Physical Model

"Model A" is made of 10 nF ceramic capacitors type CC81 with a low dissipation factor < 0.5%. The maximum voltage of the "Model A" is 30 kV and the maximum voltage drop Vp_1 between two junctions is about 3 kV in the idle state. However, the simulations showed that the impulse voltage drop $vp_1(t)$ on the capacitor at current impulse of PD can reach up to 7.5 kV. Therefore, capacitors with a nominal voltage of 10 kV were chosen. The value of the transferred charge $q_2(t)$ is dependent on the choice of PD location and the settings of the threshold voltage $vt_1(t)$ of the ignition system. The model is designed for a wide range of transferred charge $q_2(t)$ from 0.1 to 1000 pC. Low corona connection is provided by the external spherical electrodes. And this model offers an easy option of reconfiguration and setting the correct value of pC with the help of software simulators. All 60 junctions are available for connecting the ignition system. The model also allows for creation of asymmetric faults. The physical realization of the "Model A" is shown in Figure 6.

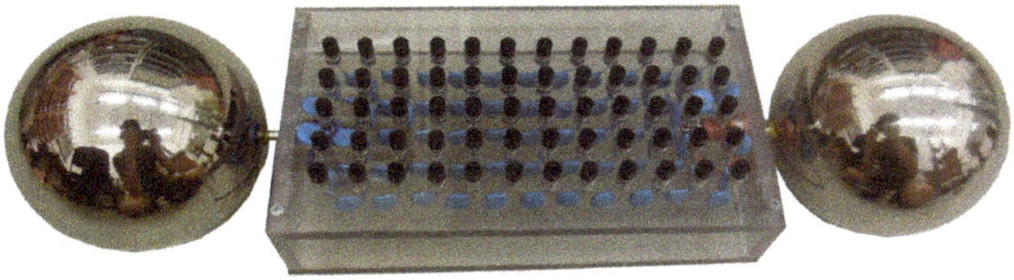

Figure 6. Realization of Physical Calibrator "Model A".

Physical Calibrator "Model B" is designed for voltages up to 50 kV AC (100 kV with external electrodes). However this model is less sensitive in comparison with "Model A" and "C". Overall, the "Model B" is designed and implemented as a rotationally symmetric capacitor with a small radial cavity in the middle. It is possible to insert test materials into the cavity (homogeneous materials, inhomogeneous, porous, with air or gas bubbles etc.). The photo of the latest version of "Model B" is presented in Figure 7.

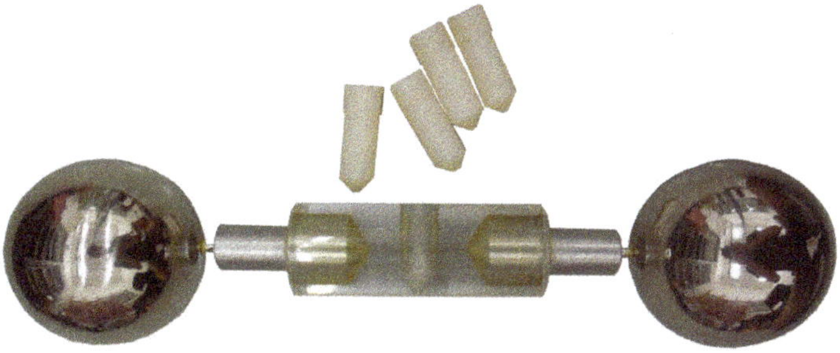

Figure 7. Realization of Calibrator Model B with Solid Dielectric Test Samples.

The input electrodes of "Model B" are made of aluminum in conical design. The construct material is a homogeneous cylinder of PMMC. Into the radial cavity located in the middle of the cylinder, it is possible to insert other kinds of solid or porous materials. Inhomogeneous electric field is produced in the vicinity of the input electrode. However, the electric field becomes uniform in the direction to the cavity. The formation of PD is minimized inside a homogeneous part of cylinder and discharged are appearing only in the cavity. Mineral oil or petroleum jelly can be used to prevent the formation of undesirable PD in the transition of the air gaps between the cylinder material and the tested material. The design of the model prevents the formation of the external corona of up to 100 kV between the external electrodes. Corona is eliminated by the smooth design of the model. For very accurate measurement with very low PD background noise of 150–200 fC at 100 kV in frequency range of 50 to 500 kHz, it is possible to use the external spherical input electrodes.

Ingnition System of the External Circuit

The ignition system is designed for the "Models A" and "C". The purpose of this ignition system is to ensure the relocation of the charge $q_1(t)$ within the test object. The charge $q_1(t)$ is determined by the ignition voltage and the value of the capacity C_1, which can be shorted. The ignition voltage can be set from the units of Volts to the hundreds of Volts. Coupling capacity can be chosen from pF to tens of nF. It is possible to drive the ignition system by fast electronic circuit. However, a simpler solution is to use same of different Zener diodes "C", transils "D" or gas surge protections "B". The disadvantage of this solution is only one set value of the ignition voltage. For the change of charge or ignition voltage values, it is necessary to change the electronic component coupling capacitors C_2 or value of working capacitor C_1. All electronic schematics of the tested ignition system are shown in Figure 8. In all cases the value of the transferred charge $q_1(t)$ it is possible to determine for one ignition of the PD (one current pulse $ip_1(t)$) or for all the impulses within one period. The charge change value $q_1(t)$ at the fault location or the charge change value $q_2(t)$ of the whole calibrator circuit can be determined by calculation or SW simulation.

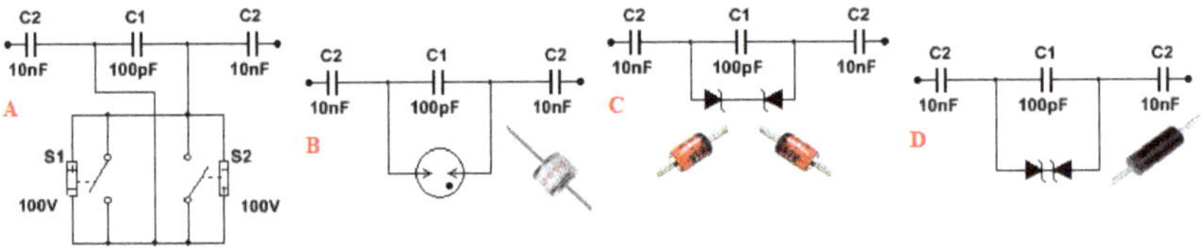

Figure 8. Different Circuits of Ignition Systems.

Capacitors C_2 are coupling and capacitor C_1 is a working one. Only this capacitor is shorted by the ignition system and the shorted differential charge $q_1(t)$ is compensated by an external circuit. The part of differential charge $q_2(t)$ is also compensated by power supply circuit with coupling capacitor and this part of charge is measured by PD meas. systems. Therefore, there are also differences between the ignition value set of charge $q_1(t)$ and the charge $q_2(t)$ measured by the PD meter in external circuit. The first Circuit "A" in Figure 8 is a theoretical circuit, which is used for the purpose of simulation. This circuit is implemented with two fast electronic relays, which are set to a threshold voltage level $vp_1(t) = vt_1(t)$ in both polarities. Unfortunately, mechanical relays are not fast enough for practical use. Gas discharge tube surge arresters or neon tubes are used in the circuit "B". This solution proves to be the most reliable, fastest, and most similar to the actual partial discharges. Alternative circuits are the circuits "C" and "D" that use symmetrical or asymmetrical fast semiconductor components with a fixed value of the ignition voltage $vt_1(t)$. In our case—mainly fast and powerful 600 W or 1.2 kW transils are used for most practical measurements.

5. Laboratory Verification of Simulators Function

Verification of functionality, parameters settings and determination of deviation from simulated parameters of all calibrator models were carried out using:

- Haefely Hipotronics DDX 7000 with internal calibrator, 1 nF coupling capacitor C_B and 100 pF injection capacitor C_I. Measurements were performed in the frequency range 20–200 kHz with a background of approx. 0.8 pC.
- Haefely Hipotronics DDX 8003 with pulse discrimination system PDS and external calibrator Tettex 9520 in frequency range 20–200 kHz with background of approx. 0.5 pC.
- Tettex DDX 9121b with coupling impedance AKV 9310 and external calibrator Tettex 9520 in frequency range 50–500 kHz with background of 0.15 pC.

Calibrator "Model A" has been validated on the accuracy of setting the level of the measured charge $q_2(t)$ by Tettex DDX 9121b. The value of pulse $q_{2IEC}(t)$ charge taken from the power supply circuit was set by the ignition system. The value of the charge has been determined by simulation. For simplicity, only one pulse per period or two pulses per period in the symmetric partial discharges have been used. The value of maximum error is 30% for DDX 8003 and 20% for DDX 9121b. The results of simulation and measurement are described in Table 1. All these values are for two pulses per one period. The generated pulses by "Model A" and "C" pulses are very similar to PD of MV cables. The pulse width at 50% amplitude is between 5 and 10 µs, thus they are easily detectable by PD detectors with a frequency range of 20 to 500 kHz.

The charge value PD is given either as a peak value Q_{peak} (Q_P) or as a Q_{IEC} value according the IEC standards, which is the largest repeatedly occurring PD magnitude. The Q_{IEC} value is the most commonly reported value in simulations or measured in measuring devices.

Table 1. Calibrator Model A—Measured Values for 4 Different Simulation.

Variant	"1"	"2"	"3"	"4"
Simulate PD level (pC)	6.5	34	185	500
C_1 value (pF), C_2 = 10nF	100	100	470	2200
Threshold Voltage vt_1(V)	18	95	110	63
Input Voltage V_1 (kV)	2.98	2.98	2.98	2.98
Meas PD Value Q_{IEC} (pC) [1]	4.98 [1]	29.5 [1]	159.3 [1]	434 [1]
Max. Error (%)[1]	23.4 [1]	13.2 [1]	13.9 [1]	13.2 [1]
Meas PD Value Q_{IEC} (pC) [2]	5.4 [2]	36.8 [2]	152 [2]	414 [2]
Max. Error (%)[2]	16.9 [2]	8.2 [2]	17.8 [2]	17.2 [2]

[1] Haefely Hipotromics DDX 8003 meas. system, [2] Tettex 9121b meas. system.

Simulated PD values are calculated for voltage range 2.3 to 10 kV of input voltage V_1. All practical measurements were performed on "Model A" and measured by DDX 8003. The result of the measurement is the verification of the correct function of the physical calibrator "Model A" in the whole range of simulated charges from pC to nC. The measurement results are shown graphically in Figure 9. The test also included a withstand voltage test and a corona test at a maximum voltage of 30 kV. Corona did not appear in practical measurements.

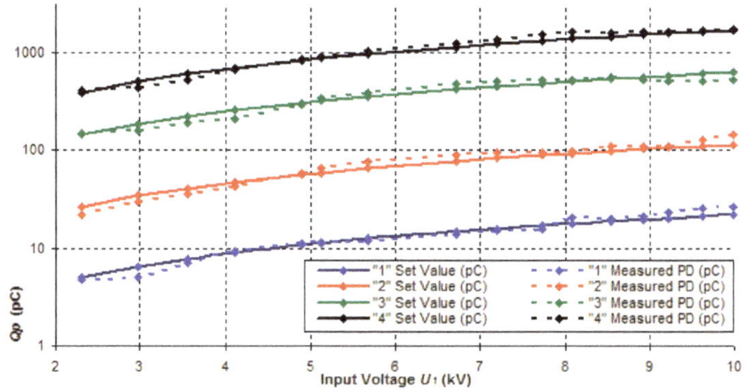

Figure 9. Verifying the Correct Operation of the "Model A" Calibrator.

The differences between the set PD values by calibrator simulation and the measured values by PD meas. systems are small (deviation max. 30% for DDX 8003). The identification of individual simulations corresponds to Table 1. The measurements made by the PD meter DDX 9121b are even more accurate. Calculated deviations are in the order of several percent or tens of percent. These laboratory measurements demonstrated the functionality of the "Model A" in a wide range of configurations and different voltage levels.

Simulation of symmetric and asymmetric faults on "Model C" has also been successful. Malfunctions of insulation condition of MV cables were simulated in the range of 0.5 to 5 pC. The deviations between the simulation and the measured values are about 10%. The data obtained from this model will be used as samples of standardized MV cable faults in the database system.

The Calibrator "Model B" is designed for the measurement of tested material properties. Four different solid materials were selected for testing in the Calibrator "Model B". The measured values for the individual materials are shown in Table 2.

Table 2. Calibrator Model B Different Solid Dielectric Materials PD Measurement.

Material	Air *	Polyethylene	Polystyrene Foam	Bakelite
Set Voltage				
3 kV (2.29 kV)	0.38 pC	0.35 pC	0.35 pC	0.38 pC
5 kV (5.02 kV)	0.35 pC	0.34 pC	0.34 pC	0.39 pC
10 kV (9.89 kV)	0.44 pC	0.35 pC	0.38 pC	0.37 pC
15 kV (15.13 kV)	0.42 pC	0.38 pC	0.37 pC	0.41 pC
20 kV (20.08 kV)	0.51 pC	0.42 pC	0.45 pC	0.39 pC
25 kV (25.21 kV)	0.58 pC	0.48 pC	0.46 pC	0.51 pC
30 kV (29.95 kV)	0.92 pC	0.48 pC	0.87 pC	0.48 pC
35 kV (34.88 kV)	1.34 pC	0.62 pC	1.21 pC	0.79 pC
40 kV (40.12 kV)	1.79 pC	0.63 pC	1.49 pC	0.82 pC
45 kV (45.23 kV)	10.2 pC	1.11 pC	7.73 pC	1.98 pC
50 kV (50.07 kV)	31.8 pC	3.21 pC	24.8 pC	5.21 pC

* Sample "Air" was measured as an empty measuring hole.

Measurements of Calibrator Model B have shown that this device is sensitive enough to measure inhomogeneities, cracks and gas cavities in homogeneous solid insulating materials. This calibrator is therefore mainly suitable as a demonstration tool for simulating partial discharges in solid dielectric materials, for example, used in MV and HV power cables. Again, data for material samples can be stored in the database system for future comparison of measured samples or mutual comparison of different samples.

6. Discussion and Conclusions

This article summarizes the basic information of our long-term research activities in the design, implementation and practical measurement of PD using non-traditional partial discharge calibrators. The purpose of these calibrators is to imitate real test objects more accurately than electronic calibrators. For the time being, the measurements of very low PD values (below 5 pC) in MV cables by these calibrators have also proved to be beneficial. Also, collecting the examples of amplitude and phase spectra for predefined faults of MV cables is of great benefit.

Above all, the possibility to use this type of calibrators for mutual relative comparison of different systems makes it a unique tool for laboratory use, further research in MV and HV cables, new dielectric materials and sensitive PD measurement systems. The simulation possibilities of individual processes in the insulation dielectric materials are a suitable tool for final student's thesis (bachelor's, master's or dissertation thesis). Also in addition, it is possible to use these calibrators for practical laboratory exercises and precise measurement for simulations and practical verification of set parameters and PD on voltage dependences.

The features and options of the Calibrators "Model A" and "Model C" allow the simulation of all possible faults as in the real objects, especially MV cables and instrument transformers. One or more symmetrical or asymmetrical partial discharges can be simulated at the same time, within one period. The "Model B" allows testing the real materials and measuring their properties. Calibrator "Model B" can be used to measure other electrical parameters such as permittivity and dissipation factor in combination with an electronic bridge. And the main measured parameter is a partial discharge depending on the applied voltage. The sensitivity of the "Model B" is sufficient for measurements from 100 fC.

Further development of these calibrators will be directed to increasing the sensitivity, the elimination of external influences and the option settings. Another goal is to reduce the errors between the software simulations and the practical settings of the calibrators.

Author Contributions: Conceptualization, M.K.; Data curation, L.P., J.S. and J.O.; Formal analysis, L.P.; Investigation, L.P.; Methodology, M.K. and J.S.; Project administration, J.O.; Validation, L.P.; Resources, I.K.; Supervision, J.O.; Writing-original draft, M.K. and L.P.; Writing-review & editing, J.O. and I.K.

Funding: This research was funded by the Ministry of Education, Youth and Sports of the Czech Republic under OP VVV Programme (project No. CZ.02.1.01/0.0/0.0/16_013/0001638 CVVOZE Power Laboratories—Modernization of Research Infrastructure).

Acknowledgments: Authors gratefully acknowledge the Centre for Research and Utilization of Renewable Energy (CVVOZE) in the research infrastructure CVVOZEPowerLab.

Conflicts of Interest: The authors declare no conflict of interest.

References

1. *IEC 60270: High-Voltage Test Techniques, Partial Discharge Measurements, Consolidated Version with Amendment 1, Ed. 3.1*; IEC: Geneva, Switzerland, 2015.
2. Lemke, E. A critical review of partial-discharge models. *Electr. Insul. Mag.* **2012**, *28*, 11–16. [CrossRef]
3. Mentlik, V.; Pihera, J.; Polansky, R.; Prosr, P.; Trnka, P. *Diagnostika elektrických zařízení*; BEN—Technicka Literatuta: Prague, Czech Republic; 439p, ISBN 978-80-7300-232-9.
4. Heller, R. *Aspects of Partial Discharge Measurement*; University of West Bohemia: Pilsen, Czech Republic, 2007.
5. Fidan, M.; Ismailoglu, H. A Novel Partial Discharge Calibrator Design via Dual Microcontroller and High Speed DAC. In Proceedings of the ELECO'2007 5th International Conference on Electrical and Electronics Engineering, Bursa, Turkey, 5–9 December 2007.
6. Gunnarsson, O.; Bergman, A.; Rydler, K.E. A Method for Calibration of Partial Discharge Calibrators. *IEEE Trans. Instrum. Meas.* **1999**, *48*, 453–456. [CrossRef]
7. Gupta, A.K.; Ray, S. Modeling of Calibration Circuit for Partial Discharge Measurement. Bacherol's Thesis, Department of Electrical Engineering, National Institute of Technology Rourkela, Odisha, India, 2013.
8. Kimt, K.H.; Yi, S.H.; Lee, H.J.; Kang, D.S. Setup of standard PD calibrator and its uncertainties. *J. Electr. Eng. Technol.* **2011**, *6*, 677–683. [CrossRef]
9. Gemat, A.; Philippoff, W. Die Funkenstrecke mit Vorkondensator. *Z. Für Tech. Phys.* **1932**, *13*, 425–430.
10. Böning, W. Luftgehalt und Luftspaltverteilung geschichteter Dielektrika I. In *Electrical Engineering (Archive fur Elektrotechnik)*; Springer: Berlin, Germany, 1963; Volume 48, ISSN 0948-7921, online ISSN 1432-0487.
11. Chen, G.; Baharudin, F. Partial Discharge Modeling Based on Cylindrical Model in Solid Dielectrics. In Proceedings of the International Conference on Condition Monitoring and Diagnosis, Beijing, China, 21–24 April 2008; IEEE: Pisctaway, NJ, USA, 2008. ISBN 978-14-2441-621-9.

energies

MDPI

Article

Modeling of Dry Band Formation and Arcing Processes on the Polluted Composite Insulator Surface

Jiahong He *, Kang He and Bingtuan Gao

School of Electrical Engineering, Southeast University, Nanjing 210096, China; 220192809@seu.edu.cn (K.H.); gaobingtuan@seu.edu.cn (B.G.)

* Correspondence: hejiahong@seu.edu.cn

Received: 26 September 2019; Accepted: 12 October 2019; Published: 15 October 2019

Abstract: This paper modeled the dry band formation and arcing processes on the composite insulator surface to investigate the mechanism of dry band arcing and optimize the insulator geometry. The model calculates the instantaneous electric and thermal fields before and after arc initialization by a generalized finite difference time domain (GFDTD) method. This method improves the field calculation accuracy at a high precision requirement area and reduces the computational complexity at a low precision requirement area. Heat transfer on the insulator surface is evaluated by a thermal energy balance equation to simulate a dry band formation process. Flashover experiments were conducted under contaminated conditions to verify the theoretical model. Both simulation and experiments results show that dry bands were initially formed close to high voltage (HV) and ground electrodes because the electric field and leakage current density around electrode are higher when compared to other locations along the insulator creepage distance. Three geometry factors (creepage factor, shed angle, and alternative shed ratio) were optimized when the insulator creepage distances remained the same. Fifty percent flashover voltage and average duration time from dry band generation moment to flashover were calculated to evaluate the insulator performance under contaminated conditions. This model analyzes the dry band arcing process on the insulator surface and provides detailed information for engineers in composite insulator design.

Keywords: composite insulator; dry band formation; heat transfer model; generalized finite difference time domain

---

1. Introduction

Composite insulators have been extensively used to provide electrical insulation and mechanical support for high voltage (HV) transmission lines [1–4]. The shank of the composite insulator is made of fiberglass or epoxy, and the sheds of the composite insulator are made of composite materials. The hydrophobic nature of composite materials discretizes water into small droplets on the insulator surface and ensures good performance of insulators under contaminated conditions [5,6]. However, the humid pollution could form a layer under the severely contaminated environment and increase the leakage current density [7,8]. The leakage current generates heat and evaporates water in the pollutant layer to create the dry band [9,10]. The arc initializes due to the significantly increased electric field close to the dry band [11–14]. Theoretical models for dry band formation and arcing processes are valuable because they contribute to the investigation of composite insulator flashover mechanism [15,16] and provide detailed information for engineers to optimize insulator geometry.

B.F. Hampton first studied the formation of dry bands in 1964 [17]. E.C. Salthouse and J.O. Löberg introduced the specific process of dry band formation in 1971. In terms of surface resistivity and electric field, E. C. Salthouse pointed out that dry band formation is caused by energy dissipation [18,19].

J.O. Löberg concluded that the width and speed of dry band formation are related to the surface temperature [20]. The distorted distribution of the field strength of dry bands also plays an important role in dry band expansion [21,22]. By analyzing a 3-D insulator model with the finite element method (FEM), J. Zhou et al. gave the opinion that the distortion field strength increases the length of the dry band. The number of dry bands also influences the electric field distribution [21]. A. Das et al. summarized that the dry band position has a significant influence on the maximum electric field strength [22]. These studies present the characteristics of the dry band and analyze the effects of dry bands on electric field distribution and flashover phenomena. However, the process of dry band formation caused by leakage current and electric arc has not been investigated in detail. This paper proposes a model to analyze instantaneous electric and thermal field variation during the dry band formation and arcing processes. The fields were calculated using a generalized finite difference time domain (GFDTD) method.

The GFDTD method consists of the generalized finite difference method (GFDM) and the finite difference time domain (FDTD) method. The GFDM is a method improved from the finite difference method (FDM). The traditional FDM depends on mesh-dividing, which is not suitable for fields with complicated boundaries, while the GFDM is a meshless method to compute the relationship of any discrete point in the field of the boundary conditions. The GFDM has an advantage over traditional FDM in the sense that the density of the calculation points could be different according to the boundary conditions and precision required in the field domain. The concept of the GFDM was first put forward by J.J. Benito in 2001 [23]. L. Gavate et al. compared the GFDM with other methods and reviewed its application in fluid and force fields [24]. J. Chen et al. then calculated electromagnetic field using the GFDM to reduce the computation time [25]. Currently, GFDM has been used in field calculation problems such as heat transfer and fluid mechanics to increase the calculation accuracy of a relatively small area in a large field domain [26,27].

This paper analyzed instantaneous electric and thermal field distributions close to composite insulators and arcs. Finite difference time domain (FDTD) was utilized to investigate the characteristics of continuously changing fields. In 1966, K.S. Yee dispersed Maxwell's equations with time variables using the method of discretization later-called Yee cell [28], which was gradually developed into FDTD. This paper investigated electric and thermal fields variation by combining the GFDM with FDTD. GFDTD is capable of increasing the calculation accuracy in a high precision requirement area and reducing the computational complexity in a low precision requirement area. The arc propagation and heat transfer processes are modeled based on the electric and thermal field distribution.

Many theories and laboratory experiments have demonstrated that the elongation of the insulator creepage distance is an effective way to increase flashover voltage, but it also increases the weight and reduces the mechanical stress endurance of the composite insulator. Therefore, recent studies have focused on insulator parameter optimization when creepage distances remain the same [29–31]. The simulation models proved that the flashover probability slightly increases with the insulator shank diameter and decreases with shed spacing [32–34]. The creepage factor (*CF*), shed angle, and the ratio of overhangs between alternating sheds are the factors that impact dry band formation and arcing processes on the composite insulator surface. These parameters were optimized in the paper to analyze the 50% flashover voltage and the average duration time from pollutant layer formation to flashover.

This paper investigated the mechanism of dry band arcing by simulating the processes of dry band formation and arcing under the influence of a heat transfer model and an arc propagation model. The simulation results were compared with the results of the experiment to verify the model. This paper optimized the composite insulator geometry when the creepage distances remain the same.

2. Model Schematic and Method

2.1. Insulator Model Schematic

The composite insulator dimension and geometry were selected according to IEC 60815. Due to the symmetric geometry of composite insulators, a two-dimension model was applied to simulate the dry band formation and arc propagation processes and reduce the computational complexity. The composite insulators were designed for a 110 kV transmission line with 15 large sheds and 14 small sheds. The insulator shed radius and the dimensions of the electrodes are show in Figure 1a. The environment temperature and air pressure were 293 K and 101.325 kPa, respectively.

In Figure 1a, the pollution distribution on the top and bottom surface of the insulator was defined as $ESDD_T$ and $ESDD_B$. The flashover voltage reduces with the increase of the ratio of $ESDD_T$ to $ESDD_B$. The range of the ratio was 0.1 to 1 [35]. In this paper, the ratio was set as 1 to simulate the dry band formation and arcing phenomena the under a severe polluted scenario with relatively low flashover voltage. Therefore, the ESDD value was 0.1 mg/cm^2 and the surface resistivity is 8.3×10^5 Ω·m under the influence of environment temperature and air humidity, and water particles in the air were not considered in the model [36,37].

In Figure 1b, θ is the shed angle. CF is defined as the ratio of the insulator creepage distance to the arcing distance.

$$CF = \frac{l_1 + l_2}{d} \quad (1)$$

where the sum of l_1 and l_2 is the total nominal creepage distance of the insulator. d is the arcing distance of the insulator.

r_1 and r_2 are the radius of large and small sheds respectively. k_{shed} is defined as the ratio of r_2 to r_1.

$$k_{shed} = \frac{r_2}{r_1}. \quad (2)$$

In order to reduce the probability of dry band arcing and arc propagation, the geometry structure of insulator was optimized under the premise that creepage distances remain the same. The optimization variables of insulator geometry were CF, k_{shed}, and θ.

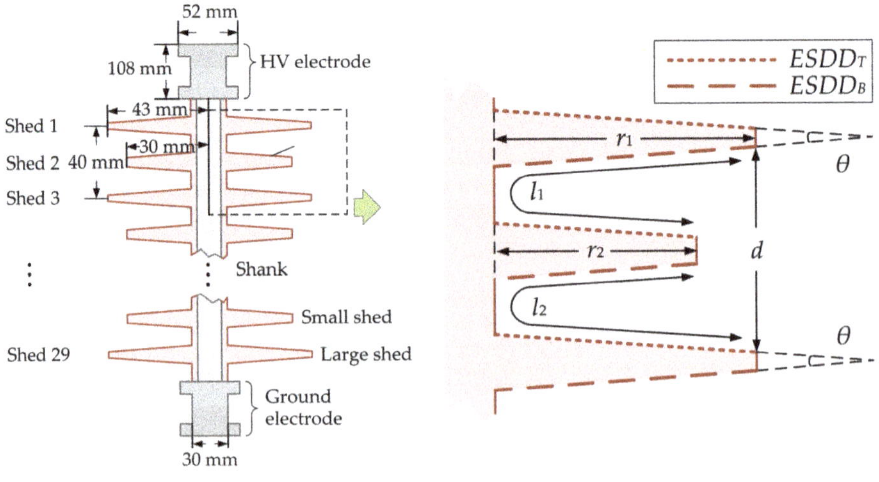

(**a**) Composite insulator schematic (**b**) Geometry parameters

Figure 1. Composite insulator model schematic.

2.2. Dry Band Formation and Arc Propagation Models

2.2.1. Electric Field and Arc Propagation Model

In the electric field close to the insulator, the Poisson is shown below:

$$\begin{cases} \nabla^2\varphi = \frac{\partial^2\varphi}{\partial x^2} + \frac{\partial^2\varphi}{\partial y^2} = -\frac{\rho_c}{\varepsilon} & \text{Possion equation} \\ \varphi(x,y)\big|_\Gamma = f_1(\Gamma) & \text{Dirichlet boundary condition} \\ \frac{\partial\varphi}{\partial n}\big|_\Gamma = f_2(\Gamma) & \text{Neumann boundary condition} \end{cases} \quad (3)$$

where φ is the electric potential, ρ_c is bulk charge density and ε is permittivity.

Before the arc ignition, the electric field calculation model computed the electric field distribution to determine the arc ignition and obtain the leakage current density on the insulator surface. After the arc ignition, the electric field and arc propagation model computed the instantaneous electric field strength around the arc leader during the propagation. The random theory was utilized to determine the arc propagation directions based on the instantaneous electric field.

The instantaneous electric field close to the composite insulator was calculated by the GFDTD method shown in Appendix A. The advantage of GFDTD is that the density of discrete calculation points could be different in the field domain according to the precision requirement and boundary conditions. To focus on the field close to the arc and reduce the computational complexity in the low precision requirement area, the distribution of points close to the arc is denser than the points distribution in other parts of the field (Figure 2a).

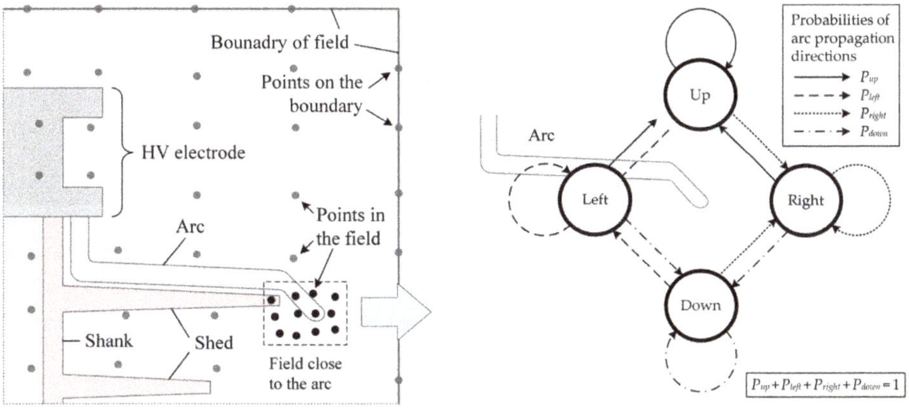

(a) Discrete points in the field domain (b) Random walk theory

Figure 2. Electric field calculation of the discretized points and arc propagation model based on the random walk theory.

Random walk theory calculated the probabilities of arc propagation in all the directions (Figure 2b). The random number was generated at each step of arc propagation to determine the exact direction of the next step. Therefore, the arc growth direction could be different even when the electric field distribution remains the same, which describes the stochastic characteristics of arc propagation [38].

$$P = \frac{E^2}{\sum E^2} a(E - E_c) \quad (4)$$

where E is the electric field strength summation of all possible directions with $E > E_c$, and E_c (2.1 kV/mm) is the RMS value of the threshold field. a is the step function. The arc propagation velocity is in proportion to the magnetite of the electric field strength.

2.2.2. Heat Transfer Model

The heat transfer model simulates the energy balance of the evaporation process, including the leakage current injection energy, heat conduction and convection energies on the insulator surface, heat radiation energy of the arc, and the water evaporation energy of phase changing.

Before the arc ignition, the source of the thermal field was the accumulated energy on the insulator surface generated by the leakage current density. After the arc ignition, the heat transfer model included the heat radiation of the arc as the dominant factor to affect the dry band formation during arc propagation.

The leakage current injection energy is calculated as follows:

$$W_{leakage} = \sum_{t=0}^{t_0} \sum_{i=1}^{l} \frac{{E_i^{t_n}}^2}{\rho_r}. \tag{5}$$

Heat conduction and convection are the main forms of heat dissipation on the composite insulator surface before arc initializes. Heat conduction partial differential equation (PDE) and boundary conditions are given as Equation (6).

$$\begin{cases} \rho c \frac{\partial T}{\partial t} = \lambda \left(\frac{\partial^2 T}{\partial x^2} + \frac{\partial^2 T}{\partial y^2} \right) + \Phi & \text{Heat conduction PDE} \\ T(x,y)\big|_{\Gamma} = f_1(\Gamma) & \text{Dirichlet boundary condition} \\ \frac{\partial T}{\partial n}\Big|_{\Gamma} = f_2(\Gamma) & \text{Neumann boundary condition} \end{cases} \tag{6}$$

where T is the thermal temperature, t is time, ρ, c and λ are the density, specific heat capacity and thermal conductivity of different insulating materials, respectively. Φ is the internal heat sources caused by dry band arcing and the leakage current density of the insulator surface.

The GFDTD method in the heat conduction calculation is similar to the electric field computation. The discretized heat conduction PDE is shown in Equation (7)

$$\rho_i c_i \left(\frac{T_i^{t_{n+1}} - T_i^{t_n}}{\Delta t} \right) = \lambda_i d_{1,1} T_i^{t_{n+1}} + \sum_{j=1}^{n} \lambda_j d_{1,(j+1)} T_j^{t_{n+1}} + \lambda_i d_{2,1} T_i^{t_{n+1}} + \sum_{j=1}^{n} \lambda_j d_{2,(j+1)} T_j^{t_{n+1}} \tag{7}$$

where the superscript "t_{n+1}" represents the next stage in the discrete time domain.

Φ is calculated below:

$$\Phi_i = E_i^{t_n} J_i^{t_n} = \frac{\left(E_i^{t_n}\right)^2}{\rho_r} \tag{8}$$

where E is the electric field strength, J is the leakage current density and ρ_r is the resistivity of the insulator surface.

Thermal conduction and convection energies on the insulator surface are calculated below:

$$W_{conduction} = \sum_{t=0}^{t_0} \sum_{i=0}^{l} \lambda \Delta T_i^{t_n} \tag{9}$$

$$W_{convention} = \sum_{t=0}^{t_0} \sum_{i=0}^{l} h\left(T_i^{t_n} - T_0\right) \tag{10}$$

where l is the length of the insulator creepage distance, t_0 is the time duration, λ is the thermal conductivity of the insulating material, T_i^{tn} is the thermal temperature on the insulator surface, ΔT is the temperature difference as a function of distance and time. T_0 is the environment temperature. h is the heat transfer coefficient of convection.

Heat radiation becomes the dominant factor to cause heat transfer on the insulator surface after arc initialization. Heat radiation is the process of arc generating radiant energy. Arc radiation energy $W_{arc\_radiation}$ is calculated below:

$$W_{arc\_radiation} = \sum_{t=0}^{t_0} \sum_{i=0}^{l} \varepsilon_{emit} \sigma \left(T_i^{t_n}\right)^4 \tag{11}$$

where ε_{emit} is the emissivity of actual objects, $\sigma = 5.67 \times 10^{-8}$ is the Stefan–Boltzmann constant, and t_0 is the time period from the radiation start to the moment of field calculation.

Water in the pollutant layer evaporates during the heat transfer process. The Clausius–Clapeyron equation describes enthalpy variation based on air pressure and thermal temperature.

$$\ln \frac{P_2}{P_1} = \frac{\Delta H_{water}^{steam}}{R} \left(\frac{1}{T_1} - \frac{1}{T_2} \right) \tag{12}$$

where ΔH_{water}^{steam} is the phase-changing enthalpy of water, $R = 8.314$ is the universal gas constant, P_1 and P_2 remain the same as the standard atmospheric pressure (101.325 kPa), and T_1 and T_2 are the thermal temperature change before and after arc initialization. Therefore, ΔH is a function of thermal temperature during the dry band formation and the arc propagation processes. The evaporation energy is calculated in Equations (13) and (14).

$$\Delta H_{water}^{steam} = \frac{RT_1}{T_2} \tag{13}$$

$$W_{water\_steam} = \Delta H_{water}^{steam} V_{water}. \tag{14}$$

The thermal balance equation of dry band formation on the insulator surface is shown below:

$$W_{water\_steam} + W_{conduction} + W_{convection} = W_{arc\_radiation} + W_{leakage}. \tag{15}$$

3. Simulation Results

The dry band formation process from the moment of the insulator energization (t = 0 s) to the moment of arc initialization was simulated, in the first place, to analyze the effects of leakage current density on dry band formation. Then, the arc propagation process was simulated after arc initialization to investigate the effects of arc energy dissipation on further dry band formation and flashover.

3.1. Dry Band Formation and Arcing Simulations

The three stages of dry band formation before arc initialization are shown in Figure 3a–c respectively, when time t equals 0 s, 0.9 s, and 1.6 s.

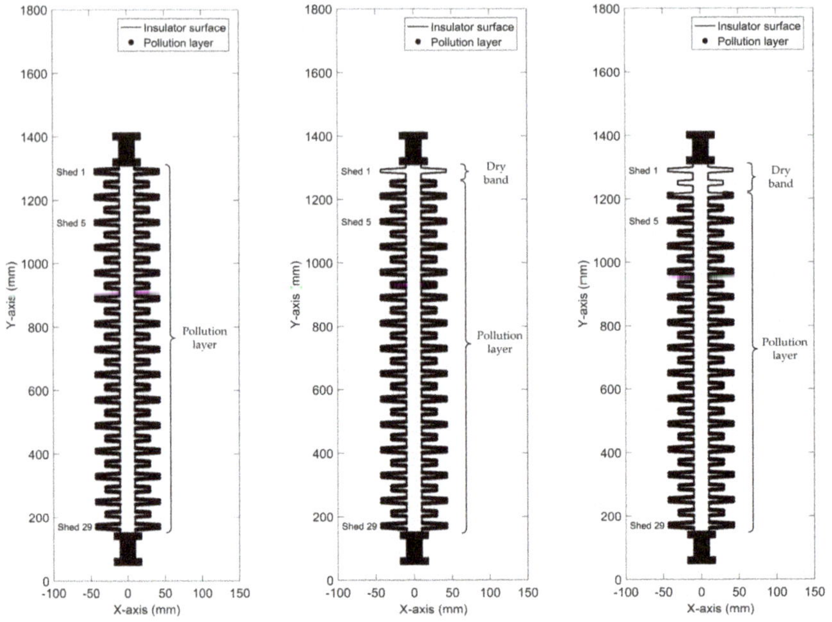

(a) $t = 0$ s (b) $t = 0.9$ s (c) $t = 1.6$ s

Figure 3. Dry band formation process on the insulator surface at different time nodes.

Figure 4 shows the electric and thermal field distributions at the initial state (t = 0 s) in Figure 3a before dry band formation. From Figures 3a and 4b, it is evident that the dry band was first generated at the location with the maximum thermal field.

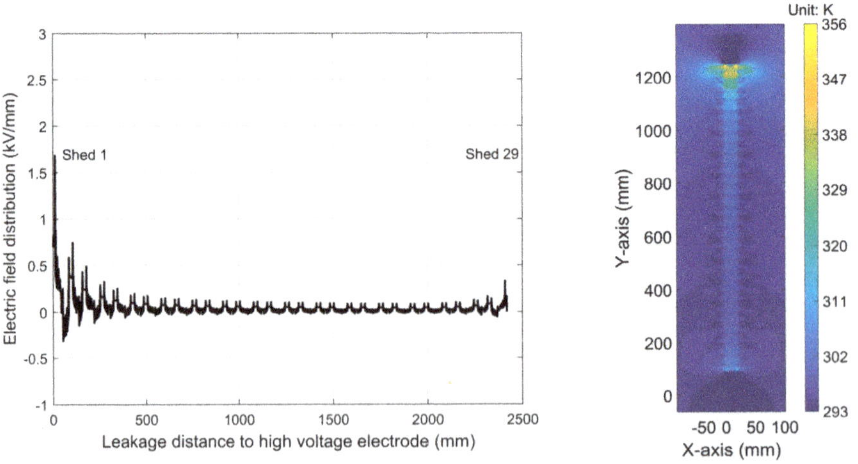

(a) Electric field distribution along the creepage distance (b) Thermal field distribution

Figure 4. Electric and thermal field distributions before the pollutant layer generation. (t = 0 s).

The dry band area expands as the water in the pollutant layer continues evaporating. The thermal field distributions on the insulator surface close to the HV electrode in Figure 3b,c are shown in Figure 5. Figure 5 indicates the mutual positive effects on thermal temperature and dry band length.

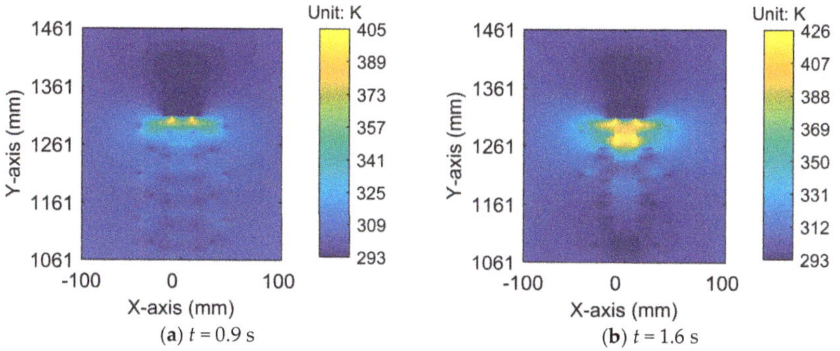

Figure 5. Thermal field distributions close to the high voltage (HV) electrode when the length of the dry band increases.

The electric field distributions on the insulator surface close to the HV electrode in Figure 3a–c are compared in Figure 6. Figure 6 shows that the maximum electric field with the dry band was higher than the maximum electric field without the dry band. The maximum electric field reduced when the dry band expanded.

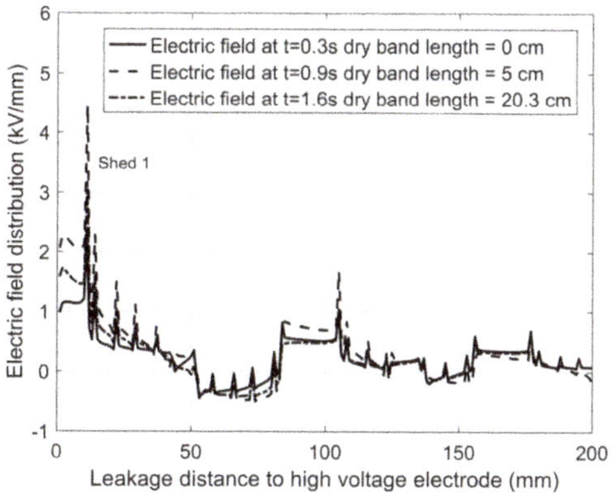

Figure 6. Electric field comparison with different lengths of the dry band.

The arc initializes when the maximum electric field exceeds the dielectric strength of air. However, the arc did not ignite immediately in Figure 3b because the water evaporation consumed the energy so that the maximum electric field could not maintain above the dielectric strength of the air. The arc initialized at $t = 1.6$ s, even though the electric field reduced slightly due to the expansion of the dry band. ($t = 5.42$ s). The first arc initialization and the thermal field distribution are shown in Figure 7. Arc initializes at the location on the insulator surface with the maximum electric field. It is observed that the thermal temperature significantly increases when the arc ignites, the arc thermal radiation dissipates energy from arc to the air and insulator surface. The dominant factor of dry band formation becomes arc energy radiation during the propagation process. However, the arc extinguishes when the length and number of dry bands increase because the leakage current reduces as the surface resistivity at the dry band is dramatically higher than the resistivity at the pollutant layer.

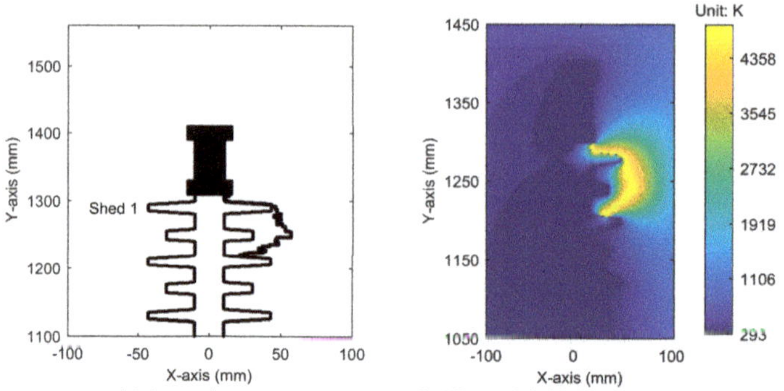

(**a**) Arc trajectory (**b**) Thermal field distribution close to the arc

Figure 7. Arc trajectory and thermal field distributions close to the HV electrode when the first arc ignites.

Two dry bands were generated during the arc propagation process. The dry band and thermal field distributions are shown in Figure 8a,b. The electric field distribution along the creepage distance is shown in Figure 9. Figure 9 shows that the maximum electric field was lower than the maximum field with one dry band in Figure 6. However, the electric field distortion along the creepage distance was more severe than the field distribution with fewer dry bands. The electric field at more than one location on the insulator surface exceeded the dielectric strength of air. Therefore, multiple arcs reignited at different places on the composite insulator surface.

The distorted electric field led to the same distribution of the leakage current density. Therefore, temperature increased more significantly close to the dry band than the other locations on the insulator surface (Figure 8b).

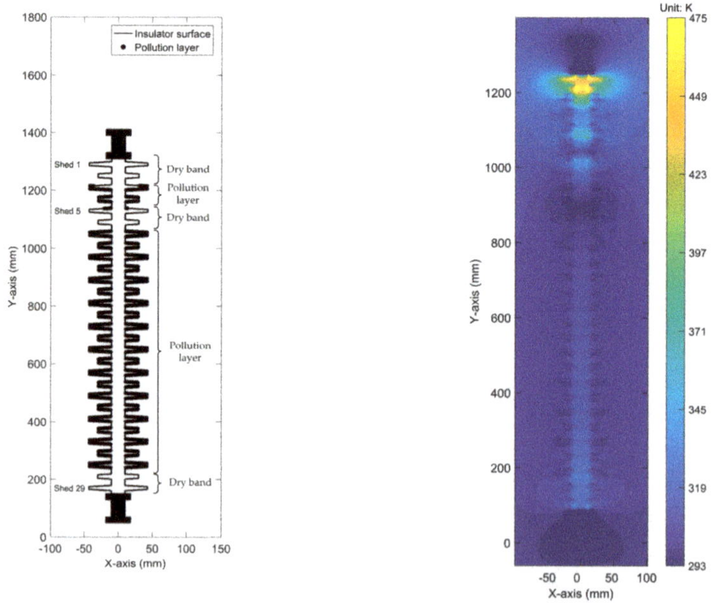

(**a**) Dry bands distribution (**b**) Thermal field distribution

Figure 8. Dry bands and thermal field distributions close to the HV electrode.

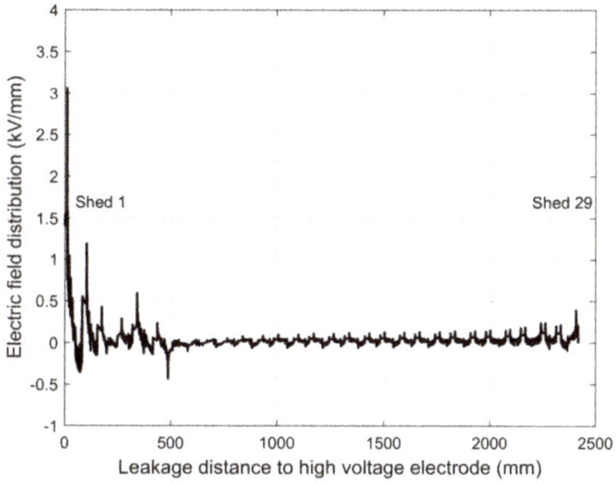

Figure 9. Electric field distribution along the creepage distance with three dry bands on the insulator surface.

Arcs ignited at different locations on the insulator surfaces after multiple dry bands generation when t was equal to 9.89 s (Figure 10a). Arcs distinguished with the expansion of dry band and ignited due to the distorted electric field close to dry bands. The iterations were repeated six times, and the number of arcs significantly increased after each iteration. The separated arcs were finally connected to a conductive path from the HV electrode to the ground electrode of the composite insulator and caused flashover (Figure 10b) when t was equal to 14.64 s.

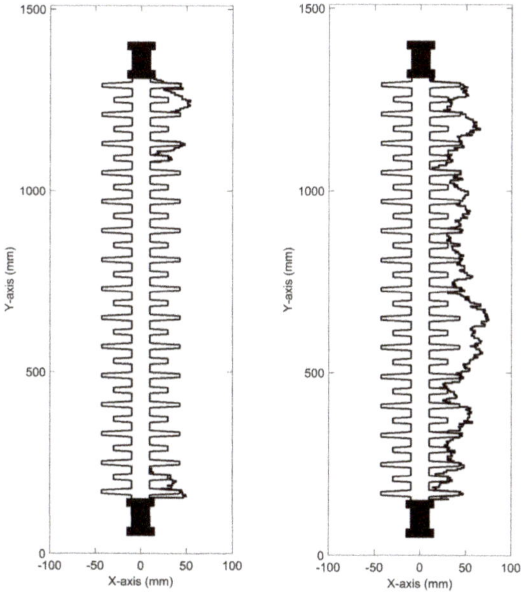

(a) $t = 9.89$ s **(b)** $t = 14.64$ s

Figure 10. Arc trajectory when multiple arcs occur and connect into a conductive path.

It was observed that the arc can jump between insulator sheds rather than traveling along the creepage distance. The arc trajectories could be slightly different due to the random walk theory. The arc jumping between sheds and stochastic characteristics in the model are consistent with the physical phenomena of the arc.

3.2. Experiment Results

The scheme of the experiment system is shown in Figure 11. Due to the hydrophobicity of the composite material, the insulator samples were coated with dry kaolin powder and NaCl to form the contamination layer [39]. The ESDD value of the contamination layer was 0.1 mg/cm^2 to evaluate the dry band formation and arcing processes under a severely polluted scenario. The samples were firstly wetted by the clean fog and then energized with 110 kV at rated voltage to observe the dry band formation and arcing processes. The surface resistivity was 8.3 × 10^5 Ω·m. The HV and ground electrodes had a diameter of 52 mm and a length of 108 mm as shown in Figure 1a. The high-speed camera (2F01) recorded 500 video frames per second at 800 pixels × 600 pixels from the start of the experiment to flashover. The videos were transmitted to the computer with 400 MB/s Ethernet.

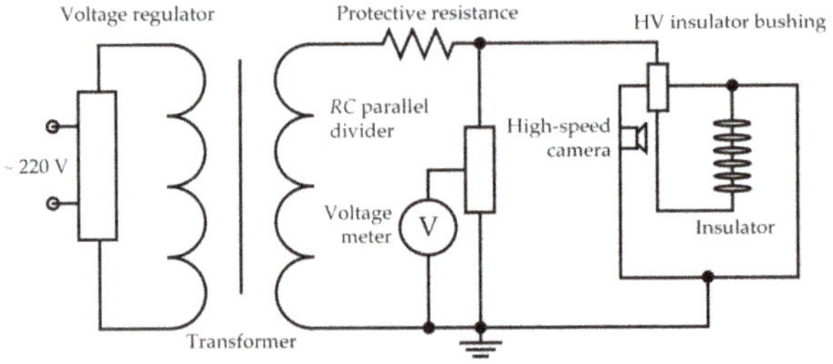

Figure 11. The schematic of the dry band formation and arcing experiment system.

Figure 12 shows the experiment results of the dry band formation and arc propagation processes. The arc propagation at different time frames was compared to analyze the dry band location and arcing phenomena. Figure 12a shows the dry band formation process before arc ignition. Figure 12b shows the first arc ignition due to the dry band close to the HV electrode. Figure 12c shows the arcs reignite at different locations due to the presence of multiple dry bands on the insulator surface. Figure 12d shows that the separated arcs were connected and led to flashover. The time frames and phenomena are consistent with the simulation results in the model.

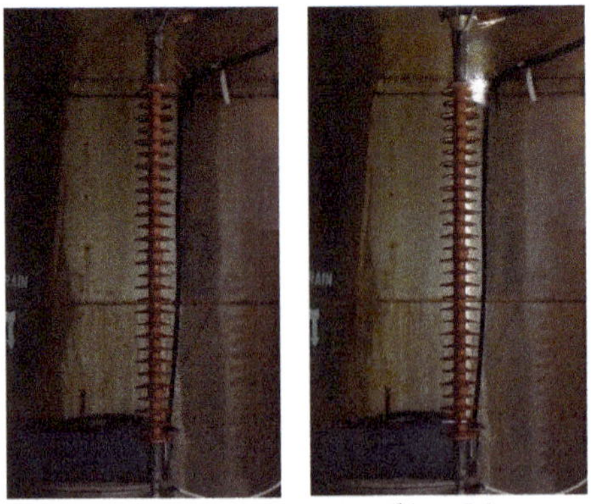

(a) $t = 1.6$ s **(b)** $t = 5.6$ s

Figure 12. *Cont.*

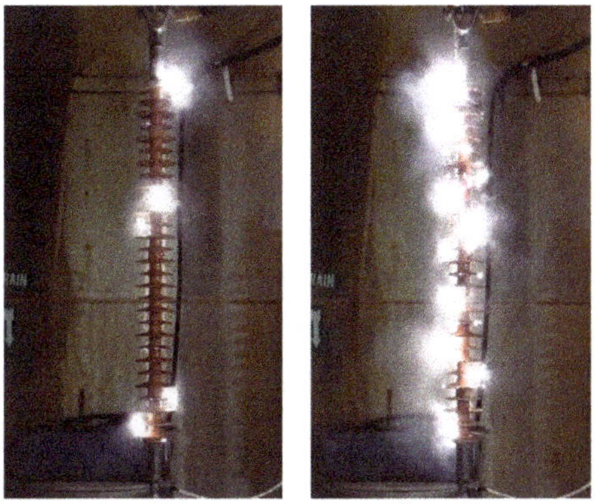

(c) $t = 10.1$ s (d) $t = 15.4$ s

Figure 12. The experiment results at different time nodes of dry bands formation and arc propagation processes.

The time nodes during arc propagation of the experimental and simulation results are compared in Table 1 to validate the simulation model.

Table 1. Comparison of time nodes between experimental and simulation results.

Time Nodes	Simulation (s)	Experiment (s)	Error (%)
Dry band formation	1.6	1.6	0
Arc igniting	5.42	5.6	3.2
Arc extinction	5.51	5.8	5
Arc reigniting	8.57	9.7	11.6
Multiple arc occurrences	9.89	10.1	2.1
Flashover	14.64	15.4	4.9

The time node errors between simulation and experimental results were caused by the stochastic characteristics of the arc propagation.

4. Insulator Geometry Optimization

Since the experiment results verify the dry band formation and arc propagation model, three factors (creepage factor, shed angle, and alternative shed ratio) of composite insulator geometry were optimized in this section, while the creepage distance of the composite insulator remained the same as 2416 mm and the ESDD value was 0.1 mg/cm^2. Due to the stochastic property of arc propagation, the dry band formation and arcing processes were repeated 187 times to calculate the 50% flashover voltage and average duration time from $t = 0$ s to the moment of flashover. The *CF* is defined as the ratio of insulator creepage distance versus the arcing distance. The shed angle is the angle of the shed surface slope. The alternative shed ratio is the ratio of the small shed radius to the large shed radius.

4.1. Creepage Factor Optimization

The *CF* was optimized when the shed angle was 10° and the alternative shed ratio was 0.8. The range of *CF* was from 2.5 to 3.5 according to IEC 60815. The insulator geometry with different *CF* values are shown in Figure 13. 50% flashover voltage and time duration as functions of *CF* value are shown in Figure 14. Figure 14 indicates that 50% flashover voltage first increased then reduced with the *CF* values. The optimal value of *CF* was 2.94 to achieve the minimum flashover voltage. The average duration time decreased with the *CF* values because the arc had a high probability to bridge sheds as the distance between sheds reduced with the increase of *CF*.

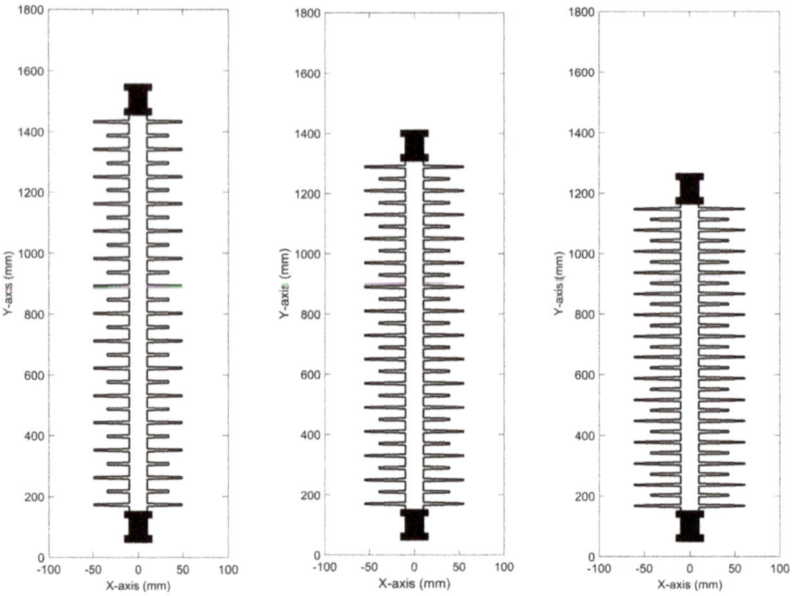

(**a**) $CF = 2.8$ (**b**) $CF = 3.0$ (**c**) $CF = 3.2$

Figure 13. Insulator geometry with different *CF* values ($\theta = 10°$ and $k_{shed} = 0.8$).

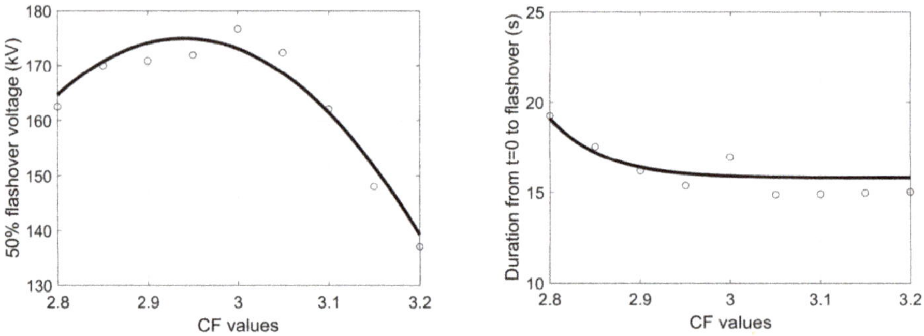

(**a**) 50% flashover voltage as a function of *CF* (**b**) Time duration as a function of *CF*

Figure 14. 50% flashover voltage and time duration as functions of the *CF* value.

4.2. Shed Angle Optimization

The shed angle was optimized when the *CF* value was 3.0 and the alternative shed ratio was 0.8. The range of shed angle was from 0° to 25° according to IEC 60815. The insulator geometry with different shed angles are shown in Figure 15. 50% flashover voltage and time duration as functions of shed angle are shown in Figure 16. Figure 16 indicates that shed angle had little impact on the 50% flashover voltage. The average duration time increased with shed angle because the average length of arc trajectories increased with shed angle.

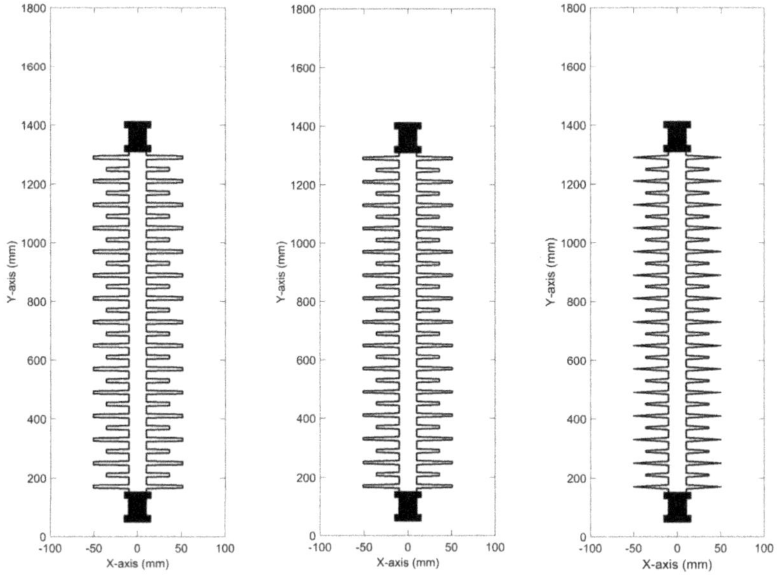

(a) $\theta = 5°$ (b) $\theta = 10°$ (c) $\theta = 20°$

Figure 15. Insulator geometry with different shed angles (CF = 3.0 and k_{shed} = 0.8).

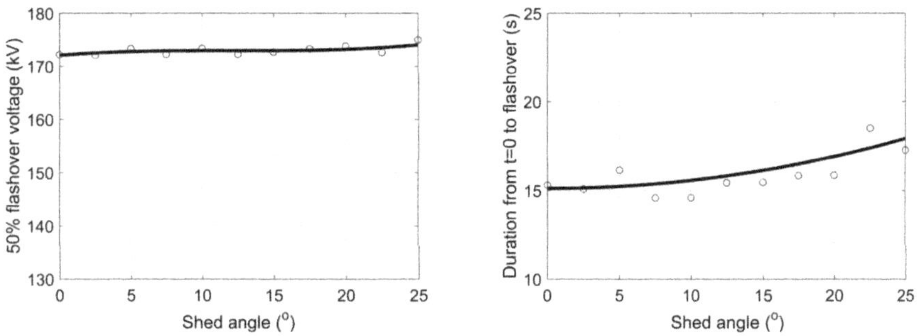

(a) 50% flashover voltage as a function of angle (b) Time duration as a function of angle

Figure 16. 50% flashover voltage and time duration as functions of the shed angle.

4.3. Alternative Shed Ratio Optimization

The alternative shed ratio was optimized when the CF value was 3.0 and the shed angle was 10°. The range of alternative shed ratio was from 0.7 to 1.0 according to IEC 60815. The insulator geometry with different alternative shed ratios are shown in Figure 17. 50% flashover voltage and time duration as functions of alternative shed ratio are shown in Figure 18. Figure 18 indicates that the 50% flashover voltage increased with alternative shed ratio because a small alternative shed ratio leads to the increasing occurrence of arc bridging between sheds. The alternative shed ratio had little impact on the average duration time.

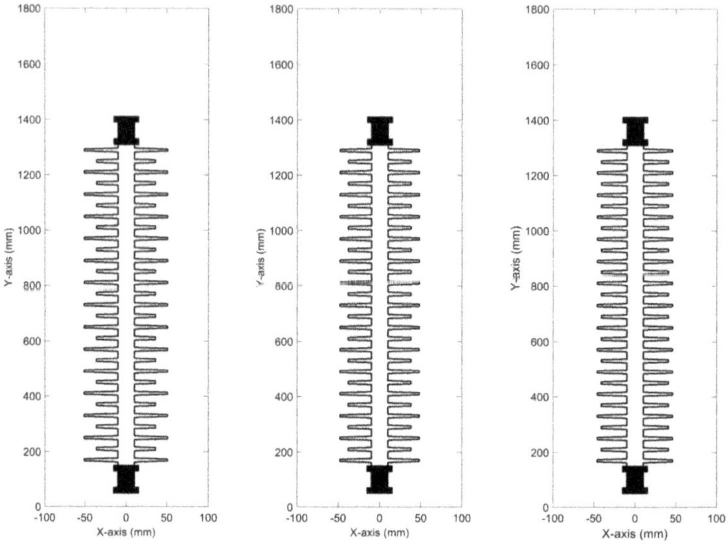

(**a**) $k_{\text{shed}} = 0.7$ (**b**) $k_{\text{shed}} = 0.8$ (**c**) $k_{\text{shed}} = 0.9$

Figure 17. Insulator geometry with different alternative shed ratios ($CF = 3.0$ and $\theta = 10°$).

(**a**) 50% flashover voltage as a function of ratio (**b**) Time duration as a function of ratio

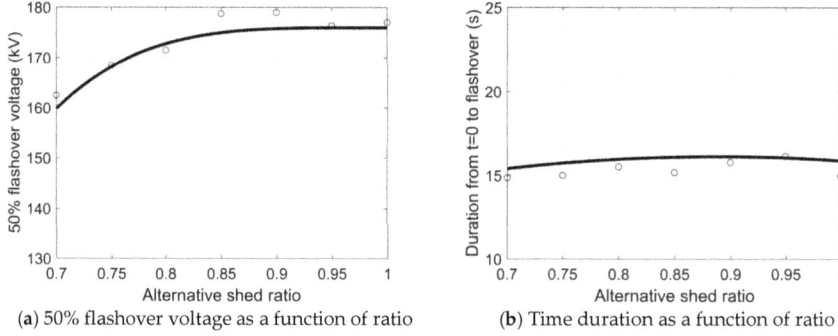

Figure 18. 50% flashover voltage and time duration as functions of the alternative shed ratio.

5. Conclusions

This paper modeled the dry band formation and arcing processes of polluted composite insulators. Instantaneous electric and thermal fields were calculated by the GFDTD method to investigate the mechanism of dry band arcing and flashover. The simulation results were verified by the laboratory experiments. Insulator dimension factors were analyzed to optimize insulator geometry when the creepage distances remained the same.

1. The GFDTD method is suitable to calculate the electric and thermal fields for the insulator geometry by improving the field calculation accuracy at the high precision requirement area and reducing the computational complexity at the low precision requirement area.
2. The stochastic characteristics and the arc trajectory jumping between insulator sheds were modelled to simulate the physical phenomena of the arc.
3. The maximum electric field decreases with the expansion of the dry band. The heat transfer model demonstrates that the leakage current density is the dominant factor to affect dry band

formation before the arc initialization, while the arc radiation becomes the dominant factor to form the dry band after the arc ignition.

4. The 50% flashover voltage of composite insulators increases with the decrease of the *CF* value and the increase of the alternative shed ratio. The duration time from the pollution layer generation moment to flashover increases with the decrease of the *CF* value and the increase of the alternative shed angle.

Author Contributions: Conceptualization: J.H.; Methodology: J.H.; Software: J.H. and K.H.; Validation: K.H. and B.G.; Formal Analysis: J.H. and K.H.; Investigation: J.H.; Experiment: B.G.; Writing—Original Draft Preparation: J.H.; Writing—Review and Editing: J.H., K.H., and B.G.

Funding: This research was supported by the National Natural Science Foundation of China (Grant No. 51807028), the Basic Research Program of Jiangsu Province (Grant No. BK20170672).

Conflicts of Interest: The authors declare no conflict of interest.

Nomenclature

A	coefficient matrix multiply with $[D_\varphi]$
$(a^{-1})_{r,c}$	element at r-th row and c-th column of matrix $[A]^{-1}$
$a(E - E_c)$	step function of random walk
B	coefficient matrix multiply with $[\varphi]$
$B(u)$	residual function of two discrete points
b	constant matrix which equals to the product of $[A]$ and $[D_\varphi]$
$b_{r,\,c}$	element at r-th row and c-th column of matrix $[B]$
c	specific heat capacity
CF	creepage factor
D	constant matrix which equals to the product of $[A]^{-1}$ and $[B]$
D_u	partial difference column matrix
d	insulator arcing distance
$d_{1,\,2,\,3\ldots}$	distance between two discrete points
$d_{r,\,c}$	element at r-th row and c-th column of matrix $[D]$
E	electric field strength
E_c	RMS value of the threshold field
E_i^{tn}	electric field strength in GFDTD form
$ESDD_B$	ESDD value of bottom part of the insulator
$ESDD_T$	ESDD value of top part of the insulator
ΔH_{water}^{steam}	phase changing enthalpy of water
h	heat transfer coefficient of convection
h_{ij}	absolute value X coordinate differences between two discrete points
J	leakage current density
J_i^{tn}	leakage current density in GFDTD form
k_{ij}	absolute value Y coordinate differences between two discrete points
k_{shed}	ratio of radii of large and small sheds
l	length of insulator leakage distance
l_1, l_2	insulator leakage distance
P	probability of random walk
$P_{1,\,2,\,3\ldots}$	discrete points
p_1, p_2	saturated vapor pressure
R	universal gas constant
r_1, r_2	radius of large and small sheds
T	thermal temperature
T_0	environment temperature
T_1, T_2	thermal temperature change before and after arc initialization
t	time
t_0	time duration of insulator current leakage

u	column matrix of discrete point values
u_i	value of field at a discrete point
V	volume
$W_{arc\_radiation}$	heat radiation energy of arcs
$W_{conduction}$	energy of heat conduction
$W_{convention}$	energy of heat convention
$W_{leakage}$	energy of leakage current
W_{water_steam}	required energy for water evaporation
$w_{1,2,3\ldots}$	weight function of discrete points in residual function
Γ	field boundary
ε	permittivity
ε_i	permittivity in GFDTD form
ε_{emit}	emissivity
θ	shed angle
λ	thermal conductivity
ρ	density
ρ_c	bulk charge density
ρ_i^{tn}	bulk charge density in GFDTD form
ρ_r	resistivity
σ	Stefan-Boltzmann constant
Φ	internal heat sources
Φ_i	internal heat sources in GFDTD form
φ	electric potential
φ_i^{tn}	electric potential in GFDTD form

Appendix A

The GFDTD method is used to calculate the electric and thermal field distributions.

$$\frac{\partial u}{\partial t} = \frac{\partial^2 u}{\partial x^2} + \frac{\partial^2 u}{\partial y^2} + C. \quad \text{(A1)}$$

The advantage of GFDTD is that the density of discrete calculation points could be different in the field domain based on the precision requirement and boundary conditions.

According to the Taylor series expansion, the value of u_j at point P_j of the function at near neighborhood of P_i is expressed as follows (Figure A1) [40]:

$$u_j = u_i + h_{ij}\frac{\partial u_i}{\partial x} + k_{ij}\frac{\partial u_i}{\partial y} + \frac{1}{2}\left(h_{ij}^2\frac{\partial^2 u_i}{\partial x^2} + k_{ij}^2\frac{\partial^2 u_i}{\partial y^2}\right) + h_{ij}k_{ij}\frac{\partial^2 u_i}{\partial x \partial y} \quad i = 1,2,\ldots,m \quad \text{(A2)}$$

where h_{ij} and k_{ij} are absolute values of X and Y coordinate differences, i.e., $h_{ij} = |x_j - x_i|$, $k_{ij} = |y_j - y_i|$.

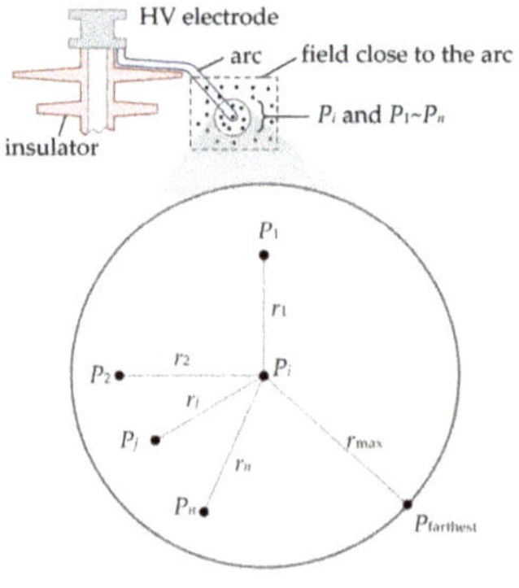

Figure A1. Generalized finite difference time domain (GFDTD) in field calculation.

P_i is the point among P_1~P_n. The value of each point P_i and P_1~P_n is u_i and u_1~u_n. The distance from each point P_1~P_n to P_i and is r_1~r_n, and the farthest distance is r_{max}.

The residual function of two points $B(u)$ is defined by Equation (A3), shown as follows:

$$B(u) = \sum_{j=1}^{n}\left[\left(u_i - u_j + h_{ij}\frac{\partial u_j}{\partial x} + k_{ij}\frac{\partial u_j}{\partial y} + \frac{h_{ij}{}^2}{2}\frac{\partial^2 u_j}{\partial x^2} + \frac{k_{ij}{}^2}{2}\frac{\partial^2 u_j}{\partial y^2}\right)w_j\right]^2. \tag{A3}$$

The weight function of the j-th point w_j is calculated below:

$$w_j = 1 - 6\left(\frac{r_j}{r_{max}}\right)^2 + 8\left(\frac{r_j}{r_{max}}\right)^3 - 3\left(\frac{r_j}{r_{max}}\right)^4. \tag{A4}$$

Derive $B(u)$ for $\partial^2 u/\partial x^2$ and $\partial^2 u/\partial y^2$, then get $[A][D_u] = [b]$, where matrixes $[A]$, $[D_u]$ and $[b]$ are shown as follows,

$$[A] = \begin{bmatrix} \sum_{j=1}^{n} \frac{h_{ij}{}^4 w_j{}^2}{4} & \sum_{j=1}^{n} \frac{h_{ij}{}^2 k_{ij}{}^2 w_j{}^2}{4} \\ \sum_{j=1}^{n} \frac{h_{ij}{}^2 k_{ij}{}^2 w_j{}^2}{4} & \sum_{j=1}^{n} \frac{k_{ij}{}^4 w_j{}^2}{4} \end{bmatrix} \tag{A5}$$

$$[D_u] = \begin{bmatrix} \frac{\partial^2 u}{\partial x^2} & \frac{\partial^2 u}{\partial y^2} \end{bmatrix}^{\mathrm{T}} \tag{A6}$$

$$[b] = \begin{bmatrix} \sum_{j=1}^{n} (u_j - u_i)\frac{h_{ij}{}^2 w_j{}^2}{2} & \sum_{j=1}^{n} (u_j - u_i)\frac{k_{ij}{}^2 w_j{}^2}{2} \end{bmatrix}^{\mathrm{T}} \tag{A7}$$

Decompose the matrix $[b]$ as $[b] = [B][u]$, where

$$[B] = \begin{bmatrix} \sum_{j=1}^{n} \frac{-h_{ij}{}^2 w_j{}^2}{2} & \frac{h_{i1}{}^2 w_1{}^2}{2} & \frac{h_{i2}{}^2 w_2{}^2}{2} & \cdots & \frac{h_{ij}{}^2 w_j{}^2}{2} & \cdots & \frac{h_{in}{}^2 w_n{}^2}{2} \\ \sum_{j=1}^{n} \frac{-k_{ij}{}^2 w_j{}^2}{2} & \frac{k_{i1}{}^2 w_1{}^2}{2} & \frac{k_{i2}{}^2 w_2{}^2}{2} & \cdots & \frac{k_{ij}{}^2 w_j{}^2}{2} & \cdots & \frac{k_{in}{}^2 w_n{}^2}{2} \end{bmatrix} \tag{A8}$$

$$[u] = \begin{bmatrix} u_i & u_1 & u_2 & \cdots & u_j & \cdots & u_n \end{bmatrix}^{\mathrm{T}} \tag{A9}$$

The matrix $[D_u]$ is written in another form, i.e., $[D_u] = [A]^{-1}[b] = [A]^{-1}[B][u] = [D][u]$, where $[D] = [A]^{-1}[B]$ is a matrix with two rows and $(n + 1)$ columns shown below:

$$[D] = [A]^{-1}[B] = \begin{bmatrix} \sum_{i=1}^{2} (a^{-1})_{1,i} b_{i,1} & \sum_{i=1}^{2} (a^{-1})_{1,i} b_{i,2} & \sum_{i=1}^{2} (a^{-1})_{1,i} b_{i,3} & \cdots & \sum_{i=1}^{2} (a^{-1})_{1,i} b_{i,n+1} \\ \sum_{i=1}^{2} (a^{-1})_{2,i} b_{i,1} & \sum_{i=1}^{2} (a^{-1})_{2,i} b_{i,2} & \sum_{i=1}^{2} (a^{-1})_{2,i} b_{i,2} & \cdots & \sum_{i=1}^{2} (a^{-1})_{2,i} b_{i,n+1} \end{bmatrix} \tag{A10}$$

where $(a^{-1})_{r,c}$ and $b_{r,c}$ are the elements at r-th row and c-th column of matrix $[A]^{-1}$ and $[B]$.

Thus, $\partial^2 u/\partial x^2$ and $\partial^2 u/\partial y^2$ are written as

$$\begin{cases} \frac{\partial^2 u}{\partial x^2} = d_{1,1} u_i + \sum_{j=1}^{n} d_{1,(j+1)} u_j \\ \frac{\partial^2 u}{\partial y^2} = d_{2,1} u_i + \sum_{j=1}^{n} d_{2,(j+1)} u_j \end{cases} \tag{A11}$$

where $d_{r,c}$ is the element at r-th row and c-th column of matrix $[D]$.

After substituting Equation (A11) into Equation (A1), the PDE is written as

$$\frac{u_i^{t_{n+1}} - u_i^{t_n}}{\Delta t} = d_{1,1} u_i^{t_n} + \sum_{j=1}^{n} d_{1,(j+1)} u_j^{t_n} + d_{2,1} u_i^{t_n} + \sum_{j=1}^{n} d_{2,(j+1)} u_j^{t_n} + C_i^{t_n} \tag{A12}$$

where the superscript "t_n" and "t_{n+1}" are the present and next stages of the point u, respectively.

References

1. Gorur, R.S.; Cherney, E.A.; Burnham, J.T. *Outdoor Insulators*; Ravi, S., Ed.; Gorur Inc.: Phoenix, AZ, USA, 1999.
2. Gorur, R.S. Status Assessment of Composite Insulators for Outdoor HV Applications. In Proceedings of the 5th International Conference on Properties and Applications of Dielectric Materials, Seoul, Korea, 25–30 May 1997.
3. Headley, P. Development and Application Experience with Composite Insulators for Overhead Lines. In Proceedings of the IEE Colloquium on Non-Ceramic Insulators for Overhead Lines, London, UK, 16 October 1992.
4. Clift, S. Composite Fiber Optic Insulators and Their Application to High Voltage Sensor Systems. In Proceedings of the IEE Colloquium on Structural Use of Composites in High Voltage Switchgear/Transmission Networks, London, UK, 16 October 1992.
5. Ilomuanya, C.S.; Nekahi, A.; Farokhi, S. Acid Rain Pollution Effect on the Electric Field Distribution of a Glass Insulator. In Proceedings of the 2018 IEEE International Conference on High Voltage Engineering and Application (ICHVE), Athens, Greece, 10–13 September 2018.
6. Liu, Y.; Wang, J.; Zhou, M.; Fang, C.; Zhou, W. Research on the Silicone Rubber Sheds Performance of Composite Insulator. In Proceedings of the 2008 International Conference on High Voltage Engineering and Application, Chongqing, China, 9–12 November 2008.
7. Hussain, M.M.; Farokhi, S.; McMeekin, S.G.; Farzaneh, M. Effect of Cold Fog on Leakage Current Characteristics of Polluted Insulators. In Proceedings of the 2015 International Conference on Condition Assessment Techniques in Electrical Systems (CATCON), Bangalore, India, 10–12 December 2015.
8. Hussain, M.M.; Farokhi, S.; McMeekin, S.G.; Farzaneh, M. Mechanism of Saline Deposition and Surface Flashover on Outdoor Insulators near Coastal Areas Part II: Impact of Various Environment Stresses. *IEEE Trans. Dielectr. Electr. Insul.* **2017**, *24*, 1068–1076. [CrossRef]
9. Hussain, M.; Farokhi, S.; McMeekin, S.G.; Farzaneh, M. Effect of Uneven Wetting on E-field Distribution along Composite Insulators. In Proceedings of the 2016 IEEE Electrical Insulation Conference (EIC), Montreal, QC, USA, 19–22 June 2016.
10. Hussain, M.M.; Farokhi, S.; McMeekin, S.G.; Farzaneh, M. Dry Band Formation on HV Insulators Polluted with Different Salt Mixtures. In Proceedings of the 2015 IEEE Conference on Electrical Insulation and Dielectric Phenomena (CEIDP), Ann Arbor, MI, USA, 18–21 October 2015.
11. Nekahi, A.; McMeekin, S.G.; Farzaneh, M. Ageing and Degradation of Silicone Rubber Insulators due to Dry Band Arcing under Contaminated Conditions. In Proceedings of the 2017 52nd International Universities Power Engineering Conference (UPEC), Heraklion, Greece, 28–31 August 2017.
12. Qiao, X.; Zhang, Z.; Jiang, X.; Li, X.; He, Y. A New Evaluation Method of Aging Properties for Silicon Rubber Material Based on Microscopic Images. *IEEE Access* **2019**, *7*, 15162–15169. [CrossRef]
13. Kim, S.H.; Cherney, E.A.; Hackam, R. Effect of Dry Band Arcing on the Surface of RTV Silicone Rubber Coatings. In Proceedings of the Record of the 1992 IEEE International Symposium on Electrical Insulation, Baltimore, MD, USA, 7–10 June 1992.
14. Qiao, X.; Zhang, Z.; Jiang, X.; Liang, T. Influence of DC Electric Fields on Pollution of HVDC Composite Insulator Short Samples with Different Environmental Parameters. *Energies* **2019**, *12*, 2304. [CrossRef]
15. Hussain, M.M.; Chaudhary, M.A.; Razaq, A. Mechanism of Saline Deposition and Surface Flashover on High-Voltage Insulators near Shoreline: Mathematical Models and Experimental Validations. *Energies* **2019**, *12*, 3685. [CrossRef]
16. Hussain, M.M.; Farokhi, S.; McMeekin, S.G.; Farzaneh, M. Observation of Surface Flashover Process on High Voltage Polluted Insulators near Shoreline. In Proceedings of the 2016 IEEE International Conference on Dielectrics (ICD), Montpellier, France, 3–7 July 2016.
17. Hampton, B.F. Flashover Mechanism of Polluted Insulation. *Proc. Inst. Electr. Eng.* **1964**, *111*, 985–990. [CrossRef]
18. Salthouse, E.C. Dry-Band Formation and Flashover in Uniform-Field Gaps. *Proc. Inst. Electr. Eng.* **1971**, *118*, 630. [CrossRef]
19. Salthouse, E.C. Initiation of dry bands on polluted insulation. *Proc. Inst. Electr. Eng.* **1968**, *115*, 1707–1712. [CrossRef]
20. Löberg, J.O.; Salthouse, E.C. Dry-Band Growth on Polluted Insulation. *IEEE Trans. Electr. Insul.* **1971**, *EI-6*, 136–141.

21. Zhou, J.; Gao, B.; Zhang, Q. Dry Band Formation and Its Influence on Electric Field Distribution along Polluted Insulator. In Proceedings of the 2010 Asia-Pacific Power and Energy Engineering Conference, Chengdu, China, 28–31 March 2010.
22. Das, A.; Ghosh, D.K.; Bose, R.; Chatterjee, S. Electric Stress Analysis of a Contaminated Polymeric Insulating Surface in Presence of Dry Bands. In Proceedings of the 2016 International Conference on Intelligent Control Power and Instrumentation (ICICPI), Kolkata, India, 21–23 October 2016.
23. Benito, J.J.; Urena, F.; Gavate, L. Influence of Several Factors in the Generalized Finite Difference Method. *Appl. Math. Model.* **2001**, *25*, 1039–1053. [CrossRef]
24. Gavete, L.; Gavate, M.L.; Benito, J.J. Improvements of Generalized Finite Difference Method and Comparison with Other Meshless Method. *Appl. Math. Model.* **2003**, *27*, 831–847. [CrossRef]
25. Chen, J.; Gu, Y.; Wang, M.; Chen, W.; Liu, L. Application of the Generalized Finite Difference Method to Three-dimensional Transient Electromagnetic Problems. *Eng. Anal. Bound. Elem.* **2018**, *92*, 257–266. [CrossRef]
26. Ureña, F.; Gavete, L.; García, A.; Benito, J.J.; Vargas, M. Solving Second Order Non-linear Parabolic PDEs Using Generalized Finite Difference Method (GFDM). *J. Comput. Appl. Math.* **2019**, *354*, 211–241. [CrossRef]
27. Suchde, P.; Kuhnert, J.; Tiwari, S. On Meshfree GFDM Solvers for the Incompressible Navier–Stokes Equations. *Comput. Fluids* **2018**, *165*, 1–12. [CrossRef]
28. Yee, K. Numerical Solution of Initial Boundary Value Problems Involving Maxwell's Equations in Isotropic Media. *IEEE Trans. Antennas Propag.* **1966**, *14*, 302–307.
29. Hou, K.; Li, W.; Ma, L.; Cheng, Y.; Jin, L. Multi-Objective Structural Optimization of UHV Composite Insulators based on Pareto Dominance. In Proceedings of the 2018 12th International Conference on the Properties and Applications of Dielectric Materials (ICPADM), Xi'an, China, 20–24 May 2018.
30. Li, L. Shed Parameters Optimization of Composite Post Insulators for UHV DC Flashover Voltages at High Altitudes. *IEEE Trans. Dielectr. Electr. Insul.* **2015**, *22*, 169–176. [CrossRef]
31. Doufene, D.; Bouazabia, S.; Ladjici, A.A. Shape Optimization of a Cap and Pin Insulator in Pollution Condition Using Particle Swarm and Neural Network. In Proceedings of the 2017 5th International Conference on Electrical Engineering-Boumerdes (ICEE-B), Boumerdes, Algeria, 29–31 October 2017.
32. Liu, L. The Influence of Electric Field Distribution on Insulator Surface Flashover. In Proceedings of the 2018 IEEE Conference on Electrical Insulation and Dielectric Phenomena (CEIDP), Cancun, Mexico, 21–24 October 2018.
33. Yu, X.; Yang, X.; Zhang, Q.; Yu, X.; Zhou, J.; Liu, B. Effect of Booster Shed on Ceramic Post Insulator Pollution Flashover Performance Improvement. In Proceedings of the 2016 IEEE International Conference on High Voltage Engineering and Application (ICHVE), Chengdu, China, 19–22 September 2016.
34. Jiang, X.; Yuan, J.; Zhang, Z.; Hu, J.; Sun, C. Study on AC Artificial-Contaminated Flashover Performance of Various Types of Insulators. *IEEE Trans. Power Deliv.* **2007**, *22*, 2567–2574. [CrossRef]
35. Jiang, X.; Zhang, Z.; Hu, J. Investigation on Flashover Voltage and Non-uniform Pollution Correction Coefficient of DC Composite Insulator. In Proceedings of the 2008 International Conference on High Voltage Engineering and Application, Chongqing, China, 9–12 November 2008.
36. He, J.; Gorur, R.S. Flashover of Insulators in a Wet Environment. *IEEE Trans. Dielectr. Electr. Insul.* **2017**, *24*, 1038–1044. [CrossRef]
37. Abd Rahman, M.S.B.; Izadi, M.; Ab Kadir, M.Z.A. Influence of Air Humidity and Contamination on Electrical Field of Polymer Insulator. In Proceedings of the 2014 IEEE International Conference on Power and Energy (PECon), Kuching, Malaysia, 1–3 December 2014.
38. He, J.; Gorur, R.S. A Probabilistic Model for Insulator Flashover under Contaminated Conditions. *IEEE Trans. Dielectr. Electr. Insul.* **2016**, *23*, 555–563. [CrossRef]
39. Pushpa, Y.G.; Vasudev, N. Artificial Pollution Testing of Polymeric Insulators by CIGRE Round Robin Method -Withstand & Flashover Characteristics. In Proceedings of the 3rd International Conference on Condition Assessment Techniques in Electrical Systems (CATCON), Chandigarh, India, 16–18 November 2017.
40. Chan, H.; Fan, C.; Kuo, C. Generalized Finite Difference Method for Solving Two-Dimensional Non-Linear Obstacle Problems. *Eng. Anal. Bound. Elem.* **2013**, *37*, 1189–1196. [CrossRef]

energies

MDPI

Article

Investigation of Surface Degradation of Aged High Temperature Vulcanized (HTV) Silicone Rubber Insulators

Mohamed Ghouse Shaik * and Vijayarekha Karuppaiyan

Department of Electrical and Electronics Engineering, Shanmuga Arts, Science, Technology and Research Academy (SASTRA) Deemed University, Thanjavur 613403, India; vijayarekha@eee.sastra.edu

* Correspondence: ghouseee@eee.sastra.edu; Tel.: +91-989-449-0687

Received: 22 August 2019; Accepted: 29 September 2019; Published: 3 October 2019

Abstract: Polymeric composite insulators are subjected to varying work conditions like rain and heat, which create an impact on degradation during their long service period. Electrical tracking under the Alternating Current (AC) field plays a predominant role in surface degradation, which can be different for fresh and aged insulations. The tracking studies on the fresh and aged polymeric insulation therefore become significant. Motivated by this, an indigenous low-cost electrical tracking setup was developed, and the tracking studies were carried out as per International Electro technical Commission standard (IEC) 60587 on fresh, thermal-aged and water-aged silicone rubber samples. Contact angles of samples were measured to analyse the effect of ageing on hydrophobicity. Further, to analyse the influence of ageing on insulation integrity, tracking tests were conducted and parameters like leakage current pattern and magnitudes, tracking length and loss of weight in the material due to tracking were examined. The physicochemical impacts of ageing on the surface degradation of the samples were also analysed using X-ray diffraction analysis and Fourier Transform Infrared Spectroscopy analysis. The investigations added insight into the degradation mechanism of polymeric insulators in terms of their electrical performance and physicochemical changes in the material. Comparison of these changes showed that ageing could influence surface degradation of samples.

Keywords: tracking; ageing; hydrophobicity; leakage current; dry band arcing; degradation; polymeric insulation; tracking test setup

---

1. Introduction

Polymeric insulators like Silicone Rubber (SR), High-Density Polyethylene (HDPE), Ethylene Propylene Diene Monomer (EPDM), Ethylene Propylene Rubber (EPR) and Poly Tetra Fluro Ethylene (PTFE) are used in power systems because of their superior electrical and mechanical properties over the inorganic insulators. They have less weight, superior insulation strength, higher flexibility, resistant to erosion/degradation, less chances of breakage and low installation cost.

However, there are also some disadvantages associated with polymeric insulators. Being organic in nature, they have a tendency to degrade due to environmental effects. In the presence of temperature, Ultra Violet (UV) radiation, mist, fog, and rain, the surface of the insulator in service will be affected, and it leads to the settlement of contaminants on the surface. These conducting contaminants promote leakage current, scintillation and dry-band arcing over the surface and lead to the formation of conducting tracks and permanent erosion of the materials.

As electrical discharges on the surface are more predominant in polymeric insulators due to applied electrical stress and changes in environmental conditions such as heat, rain, etc., tracking studies on them has become significant.

So far, various tracking studies have been reported in the literature on solid insulation under normal and contaminated conditions. Mathes et al. had proposed the inclined plane liquid- contaminant test [1]. It was standardised further and IEC 60587 and American Society for Testing and Materials standard (ASTM) D2303 were developed to evaluate the resistance due to tracking and erosion of insulating material under severe ambient conditions [2,3]. These standards were revised in 2007 and 2013, respectively. Researchers are implementing them to study the tracking behaviour of various solid insulating materials.

The tracking effects were investigated based on the tracking length, erosion depth due to the leakage current flow and dry-band arcing over the surface of the material. The variation of leakage current magnitude, its pattern and dry-band arcing periods were considered as the parameters in the investigation. Approaches like the tracking wheel test were also employed for investigating polymeric insulation degradation under an accelerated ageing laboratory environment [4]. Partial discharge analysis and flashover voltage tests were also carried out on polymeric insulators under clean and salt fog environments to analyse the failure mechanism with environmental stresses and were reported in literature [5]. Multi-stress ageing effects of different types of polymeric insulators [6–9], ageing effects of HTV silicone rubber/silica hybrid composites for high-voltage insulation [10], accelerated aging effects of silicone rubber insulators and silicone rubber-coated insulators [11,12], ageing of polymeric insulators under acid rain conditions [13] and arid climatic conditions [14] have also been reported in recent literature.

With the recent development of the High Voltage Direct Current (HVDC) transmission system, researchers have also started focussing on the DC tracking studies adapting the conventional IEC 60587 and ASTM D2303 standards of AC tracking studies. Tracking studies were also carried out on polymeric insulators under AC and DC fields and for developing some standard procedures for tracking studies on polymeric insulators. The influence of DC and AC fields on tracking was investigated based on leakage current magnitude and the erosion severity [15–19].

Many methods of degradation of different types of polymeric materials have been investigated in the literature. The effect of gamma radiation ageing on silicone rubber decomposition and its electromechanical properties was investigated by Rajini et al. [20]. The effect of gamma radiation on EPDM was investigated by Pourmand et al. [21]. The ageing effect of radio frequency and microwave oxygen plasma on polydimethylsiloxane (PDMS) was investigated by Hillborg and Gedde [22]. The effect of combined UV-thermal and Hydrolytic ageing on hardness, contact pressure and contact area of silicone elastomers was investigated by Wu et al. [23]. The ageing effect in the form of algae growth over the surface of the polymeric insulators and the hydrophobicity changes have been studied by Yang et al. [24]. The effect of thermal ageing on the formation of cross-linked structures, gamma irradiation on the oxidation of cyclic siloxane and steam exposure on molecular weight on silicone rubber was explained by Kaneko et al. [8]. Apart from these works on polymeric insulating materials, the ageing studies on nano-composites polymeric insulators were also carried out. The effect of thermal ageing on Nano SiO_2-filled HTV silicone rubber was investigated by Loganathan et al. [25]. The water ageing of Epoxy nanocomposite was investigated by Suchitra et al. [26].

To get information about the degradation of the polymeric insulators, material characterization studies from the molecular level to the macro level was done using Fourier Transform Infrared (FTIR) spectroscopy, Scanning Electron Microscopy with Energy Dispersive X-ray (SEM-EDX) analysis, Differential Scanning Calorimetry (DSC), X-Ray Diffraction (XRD), Nuclear Magnetic Resonance (NMR) spectroscopy, Dielectric spectroscopy, Atomic Force Microscopy (AFM) and X-ray Photoelectron spectroscopy (XPS) [27–32].

It could be well seen from the literature that the focus was on a prerequisite for conducting tracking tests with good reliability. The requirement of a reliable test setup comprises of a regulated power supply, a current limiting resistor to control leakage current, dry-band arcing and erosion below the endpoint criteria during the tracking test and a controlled contaminant flow system.

The conventional high-voltage setups were found to be more costly and contained a high-voltage test transformer, high-voltage current limiting resistors and a data acquisition system. With the traditional test transformer of the high-voltage transformation ratio, it was challenging to adjust the test voltage to the required level and maintain the same throughout the test duration with proper regulation. The flow rate of the contaminant and the voltage fluctuations will significantly affect the tracking performance during the entire tracking process [18].

The focus of this research work is to develop an indigenous low-cost electrical tracking setup with which we can perform tracking studies without voltage fluctuations. This has provided great voltage regulation compared to other tracking setups in which the conventionally available high cost, high-voltage test transformers with a high-voltage rating are used. Thus, the proposed setup addressed one of the major issues (maintaining minimum Voltage fluctuation on the High Voltage (HV) side per one voltage change on the Low Voltage (LV) side) in conducting experimentation.

In the present work, the initial focus was taken to sort out the reliability issues in carrying out high-voltage tracking tests. A simple, cost-effective, easily reproducible test circuitry for conducting the tracking test was successfully assembled and implemented. The uniform flow rate of the contaminant was achieved through the use of a simple, commercially available Intravenous (IV) system. Use of a low-voltage ratio test transformer as a source ensured a continuous regulated power supply with minimum fluctuations throughout the test duration. The resistive ladder network employed served to be effective in limiting the current below the threshold value of 60 mA, and the Digital Storage Oscilloscope (DSO) was useful in leakage current data acquisition.

Secondly, in literature, the tracking performance of insulating materials were highlighted and compared through electrical and physiochemical changes. Some literature referred to above also addressed the tracking performance of aged samples. All these research works highlight the significance of investigating the effects of ageing on hydrophobicity recovery of silicone rubber and its influence on the dynamics of degradation of the surface during the tracking period. The focus was on understanding the dynamics of the degradation process and the influence of ageing on insulation integrity.

The research focus is to mainly study the effect of environmental ageing during the tracking process of HTV silicone rubber used in outdoor insulation applications. The leakage current magnitude change for fresh and artificially aged samples were analysed to understand the effect of aging in promoting conduction over the surface of the polymeric insulators. The arcing patterns were observed for fresh and aged samples. The intensity and the frequency of arcing were observed during the tracking process. After tracking, the physicochemical properties were studied to understand the change in the crystalline nature and crystal size at the most eroded region of the sample. This gives a clear inference regarding the role of Aluminium trihydrate (ATH) filler degradation during tracking and the clustering of ATH fillers at the surface of the eroded polymer during electrical tracking.

To understand the effect of ageing on surface degradation of tracked silicone rubber samples, XRD analysis and FTIR spectroscopy studies were carried out. XRD analysis was used to gain insight about the change in the crystalline structure of the polymeric material, and FTIR analysis was used to analyse the physicochemical modifications at the surface of the tracked samples [27–32].

FTIR spectroscopy was mainly used to understand the changes in the fingerprint region of the spectrum due to the ageing effect. The changes in the different functional group of polymeric chain in the aged samples due to the electrical tracking failure were analysed.

The tracking studies were carried out on HTV silicone rubber samples obtained from an outdoor transmission tower polymeric composite insulator manufacturer. These HTV silicone rubbers are made up of Aluminium trihydrate (ATH)-filled polydimethylsiloxane (PDMS) to enhance mechanical, thermal and electrical erosion/tracking properties. Fresh, thermal-aged and water-aged samples were prepared and used for this study.

The samples were water-aged, and thermal-aged to emulate the effect of environmental stresses such as rain and heat on them, as per the recommendations available in the literature [25,26].

The investigation carried out before, during and after tracking was grouped into three stages: First, the effect of ageing on hydrophobicity was investigated. Then, the impact of ageing was investigated based on electrical and physical responses during tracking, such as leakage current, dry-band arcing, rewetting, etc. Finally, the investigation was done on the effect of ageing on physicochemical changes developed during tracking.

To assess the effect of water and thermal ageing on the hydrophobicity of samples, the contact angle measurement was carried out. A comparative study was done to understand the changes in the hydrophobicity of the samples due to ageing.

After hydrophobicity studies, all the samples were subjected to tracking investigations, and the leakage current magnitude and leakage current patterns were continuously recorded using DSO. To understand the effect of ageing on the electrical characteristics of polymeric material, the leakage current pattern and its magnitude for the water-aged and thermal-aged samples were observed and compared with that of the fresh samples.

To ensure reliability and repeatability of the results, tests were conducted on fifteen samples in total with the same experimental setup and measurement procedures, i.e., five samples each from the fresh, water-aged and thermal-aged groups were tested. The tracking tests were carried out continuously, at a constant voltage of 4.5 kV (RMS), without any power cut and fluctuations.

2. Methodology

2.1. Sample Preparation

For this study, the HTV silicone rubber sheets were procured and cut to meet the required dimensions of 120 mm × 50 mm, with a thickness of 6 mm, as per IEC60587 [2]. Tracking studies were carried out with fresh, thermal-aged and water-aged samples.

Fresh samples were tested to analyse the performance of the silicone rubber under normal conditions. Silicone rubber insulation gets subjected to the varying working environment and undergoes some long-term degradation during service due to these accumulated stresses. To emulate such situations and to investigate their effects on degradation of insulation, thermal-ageing and water-ageing of some samples were done and tracking studies were carried out. Five samples were tested in each case, to ensure the reliability of the results.

For the preparation of aged samples, continuous stress for a specified period was followed and reported by researchers in the literature. Continuous stressing is difficult due to operational restrictions of the laboratory. Moreover, environmental conditions are not continuously severe during natural ageing processes and insulation undergoes stresses and recovery periods, which are cyclic and repetitive. These aspects were considered for the present work and ageing of samples were carried out by subjecting them cumulatively to the specified stress level within a period [25].

Thermal-aged samples were thus prepared by subjecting the samples at 150 °C in a hot air oven for a cumulative period of 240 h, spanned at the rate of 8 h per day for 30 days. Water-aged samples were prepared by immersing the samples in distilled water for about 500 h continuously [26].

0.1% by weight of Ammonium chloride salt was added to the distilled water to prepare the contaminant solution. 0.02% by weight Triton X-100, a non-ionic wetting agent, was also added to the contaminant solution to improve the wettability over the surface of the sample during testing. The conductivity of the contaminant solution was 2.5 mS/cm, created an effect similar to the deposition of dirt, and condensed moisture from the atmosphere [2].

2.2. Experimental Setup

The experimental arrangement was designed as per IEC 60,587 [2]. The inclined plane test method with constant tracking voltage was used to investigate the surface tracking and erosion resistance of HTV silicone rubber samples. The experimental setup is shown in Figure 1.

The sample was mounted on an inclined plane at an angle of 45° with the horizontal plane. The top and bottom electrodes were made up of stainless-steel sheets of 0.5 mm thickness. The two electrodes were mounted such that the spacing between them was 50 mm.

In this study, a low-cost 230V/5 kV, 500 VA single-phase transformer commercially available in the market was preferred to get the test voltage. This is in line with the requirements of the IEC standard, as the maximum voltage to be applied as per the Inclined Plane Test (IPT) test and the maximum allowable leakage current during tracking is 4.5 kV and 60 mA, respectively [2]. The transformer has a maximum full load current of 0.1 A on the 5 kV side, and the low transformation ratio (230/5000) ensures proper regulation in the test voltage, with the output voltage magnifying by just 21.74 times the input voltage. Supply to the test transformer was availed through a single-phase autotransformer so that the input to the test transformer can be adjusted to get the required test voltage. An isolation transformer was included between the test transformer and the autotransformer to isolate the IPT circuitry from any other circuits in the test yard and this was not mandatory.

The literature suggests the use of a peristaltic pump to control the flow of contaminant [2,3]. In this work, a low-cost controllable intravenous (IV) system was used to precisely control the contamination flow to ensure the desired flow rate as per standard.

Initially, the quantity of a single drop was measured using a small syringe to assess the volume/drop. The discharge from the IV system was adjusted based on this to meet out the required flow rate in mL/min. The contaminant solution was made to flow through the filter paper, from the high voltage top electrode to the bottom ground electrode through the lower face of the sample, as shown in Figure 1.

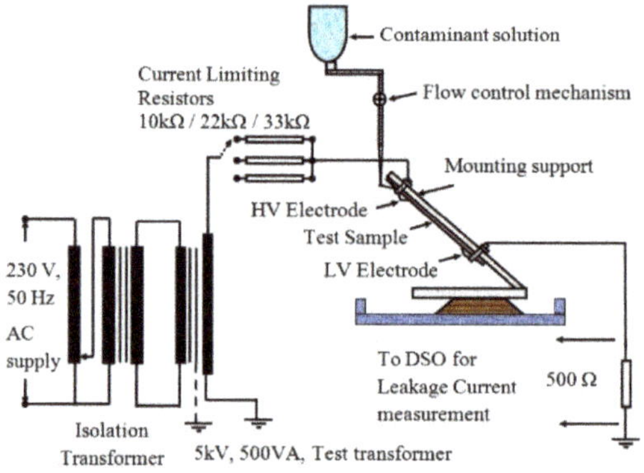

Figure 1. Experimental setup for the Inclined Plane Test.

The flow rate of the ammonium chloride contaminant solution was adjusted to 0.6 mL per minute. To precondition the sample, the flow was allowed, and the sample was wetted for 30 min.

As per IPT standards, at different test voltage levels, different series resistors are needed between the test transformer and the top electrode of the IPT setup, to limit the leakage current during tracking. In the present work, a resistive divider arrangement fabricated in the form of a ladder series connection of several 1 kΩ, 10 Watts resistors were used. It has tapping provisions for availing the required series resistance. Additionally, for the leakage current measurement purpose, a low ohmic resistor can be included in the grounding path of the bottom electrode of the IPT setup. The power dissipation capacity of these current limiting and current measurement resistors have to be decided in terms of the maximum allowable current during tracking.

The constant tracking voltage method with an AC test voltage of 4.5 kV (RMS) between the electrodes was adopted in the present experimental investigations. Accordingly, 33 kΩ was availed as the current limiting resistor and a 500 Ω, 10 Watts resistor was availed to measure the leakage current. Throughout the experimental work, the leakage current was observed and measured with the help of a digital storage oscilloscope [2]. The experimental setup of the inclined plane test is shown in Figure 1.

The experiments were conducted for a 6 h duration for all the fresh and aged samples. The IPT procedures were repeated, maintaining the same voltage level, flow rate and environmental conditions. The leakage current patterns were recorded in a Digital Storage Oscilloscope (DSO) at a regular interval of 10 min. The RMS value of the leakage current was also noted manually. The photograph of the experimental test setup and mounting support is shown in Figure 2.

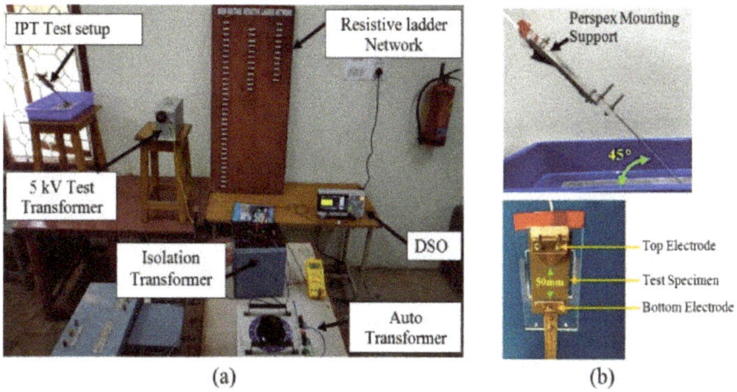

Figure 2. Photograph of the experimental setup and mounting support: (**a**) Experimental set up, (**b**) Mounting support.

Figure 2a presents the experimental setup and Figure 2b presents the mounting support with electrodes and the test sample. The mounting support was made up of perspex sheet of 8 mm thick and was used to hold the samples during the test.

As per standard, before six hours, if the leakage current reaches a value of 60 mA or if the track reaches a length of 25 mm from the ground electrode or if a hole is observed due to intensive arcing, the test has to be terminated. Otherwise, the test has to be continued and observed up to 6 h.

During the experimentation, the maximum current observed throughout the test was only 45 mA. The samples employed were industrially tested ones for tracking studies. They didn't show any deep erosion during testing but the tracking length and the dry-band arcing area was different for different samples. After testing, all the samples were cleaned with adequate care to avoid any physical damage to the tracked path and dry-band regions. The length of the tracked portion was measured for all the samples. The samples with the largest tracked area in each case were used to carry out XRD and FTIR analysis.

2.3. Research Design

Before the tracking test, the contact angle of fresh, thermal-aged and water-aged samples was measured and compared, to investigate the influence of ageing on the hydrophobicity of HTV silicone rubber.

During tracking, the focus was given to analyse the effect of thermal and water ageing of samples on variation in leakage current magnitude, its pattern and the dry-band arcing.

After carrying out the tracking test, further investigations were carried out to study the effect of thermal and water ageing on the length and depth of tracking and loss of weight of tracked samples. For this, the highly eroded sample was picked up from each of the fresh, thermal-aged and water- aged sample group and XRD and FTIR spectroscopy was carried out on them to analyse changes in the physical structure of samples.

3. Results and Discussions

This section is divided by subheadings. It should provide a concise and precise description of the experimental results, their interpretation, as well as the experimental conclusions that can be drawn.

3.1. Influence of Ageing on Hydrophobicity

Silicone rubber possesses a good hydrophobicity and is expected to recover the same after ageing. Researchers used contact angle measurement through goniometer to measure the hydrophobicity of silicone rubber insulators before and immediately after ageing [32]. If the contact angle is measured after allowing some time for recovery of hydrophobicity, it can give a further understanding on the hydrophobicity recovery nature of the aged silicone rubber samples and subsequently, its influence on the tracking.

In the present work, the contact angle of the fresh sample was considered as the base or reference value. The contact angle of all the five fresh samples was measured, and their average was used as a base. All the water-aged and thermal-aged samples were allowed to recover the hydrophobicity for about 24 h, and their contact angles were measured and compared with the base value for assessing the change in hydrophobicity due to ageing. The contact angles for all the fresh, thermal-aged and water- aged samples are shown in Figure 3.

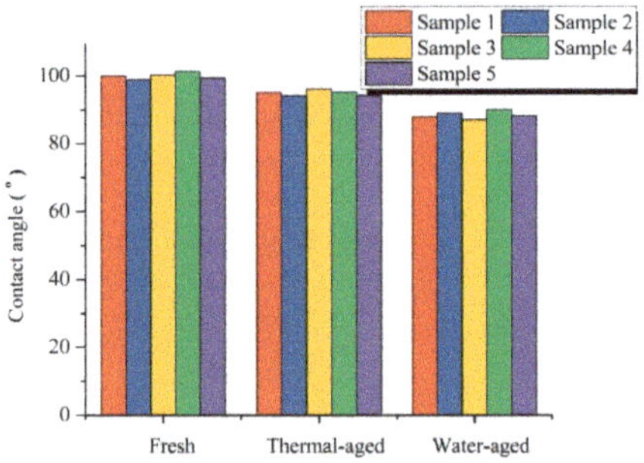

Figure 3. Contact angles of fresh, thermal- and water-aged samples.

The average values of contact angle for each case were used for comparison. For the fresh samples, it was 99.91°. The fresh samples were found to be highly hydrophobic, and their average contact angle was subsequently considered as the base value during the comparison.

For thermal-aged samples, the average contact angle was 95.14°. It decreased due to thermal ageing, indicating the change from a high hydrophobic state to a low hydrophobic state. However, when the samples were aged with water, the average contact angle was only 88.00°. With the contact angle value reaching below 90.00°, it indicated that the sample had become hydrophilic. Thus, the changes in contact angle indicated that the moisture absorbing capability of silicone rubber got altered due to thermal and water ageing. This makes the investigations on the effect of ageing on the tracking of silicone rubber insulation significant.

Motivated by this, further investigations were carried out to understand the influence of thermal and water ageing of silicone rubber on various electrical aspects of tracking, such as variation in magnitude and pattern of leakage current, dry-band arcing, tracking depth and physicochemical properties such as loss of material, change in crystallinity and functional groups of silicone rubber [27–32].

3.2. Analysis of Leakage Current Variations During Tracking

After hydrophobicity studies, all the samples were subjected to electrical tracking studies as per the standard IEC60587. For understanding the tracking mechanism, the leakage current patterns were recorded at regular intervals. The various processes observed during tracking were summarised in three states:

Initially, during the first state, a continuous conductive moisture filament was developing over the surface of the sample. Then, Joule heating took place, evaporated the contaminant and initiated dry bands. During the second state, dry-band-stretching and elongation were observed due to the large potential difference across it. Small arcs appeared over these dry bands in the conducting filament path, thus bridging the less conductive regions of the silicone rubber surface between the electrodes. As the dry band arc path offered higher resistance than the conductive filament path, the leakage current was reduced. These reductions in leakage currents were represented by discontinuities in the leakage current patterns. Figure 4a shows the initiation of arcing during the application of voltage. Figure 4b shows the dry-band elongation, promoting discontinuity in arcing and leakage current. Subsequently, the voltage-drop across the dry band was increased and led to the formation of long and intensive arcs. This is shown in Figure 4c. In the meantime, due to the continuous flow of the contaminant solution from the IV set, the surface of the specimen was rewetted, the conductive contaminant path was re-established, and the arc was quenched. Thus, in the second state, repeated arc formation and quenching though rewetting of the surface was observed. This led to the third state, which is the final stage in the tracking. During the third state, the leakage current magnitude increased slowly over time and led to carbonaceous track formation on the surface of the sample. Subsequently, the material was eroded. Continuous intense discharges were observed at this final stage with several such repetitions of rewetting and arcing, as shown in Figure 4d.

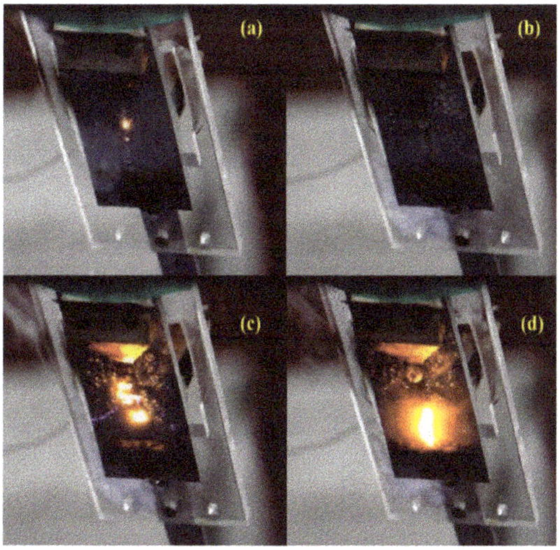

Figure 4. (**a**) Initial dry-band arcing, (**b**) Dry-band elongation promoting discontinuity in arcing and leakage current, (**c**) Repeated arcing due to rewetting and dry-band formations, (**d**) Intense discharge during tracking.

Following this, physical observations were also made during the development of tracks: First, at initial stages, the location of dry-band arcs was slowly and continuously changing from the top electrode towards the bottom electrode and looked like a moving arc. Secondly, the initiation of the track was first observed near the bottom electrode and, as time progressed, this tracking length was increased. Thirdly, a bright localised spot appeared near the ground electrode and began to extend towards the top electrode. Finally, physicochemical changes like localised erosion and carbonisation on the surface of the material happened in the tracking path.

Further investigations on these processes and the role of potential redistributions due to rewetting and dry-band arcing during tracking can lead to a better understanding of the tracking phenomena. However, they were not considered in the scope of the present work, as the main focus is only on the role of ageing on electrical tracking of silicone rubber.

From the initial stage to the final stage, at regular intervals of time, leakage current patterns were recorded. Figure 5 manifests the variations in the leakage current patterns of a fresh sample during the tracking test duration.

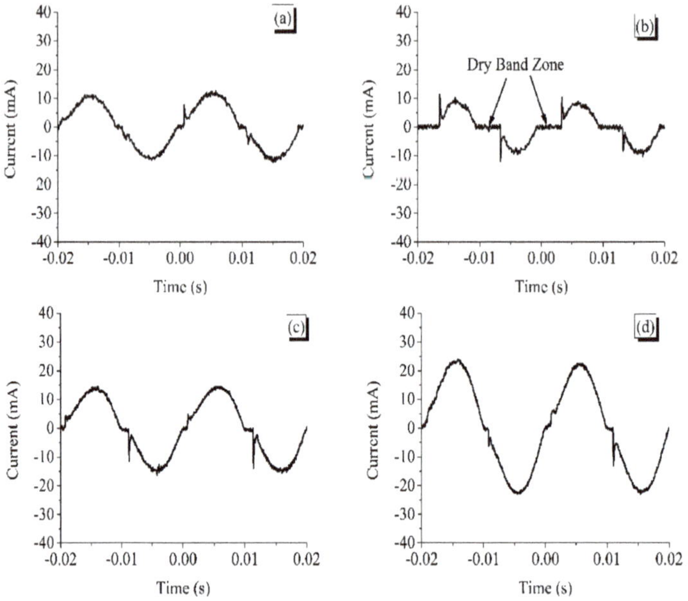

Figure 5. Leakage current through the fresh sample at different stages of tracking: (**a**)leakage current under normal condition, continuous and lower in magnitude, (**b**) discontinuous current, for smaller dry-band length, (**c**) magnitude of current and spikes increases with lesser discontinuities, for higher dry-band lengths, (**d**) increased current, after the initiation of tracking.

Figure 5a shows the pattern of the leakage current at the starting point. The magnitude of the leakage current was minimum, and the current was continuous. After some time, with the flow of this leakage current, the contaminant solution at the surface of the insulation was evaporated, and dry-band regions were formed. With this formation of the dry band, the leakage current pattern changed and there were discontinuities, as shown in Figure 5b. Repeated rewetting and heavy dry-band arcing were then observed, and the current magnitude again started to increase, as shown in Figure 5c. Subsequently, with this heavy arcing, a hotspot was witnessed near the bottom electrode, initiating erosion of the test sample near its bottom end. With time, this erosion aggravated further with simultaneous carbonisation and tracking on the surface of the material. The leakage current waveform during this continuous intense arc period is shown in Figure 5d.

During experimentation, the leakage current patterns are almost similar for fresh, thermal-aged and water-aged samples, but the variation in the magnitude of the leakage current with time was different.

Also, the total arcing and non-arcing periods were different for fresh, thermal-aged and water-aged samples: Water-aged samples showed less non-arcing periods compared to thermal-aged and fresh samples. The intensity of arcing was higher in the case of thermal-aged samples when compared to fresh and water-aged samples.

The variation of the RMS value of the leakage current during tracking was observed continuously until the end of the test and noted every five minutes, for all the fresh, thermal-aged and water-aged samples. The RMS value for the five samples of each group was averaged and considered to represent the performance of that group.

The variation of the leakage current of fresh, thermal-aged and water-aged samples with time, during the tracking period, is shown in Figure 6.

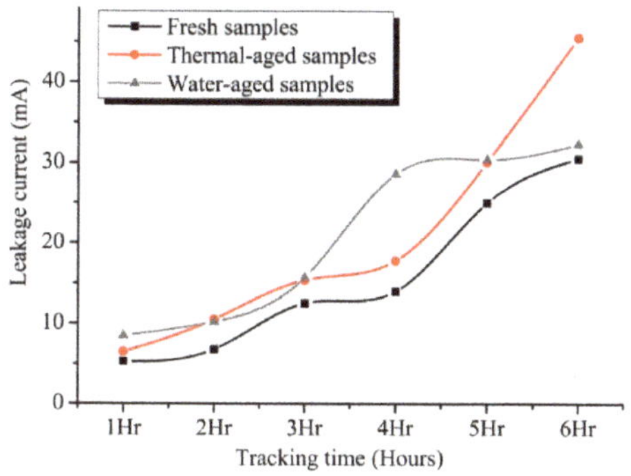

Figure 6. Variation of leakage current of fresh, water-aged and thermal-aged samples during tracking.

The plot shows an increase in the RMS values of leakage current for all three cases with respect to time. However, the following interesting observations were made regarding their dynamism, which could be significant in further understanding the effects of ageing and associated hydrophobicity changes of silicone rubber on the tracking.

For fresh samples, the leakage current magnitude increased steadily. For thermal-aged samples, the current was slightly higher than the fresh sample at the initial stage, but, at the end of the test, the highest leakage current with intense arcing was observed. For water-aged samples, the leakage current was higher from the initiation of the test and became almost constant during the last 3 h of the test.

A continuous leakage current with less arcing was observed in the case of the water-aged samples. It could be noted that the water-aged samples were found to be hydrophilic through contact angle measurement. Hence, the continuous leakage current might be due to the retained conductive contaminant layer on the surface, for an extended period.

The arcing period of the thermal-aged samples during the initial stage was more, compared to water-aged samples. Contact angles, measured earlier for these thermal-aged samples, also indicated that thermal ageing had made samples more hydrophobic. Hence, the conductive contaminant might be vaporising quickly, thereby promoting dry-band arcing.

3.3. *Analysis of Tracking Length and Loss of Weight after Tracking*

After six hours, which is an endpoint criterion for the tracking test, as per standard, the experiment was stopped. The specimens were removed from the sample holder and then cleaned with adequate care, to avoid any physical damage to the tested samples. The length of the tracked region was measured for all the fresh, thermal-aged and water-aged samples. One highly eroded sample from each group was selected for further investigation. Figure 7 presents the images of such highly eroded and tracked samples from fresh, thermal-aged and water-aged groups.

The samples were weighed before and after tracking, and the percentage weight loss due to tracking was calculated. Table 1 illustrates the comparison of the average tracking length and the average weight loss for all of the samples.

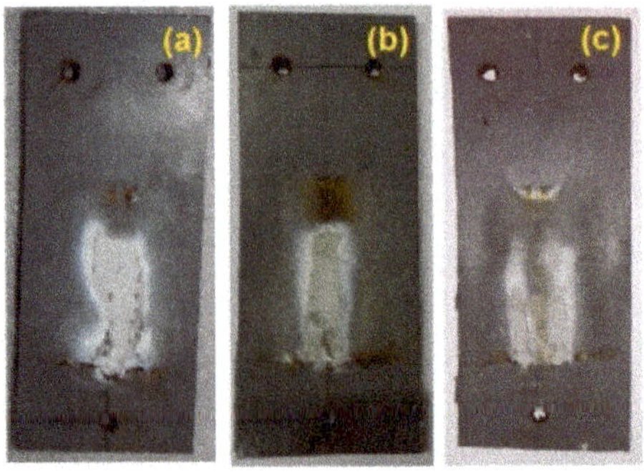

Figure 7. Images of the highly eroded and tracked samples: (**a**) Fresh, (**b**) Thermal-aged, (**c**) Water-aged.

Table 1. Comparison of tracking length and weight loss.

Sample Type	Average Tracking Length (cm)	Average Weight Loss (%)
Fresh samples	2.8	1.39
Thermal-aged samples	3.2	2.95
Water-aged samples	3.6	1.62

The average tracking length and average percentage weight loss of the thermal-aged and water-aged samples were comparatively higher than those of the fresh samples. The water-aged samples showed the maximum tracking length among all of the samples but the average percentage weight loss of the water-aged sample was less compared to the thermal-aged sample. The highest average tracking length was observed in the water-aged samples, and the highest average weight loss in percentage was observed in the thermal-aged samples.

With thermal stress, the bond of the polymer chain became weak. This might be the reason for the deep erosion and higher weight loss in thermal-aged samples during the tracking test.

Water-aged samples showed higher tracking length and the surface condition was affected due to loss of hydrophobicity. The erosion effect and weight loss were minimal. The role of ageing on tracked samples was further investigated through XRD and FTIR analysis. The focus was given to physicochemical changes happening in aged samples during tracking.

3.4. *Analysis of Crystallinity by X-Ray Diffraction (XRD) Analysis*

XRD analysis was used to explain the effect of ageing on the physical structure of the eroded polymeric insulation samples and to understand the clustering of ATH filler particles over the surface of the polymeric material during the tracking process. The highest eroded regions of the samples were cut carefully to the required dimension of 10 mm × 10 mm × 0.5 mm, and XRD analysis was carried out for the sample within the 2θ range of 10° to 90°.

Figure 8a shows the normalized XRD pattern of the fresh sample. The very low-intensity broad region from 10° to 15° indicates the amorphous nature of the silicone rubber with a low degree of crystallinity. The sharp peaks around 20° indicate the crystalline nature of the Aluminium trihydrate (ATH) filler in the polymer.

XRD analysis was then conducted on thermal-aged and water-aged silicone rubber samples. Sharpening of the amorphous hump was observed in the thermal-aged and water-aged samples. This clearly shows that the crystallinity increased due to degradation.

Figure 8b,c show the corresponding normalized XRD patterns for thermal-aged and water-aged samples.

The intensity of the peak around 18° increased considerably in the thermal-aged sample. This difference indicated that silicone rubber had become more crystalline with thermal ageing.

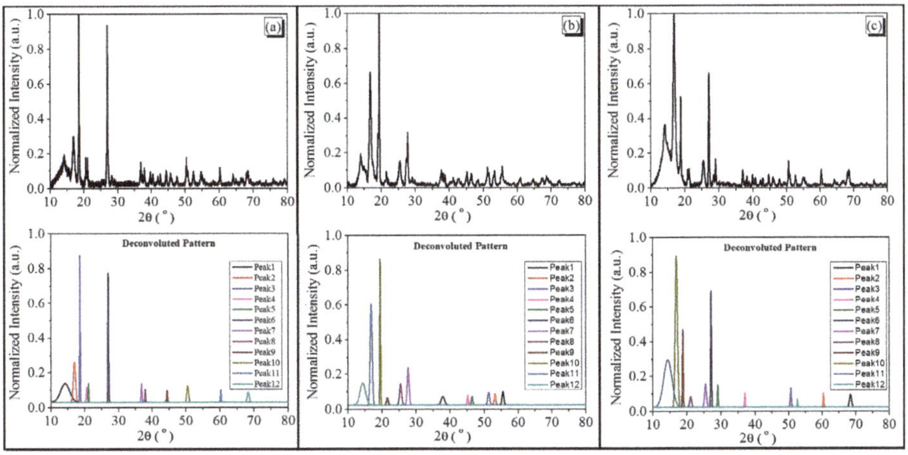

Figure 8. Comparison between the X-ray diffraction (XRD) pattern of fresh and aged samples: (**a**) Fresh, (**b**) Thermal-aged, (**c**) Water-aged.

For the water-aged sample, higher peaks were observed at 17°, but this was comparatively less than the thermal-aged sample. These peaks confirmed that chemical changes had taken place in the samples permanently due to ageing.

In order to quantify the crystallinity, the XRD characteristics were normalized and de-convolution of XRD patterns were carried out. The percentage crystallinity was calculated using 'Origin Pro' software. For this, first 'fitting of XRD peaks' were done using the Gaussian function. Then area under crystalline peaks and the area under amorphous peaks were separately determined. The percentage crystallinity was then calculated by dividing the 'area of crystalline peaks' by the 'total area of crystalline and amorphous peaks.'

The crystallinity of the fresh sample was found to be 66.04%, but for the thermal-aged sample, the crystallinity was increased to 82.82%. For the water-aged samples, it was increased to 77.05%. The variation in the percentage of crystallinity showed that the thermal-aged samples became more crystalline when compared to water-aged samples and fresh samples. Due to heat and the oxidation of the surface of the polymer, the surface material changed from amorphous to crystalline.

Using the Williamson–Hall analysis, crystal size and the strain of fresh, thermal-aged and water-aged samples were determined. The crystal size of the fresh sample was 18.1 nm with 0.1% strain. The crystal size of the thermal-aged sample was 11.6 nm with 0.1% strain. The crystal size of the water-aged sample was 11.1 nm with 0.4% strain. It is inferred that the crystal size of the eroded region of the fresh sample is higher compared to thermal-aged and water-aged samples. The reduction in the crystal size is due to the degradation of the surface layer prior to tracking studies due to accelerated ageing effects.

3.5. Fourier Transform Infrared (FTIR) Spectroscopy Analysis

FTIR spectroscopy is used to study the physicochemical changes in the material by investigating the infrared absorption spectrum. It gives the information about the presence or absence of a specific functional group of the polymeric insulation. Information about absorption or transmittance due to the O-H bond in the ATH filler, the Si-O-Si bond in the polymer chain and Si-CH_3 in the side chain of the polymer etc., can be obtained from the variations noted at different wavelengths in the IR spectra [27–32].

In the present work, FTIR spectroscopy of the fresh, thermal-aged and water-aged samples was carried out. The focus was on identifying the presence and absence of specific functional groups in samples when they were subjected to tracking after ageing. FTIR results were interpreted based on the similar discussion in various literatures [27–32].

The spectrum of the tracked fresh sample was considered as the reference. Figure 9 depicts the FTIR spectra for tracked fresh, thermal-aged and water-aged silicone rubber samples.

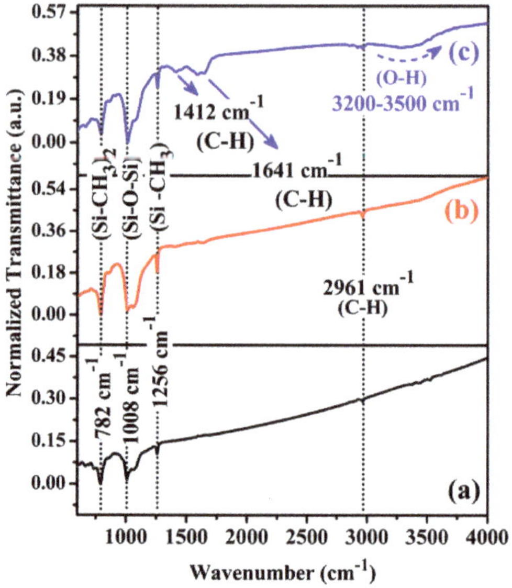

Figure 9. Fourier Transform Infrared (FTIR) spectra for samples after tracking: (**a**) Fresh, (**b**) Thermal-aged, (**c**) Water-aged.

In the FTIR spectrum, the region between 500 cm^{-1} and 1500 cm^{-1} is called the fingerprint region and the region between 1500 cm^{-1} and 4000 cm^{-1} is called the functional group region. In order to understand the change in the transmittance value, the FTIR spectrum was normalised, and the stacked patterns were used for better clarity.

There were distinct differences in the peaks of the spectrum of water-aged and thermal-aged samples. The dips observed in transmittance plots of all the three types of samples at 2961 and 1256 cm^{-1} showed an increase in the CH_3 group. Corresponding transmittance values indicated that the hydrophobic recovery levels of silicone rubber samples were different for fresh and aged samples.

The water-aged sample showed a dip in the range 3500–3200 cm^{-1} and indicated the decrease in the O-H stretch intensity. The continuous conduction through the contaminant had modified the behaviour surface of water-aged samples and decreased the vibration of the O-H bond of the polymeric material as well as the O-H bond of ATH fillers. This also indicated the dehydration of ATH fillers. Due to the tracking, the ATH fillers might have lost their bondage and been converted into aluminium oxide, with the evaporation of water molecules from the ATH filler, preventing surface oxidation [31,32]. The degree of dehydration of the water-aged sample could be inferred from the wide dip of the spectrum in this range. Carbonyl formation was also observed in this region.

Around 2961 cm^{-1} there was a decrease in the intensity of C-H stretch for both thermal-aged and water-aged samples. This could shrink the polymer chain, and the surface of the polymeric sample could become harder. The presence of O-H bond in water was indicated at 1641 cm^{-1}.

In the region 1641–1412 cm^{-1}, there was a decrease in the peak for the water-aged sample. This could result in bending of C-H and O-H bonds and produce carbonyl degradation products. This can be attributed to the continuous leakage current on the surface of water-aged samples. After water ageing, the surface hydrophobicity decreases, and it leads to an increase in the leakage current magnitude, as discussed in Section 3.2. This could result in the production of carbonyl degradation products.

At 1256 and 782 cm^{-1}, the changes in the peak for the thermal-aged and water-aged sample were almost the same, and there was not much change in the intensity of the Si-C bond, the base skeleton of the polymer matrix.

In the region 1008 to 782 cm^{-1}, there was a difference in the peak intensity of thermal-aged and water-aged samples, due to the out of plane bending of the Si-CH_3 and Si-O stretch. This could reduce the Si-O rocking vibrations in the aged samples. This clearly indicates that the aging of the polymeric sample before the tracking test considerably affects the Si-CH_3 and Si-O bond in the polymeric chain.

The differences observed between the IR spectra of fresh and aged samples confirmed that depolymerisation at the surface of the samples had increased with ageing.

Investigations also revealed that the formation of the hydroxyl group due to water ageing could accelerate the deterioration of polymeric material. Tracking studies on thermal-aged samples showed that the deformation in the Si-O and Si-CH_3 bond had resulted in higher erosion of polymeric material.

4. Conclusions

In this research work, first, a low-cost indigenous IPT setup was assembled and successfully implemented for conducting tracking studies on insulating materials. The setup implemented was simple, user-friendly in regulating voltage and contaminant flow rate, and found to be reliable.

Comparison of contact angles of fresh and aged samples showed that samples had become hydrophilic due to water ageing. To understand the effect of ageing on the electrical and physicochemical properties, tracking studies were performed on fresh, thermal-aged and water-aged silicone rubber insulation samples.

The leakage current magnitude increased due to ageing. In water-aged samples, the leakage current was initially increasing, and later, became steady and continuous with less dry-band arcing. In the thermal-aged sample, the leakage current increased steadily and, in the end, it was high with intense arcing. Water-ageing led to larger tracking length and thermal-ageing led to deeper erosion and maximum degradation. Variations in the intensity of the peaks of XRD patterns showed that thermal-aged samples had become more crystalline and offered a higher degree of depolymerisation than water-aged samples. FTIR spectra of water-aged and thermal-aged samples showed more physiochemical changes in the various functional groups, which could influence the electrical discharge activity and affect the dynamics of degradation.

Ageing was thus found to change the electrical and physicochemical properties and offered significant influence on the degradation of the polymer under sustained alternating voltage.

This work can be extended in the future for studying the effect of ageing on the tracking of insulations under sustained direct voltages.

Author Contributions: M.G.S. conceived, designed, performed the experiments, analyzed the test results and wrote the manuscript. V.K. supervised the research work. Both the authors revised and proofread the manuscript.

Funding: This research received no external funding.

Acknowledgments: The authors thank the management of SASTRA Deemed University and Department of Science and Technology (DST-FIST) (Sanction order ref: SR/FST/ETI-338/2013(C) dated 10/09/2014) for their motivation and support for establishing the facilities at HV lab of SASTRA, with which the experimental works were carried out. The authors thank Vikas Elastochem Agencies Pvt. Ltd. Chennai, Tamil Nadu, India for providing HTV silicone rubber samples. The authors thank K. Jeyadheepan, SASTRA, for helping in XRD and FTIR analysis and P. Karthik, SASTRA, for his support. The authors thank R. Rajesh and R. Shriram Srinivasarangan for helping in the preparation of the manuscript.

Conflicts of Interest: The authors declare no conflict of interest.

References

1. Mathes, K.N.; Mcgowan, E.G. Surface electrical failure in the presence of contaminants: The inclined-plane liquid contaminant test. *Trans. Am. Inst. Electr. Eng. Part 1* **1961**, *80*, 281–289. [CrossRef]
2. IEC. *Electrical Insulating Materials Used Under Severe Ambient Conditions-Test Methods for Evaluating Resistance to Tracking and Erosion*, 3rd ed.; Publications 60587; IEC: Geneva, Switzerland, 2007.
3. ASTM International. *ASTM D2303-13 Standard Test Methods for Liquid-Contaminant, Inclined-Plane Tracking and Erosion of Insulating Materials*; ASTM International: West Conshohocken, PA, USA, 2013. [CrossRef]
4. Moghadam, M.K.; Taheri, M.; Gharazi, S.; Keramati, M.; Bahrami, M.; Riahi, N. A study of composite insulator ageing using the tracking wheel test. *IEEE Trans. Dielectr. Electr. Insul.* **2016**, *23*, 1805–1811. [CrossRef]
5. Douar, M.A.; Beroual, A.; Souche, X. Assessment of the resistance to tracking of polymers in clean and salt fogs due to flashover arcs and partial discharge degrading conditions on one insulator model. *IET Gener. Transm. Dis.* **2016**, *10*, 986–994. [CrossRef]
6. Khan, H.; Amin, M.; Ahmad, A. Multistress accelerated aging and tracking/erosion-resistance investigation of high voltage polymeric insulators. *Arabian J. Sci. Eng.* **2017**, *42*, 5101–5120. [CrossRef]
7. Verma, A.R.; Reddy, G.S.; Chakraborty, R. Multistress aging studies on polymeric insulators. *IEEE Trans. Dielect. Electr Insul.* **2018**, *25*, 524–532. [CrossRef]
8. Kaneko, T.; Ito, S.; Minakawa, T.; Hirai, N.; Ohki, Y. Degradation mechanisms of silicone rubber under different aging conditions. *Polym. Degrad. Stab.* **2019**, *168*, 108936. [CrossRef]
9. Akbar, M.; Ullah, R.; Alam, S. Aging of silicone rubber-based composite insulators under multi-stressed conditions: An overview. *Mater. Res. Express* 2019. [CrossRef]
10. Rashid, A.; Amin, M.; Ali, M.; Khattak, A.; Saleem, J. Fabrication, characterization and aging influence on characteristics of high temperature vulcanized silicone rubber/silica hybrid composites for high voltage insulation. *Mater. Res. Express* **2019**, *6*, 105327. [CrossRef]
11. Verma, A.R.; Reddy, B.S. Accelerated aging studies of silicon-rubber based polymeric insulators used for HV transmission lines. *Polym. Test.* **2017**, *62*, 124–131. [CrossRef]
12. Amin, M.; Khattak, A.; Ali, M. Accelerated aging investigation of silicone rubber/silica composites for coating of high-voltage insulators. *Electr. Eng.* **2018**, *100*, 217–230. [CrossRef]
13. Verma, A.R.; Reddy, B.S. Tracking and erosion resistance of LSR and HTV silicone rubber samples under acid rain conditions. *IEEE Trans. Dielectr. Electr. Insul.* **2018**, *25*, 46–52. [CrossRef]
14. Chakraborty, R.; Reddy, B.S. Studies on high temperature vulcanized silicone rubber insulators under arid climatic aging. *IEEE Trans. Dielectr. Electr. Insul.* **2017**, *24*, 1751–1760. [CrossRef]
15. Ghunem, R.A.; Jayaram, S.H.; Cherney, E.A. Investigation into the eroded dry-band arcing of filled silicone rubber under DC using wavelet-based multi resolution analysis. *IEEE Trans. Dielectr. Electr. Insul.* **2014**, *21*, 713–720. [CrossRef]
16. Cherney, E.A.; Gorur, R.S.; Krivda, A.; Jayaram, S.H.; Rowland, S.M.; Li, S.; Marzinotto, M.; Ghunem, R.A.; Ramirez, I. DC inclined-plane tracking and erosion test of insulating materials. *IEEE Trans. Dielectr. Electr. Insul.* **2015**, *22*, 211–217. [CrossRef]
17. Ghunem, R.A.; Jayram, S.H.; Cherney, E.A. Erosion of silicone rubber composites in the AC and DC inclined plane tests. *IEEE Trans. Dielectr. Electr. Insul.* **2013**, *20*, 229–236. [CrossRef]
18. Kaaiye, S.F.; Nyamupangedengu, C. Comparative study of AC and DC inclined plane tests on silicone rubber (SiR) insulation. *IET High Volt.* **2017**, *2*, 119–128. [CrossRef]
19. Verma, A.R.; Reddy, B.S. Aging studies on polymeric insulators under DC stress with controlled climatic conditions. *Polym. Test.* **2018**, *68*, 185–192. [CrossRef]
20. Rajini, V.; Udayakumar, K. Degradation of silicone rubber under AC or DC voltages in a radiation environment. *IEEE Trans. Dielectr. Electr. Insul.* **2009**, *16*, 834–841. [CrossRef]
21. Pourmand, P.; Hedenqvist, L.; Pourrahimi, A.M.; Furó, I.; Reitberger, T.; Gedde, U.W.; Hedenqvist, M.S. Effect of gamma radiation on carbon-black-filled EPDM seals in water and air. *Polym. Degrad. Stab.* **2017**, *146*, 184–191. [CrossRef]
22. Hillborg, H.; Ankner, J.F.; Gedde, U.W.; Smith, G.D.; Yasuda, H.K.; Wikström, K. Crosslinked polydimethylsiloxane exposed to oxygen plasma studied by neutron reflectometry and other surface specific techniques. *Polymer* **2000**, *41*, 6851–6863. [CrossRef]

23. Wu, J.; Niu, K.; Su, B.; Wang, Y. Effect of combined UV thermal and hydrolytic aging on micro-contact properties of silicone elastomer. *Polym. Degrad. Stab.* **2018**, *151*, 126–135. [CrossRef]
24. Yang, S.; Jia, Z.; Ouyang, X.; Bai, H.; Liu, R. Hydrophobicity characteristics of algae-fouled HVDC insulators in subtropical climates. *Electr. Power Syst. Res.* **2018**, *163*, 626–637. [CrossRef]
25. Loganathan, N.; Muniraj, C.; Chandrasekar, S. Tracking and erosion resistance performance investigation on nano-sized SiO_2 filled silicone rubber for outdoor insulation applications. *IEEE Trans. Dielectr. Electr. Insul.* **2014**, *21*, 2172–2180. [CrossRef]
26. Suchitra, M.; Renukappa, N.M.; Ranganathaiah, C.; Rajan, J.S. Correlation of free space length and surface energy of epoxy nanocomposites to surface tracking. *IEEE Trans. Dielectr. Electr. Insul.* **2018**, *25*, 2129–2138. [CrossRef]
27. Hillborg, H.; Gedde, U.W. Hydrophobicity recovery of polydimethylsiloxane after exposure to corona discharges. *Polymer* **1998**, *39*, 1991–1998. [CrossRef]
28. Hillborg, H.; Gedde, U.W. Hydrophobicity changes in silicone rubbers. *IEEE Trans. Dielectr. Electr. Insul.* **1999**, *6*, 703–717. [CrossRef]
29. Akhlaghi, S.; Pourrahimi, A.M.; Sjöstedt, C.; Bellander, M.; Hedenqvist, M.S.; Gedde, U.W. Degradation of fluoroelastomers in rapeseed biodiesel at different oxygen concentrations. *Polym. Degrad. Stab.* **2017**, *136*, 10–19. [CrossRef]
30. Zhou, Y.; Zhang, Y.; Zhang, L.; Guo, D.; Zhang, X.; Wang, M. Electrical tree initiation of silicone rubber after thermal aging. *IEEE Trans. Dielectr. Electr. Insul.* **2016**, *23*, 748–756. [CrossRef]
31. Gao, Y.; Liang, X.; Bao, W.; Li, S.; Wu, C.; Liu, Y.; Cai, Y. Effects of liquids immersion and drying on the surface properties of HTV silicone rubber: Part I-contact angle and surface chemical properties. *IEEE Trans. Dielectr. Electr. Insul.* **2017**, *24*, 3594–3602. [CrossRef]
32. Liu, H.; Cash, G.; Birtwhistle, D.; George, G. Characterization of a severely degraded silicone elastomer HV insulator-An aid to development of lifetime assessment techniques. *IEEE Trans. Dielectr. Electr. Insul.* **2005**, *12*, 478–486. [CrossRef]

energies

MDPI

Article

Mechanism of Saline Deposition and Surface Flashover on High-Voltage Insulators near Shoreline: Mathematical Models and Experimental Validations

Muhammad Majid Hussain [1,*], Muhammad Akmal Chaudhary [2] and Abdul Razaq [3]

1 Faculty of Computing, Engineering and Science, University of South Wales, Treforest, Cardiff CF37 1DL, UK
2 Department of Electrical and Computer Engineering, Ajman University, Ajman P.O. Box 346, UAE; m.akmal@ajman.ac.ae
3 School of Design and Informatics, Abertay University, Dundee DD1 1HG, UK; a.razaq@abertay.ac.uk
* Correspondence: muhammad.hussain@southwales.ac.uk

Received: 17 July 2019; Accepted: 24 September 2019; Published: 26 September 2019

Abstract: This paper deals with sea salt transportation and deposition mechanisms and discusses the serious issue of degradation of outdoor insulators resulting from various environmental stresses and severe saline contaminant accumulation near the shoreline. The deterioration rate of outdoor insulators near the shoreline depends on the concentration of saline in the atmosphere, the influence of wind speed on the production of saline water droplets, moisture diffusion and saline penetration on the insulator surface. This paper consists of three parts: first a model of saline transportation and deposition, as well as saline penetration and moisture diffusion on outdoor insulators, is presented; second, dry-band initiation and formation modelling and characterization under various types of contamination distribution are proposed; finally, modelling of dry-band arcing validated by experimental investigation was carried out. The tests were performed on a rectangular surface of silicone rubber specimens (12 cm × 4 cm × 8 cm). The visualization of the dry-band formation and arcing was performed by an infrared camera. The experimental results show that the surface strength and arc length mainly depend upon the leakage distance and contamination distribution. Therefore, the model can be used to investigate insulator flashover near coastal areas and for mitigating saline flashover incidents.

Keywords: saline mechanism; shoreline; wind speed; outdoor insulators; dry band arcing; flashover

1. Introduction

The performance of outdoor high-voltage insulators near the shoreline is a key factor in the determination of power network systems' stability and reliability. It is well known that contamination is considered a major critical factor responsible for surface flashovers [1]. The process of saline deposition on an insulator surface, associated with flashover and consequent power outages has been a major problem for power network systems since the early 1900s [2]. The time to surface flashover initiation depends on (i) deposition of saline contamination, and (ii) how salt particles penetrate and are diffused on the insulator surface through various wetting agents such as rain, fog, snow, dew or drizzle.

It is recognized that saline deposition and diffusion are affected by various natural processes such as temperature exposure, relatively humidity or moisture level, counter-diffusion of hydroxide ions and environmental load of salts and other adverse weather conditions. Types of contamination deposition on the insulator surface influence surface flashover, which has been extensively studied by several researchers [3–8]. The use of non-ceramic insulators increased significantly in last five decades. Silicone rubber insulators both in service [9] and high-voltage laboratory tests [10] demonstrated better performance than ceramic insulators in contaminated environmental conditions. Initially, non-ceramic insulators prevent water filming on the surface due to their hydrophobic properties, but this resistance gradually decreases due to physical and chemical changes in the silicone materials which can lead to dry-band arcing and surface discharges [11]. The combination of high-voltage stress and a contaminated water film produces dry-band arcing, and the resulting heat can lead to erosion of the insulator surface. The surface is damaged by physico-chemical changes caused by dry-band arcing [12,13]. Most of the previous work on surface flashover of contaminated insulators mainly focused on laboratory or onsite experiments based on alternating current (AC) and direct current (DC) voltage [14–16].

Kim et al. [17] studied chemical changes on the silicone rubber insulators during dry-band arcing but did not investigate the physical changes, for example the behavior of arc lengths with uniform and non-uniform pollution levels, as well as arc resistivity, power and energy. Therefore, there is a need to investigate the effects of dry-band arcing for a better understanding of the physical changes on the surface of silicone rubber insulators.

This paper presents a model, which is based on mechanism of sea salt transportation, deposition and diffusion on outdoor insulators near a shoreline, taking into account the saline concentration and the distance from the shoreline. It also introduces a new mathematical model to investigate the development of dry bands for different types of pollution layers on silicone rubber. A series of simulations and experiments were performed on the model to verify the theoretical results.

2. Mechanism of Salt Transportation and Deposition

The following three sub-models simulating three different processes were combined into one theoretical model of sea salt production, transportation and deposition:

2.1. Production of Saline

There are two major regimes where saline ions and particles are generated and scattered from shoreline to inland. Sea salt particles originate from breaking sea waves, a phenomenon that is followed by a high rate of wave motion and turbulence, air entrainment and surf formation. At high-level oceans, this breaking phenomenon is encountered under higher wind action with the formation of whitecap bubbles. As these bubbles rise, they are forced into the air where they scatter, thus producing saline particles. These particles can be routed to shoreline areas by oceanic wind speeds that exceed 4 to 12 ms^{-1} [18,19], where they tend to settle on outdoor insulators after a certain time and after having covered a certain distance. This mechanism is important in the generation of saline particles at intermediate to high wind speeds. In fact, wind speed is not the only factor to be taken into consideration. However, any factor favouring wave breaking and turbulence in the sea near the coast line must contribute to the formation of saline particles.

2.2. Saline Transportation and Deposition

Pollution near coastal regions is a major source of degradation of power network system equipment. In particular, saline attack is an important and major factor in the deterioration process of high voltage outdoor insulators near the coast. Feliu et al. proposed a complex theoretical model to represent transfer and deposition of saline on testing equipment near coastal areas [20]. However, this was based on constant spray of artificial aerosol, such as haze, dust and smoke, which do not represent natural climate conditions. In the present paper, a sea salt mechanism model representing saline transportation and deposition on outdoor insulators near coastal areas is proposed. The model represents the relationship between sea salt deposition on outdoor insulators and distance from the shoreline. Saline ions and particle changes are also taken into account driven on outdoor insulators by the wind from the sea. The study and experimental implementation of this new model is particularly useful for the investigation of surface degradation and surface flashover of outdoor insulators and substation components near shoreline based on salt concentration, wind speed and direction, and distance from coastal areas. The exposure of this model and experimental work is very similar to that of high-voltage transmission lines running along the seashore of Scotland, where outdoor insulators are exposed to wind, fog and rain, but not to direct saline spray. From various studies [21,22], it was found that near the shoreline salt particles mobilization was based on gravitational settlement and wind speed, and that these salt particles can travel longer distances before deposition. Based on that, a model and mechanism of sea salt transportation and deposition is presented in Figure 1.

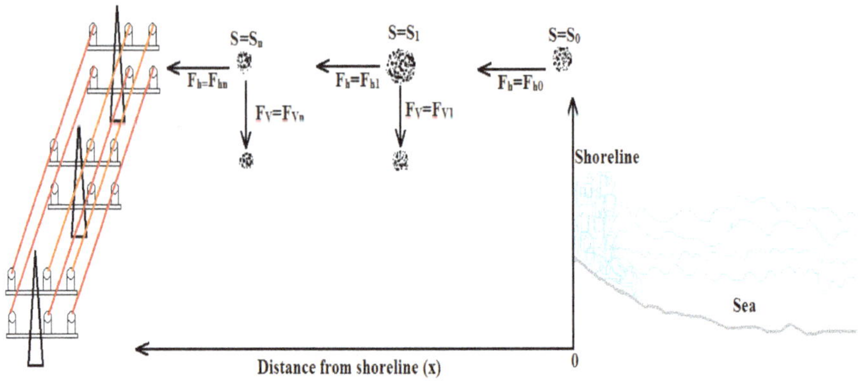

Figure 1. Schematic diagram of sea salt transport and deposition phenomenon.

Figure 1 represents a schematic mechanism of sea salt transportation and deposition, which shows the saline concentration (S) and its variation (S_0, S_1,... and S_n). In this model, oceanic winds, distance from the sea, diffusion and penetration of saline and gravitational settlement of saline on outdoor insulators are taken into account. The resultant vertical settlement flux (F_v) of saline due to gravitational effect and the saline transport near the shoreline on the outdoor insulators' surface is represented by a horizontal flux (F_h), as a significance of saline concentration, saline variation and wind regime. The relationship between saline concentration (S) and its variation (S_0, S_1, ... and S_n) from shoreline to the surface of insulator, vertical resultant deposition flux (F_v) and deposition rate (V_{dep}) is represented in Equation (1) by means of a mathematical simplification of flow velocity fluid mechanics equations:

$$F_v = V_{dep}\, S \tag{1}$$

From Equation (1), it is possible to determine the saline concentration (S) with variation (S_0, S_1 ..., and S_n) from shoreline to the surface of insulator, and deposition variation with time as a function of deposition rate (V_{dep}). It follows that the mass of saline deposited per unit of time is a negative function of the resultant vertical deposition flux. This is represented by Equation (2), where dt is the time variation, h is the thickness of the saline contamination layer and where the negative sign represents the reduction of saline concentration (S) deposition on the insulator surface with time and distance from the sea.

$$\frac{dS}{dt} = -\frac{SV_{dep}}{h} \tag{2}$$

Equation (3) is the solution of Equation (2) on the basis of an environmental natural phenomenon by which saline characteristics change when transported and deposited on outdoor insulators and substations equipment from sea to near shoreline. The saline decreases exponentially as shown in Equation (3), where S_0 is the saline concentration at shoreline, x is the distance from the sea and α is a constant represented by "$\alpha = V_{dep}/v_h$", where v as the wind speed. However, to solve Equation (3) it should be assumed that the deposition rate is constant with time and for any distance from the shoreline and its decay function may be estimated as:

$$S = S_0 e^{-\alpha x} \tag{3}$$

The unit of saline concentration is (mg/cm^2), taking into account the exponential decrease of saline deposition rate with time (saline concentration and deposition velocity has a proportional relation, thus Equation (3) can be rewritten as $V_{dep} = V_{dep0}e^{-\alpha t}$), and that, during a time period (t) as some saline particles are deposited on the surface of insulators installed at certain distance from the sea, while the remaining particles travel from the shoreline to a subsequent distance (x) driven by wind speed (v). As a result, integration of Equation (3) can be expressed by Equation (4), where S is the saline concentration at a distance x from seashore, S_0 is the saline concentration at seashore, V_{dep0} is the initial deposition rate of a saline at shoreline, h is the thickness of the contamination layer on the insulator surface and α is a coefficient of the deposition rate reduction that characterizes saline contamination distribution and its influence on deposition rate.

$$S = S_0 e^{(\frac{v_{dep0}}{\alpha h})[e(\frac{-\alpha x}{v})-1]} \tag{4}$$

where $t = \frac{x}{v}$.

The coefficient of deposition rate reduction is directly influenced by saline particles that are distributed near the shoreline. Considering that saline deposition on an insulator surface corresponds to saline particles that influence and remain on its surface during migration of saline from shoreline to inland and assuming that there is a proportional relation between saline flux (Φ), which depends upon insulator weather sheds and environmental characteristics and saline deposition (D) on the insulator surface. This can be expressed by a simplified relationship:

$$D = \Phi(VS) \tag{5}$$

where Φ is capture efficiency constant for the saline, which depends upon climate conditions and outdoor insulators, V is the wind speed and S is the saline concentration. On the basis of these assumptions and as a consequence, Equation (4) can be integrated, simplified and rewritten as:

$$D = D_0 e^{(\frac{v_{dep0}}{\alpha h})[e(-\frac{\alpha x}{v})-1]} \tag{6}$$

where D_0 is the saline deposition on the insulator surface at a certain distance from the shoreline and D is the saline deposition on the insulator surface from shoreline to inland. Equations (4) and (6) combine the proposed model of saline transportation and deposition from shoreline to inland on the insulator surface. Thus, Equation (4) represents the saline concentration variation across the seashore regime and Equation (6) represents the saline deposition rate on the insulator surface across the same regime. Contamination deposition mechanisms rather than gravitational settlement, like high speed wind and the scavenging effects due to heavy rainfall or dense fog precipitation were also considered at this stage in the model and related experimental investigations are presented.

The proposed model, which is based on a physical phenomenon, can be used to allow for the action of saline particles generated at sea and to account for the saline concentrations and deposition on the insulator surface near the shoreline. The aim of the proposed model is to observe the action of different input variables individually. Thus, simulations were done taking as a reference Equation (6) to find the saline deposition on insulator surface at different wind speeds and distances from shoreline. Saline deposition on the insulator surface represents the severity of saline transportation from the shoreline. Figure 2 indicates the saline deposition behaviour, when average wind speed varies in a range between 4.0 to 12.0 ms^{-1}. Figure 2 clearly shows that increase in wind speed contributes to migrate more saline ions further from shoreline, however, when wind speed was higher than 12 ms^{-1}, the migration of saline ions tended to be constant. Consequently, saline deposition on the insulator surface increases at fixed distance from shoreline. Moreover, insulator surface deposition significantly decreases with increasing the distance from shoreline to inland. For the simulations displayed in Figure 2, the conditions used were with variable wind speed and initial deposition rate, assuming the constant of coefficient of deposition rate reduction α (0.01 s^{-1}), which depends on insulator surface and environmental characteristics, derived from $\alpha = e^{A.T} \times \frac{V_{dep}}{vh}$, where A is total area of the insulator sheds and T is ambient temperature.

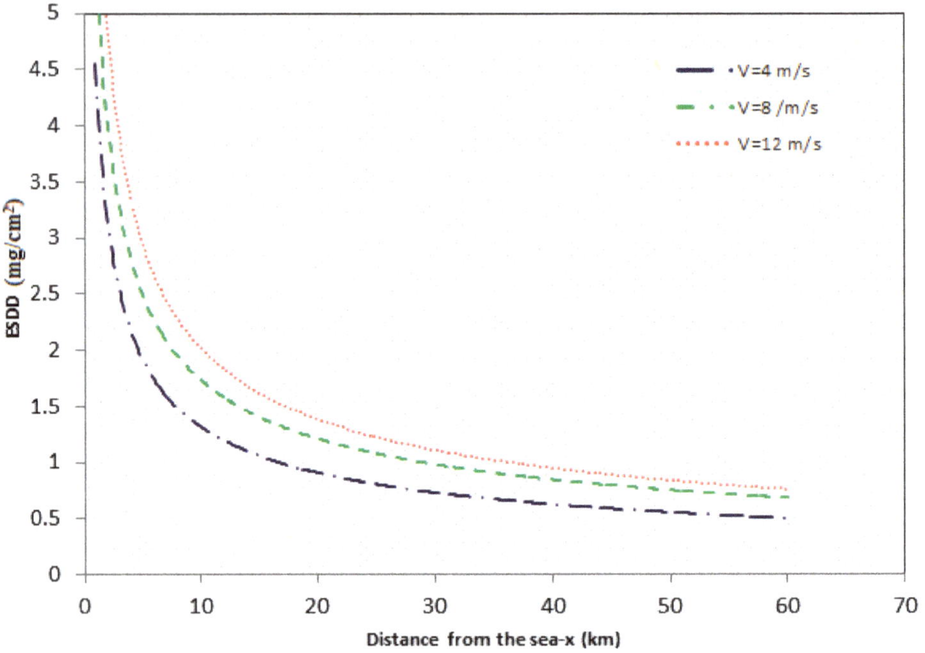

Figure 2. Simulated analysis of the model.

2.3. Saline Penetration and Moisture Diffusion

Saline and moisture migration are two important processes for studying the long-term durability of high-voltage insulators. The interaction between saline penetration and moisture diffusion become more important for high voltage insulator materials subjected to repeated wetting and drying cycles near the shoreline. Various other environmental stresses can also influence these. They will vary depending upon airborne saline particles and moisture levels in a marine environment. The saline penetration and moisture absorption on the insulator surface near the shoreline is usually caused by saline water spray carried by the prevailing winds coming from the coast [23]. It is a complex process that involves heat and mass transfer. Due to constant wetting rate, heat and mass transfer processes are much less important than diffusion and penetration processes. In this case, diffusion and penetration can be considered as one-dimensional, and numerical methods are required to solve it. The simulation was performed using COMSOL Multiphysics software. A 2-dimentional model was developed in COMSOL Multiphysics and two boundary conditions were considered for the simulation, as shown in Figure 3. The technical specifications of the 2-D model are summarized in Table 1. The first condition related to a fiber reinforced plastic (FRP) rod, was impermeability, so that saline concentration and moisture content were considered to be equal to zero. The second boundary was related to the insulator surface, for direct saline penetration and moisture diffusion. During the simulation work, saline ions and moisture struck and penetrated the silicone rubber (SiR), a process which may be described by Fick's Second Law of Diffusion, assuming constant diffusivity and direct saline binding. This law is expressed by Equation (7):

$$\frac{\partial S}{\partial t} = D\frac{\partial^2 S}{\partial^2 X} \tag{7}$$

where S is the saline penetration (mg/cm^2) as a function of time (t) at a distance X from the shoreline and D is the diffusion coefficient of moisture. The solution for saline penetration and moisture diffusion on insulator surface is given by Equation (8). The error function (*erf*) may be determined from the standard table of Fick's law.

$$S(x,t) = S_C(1 - erf\frac{X}{2\sqrt{D}}) \quad (8)$$

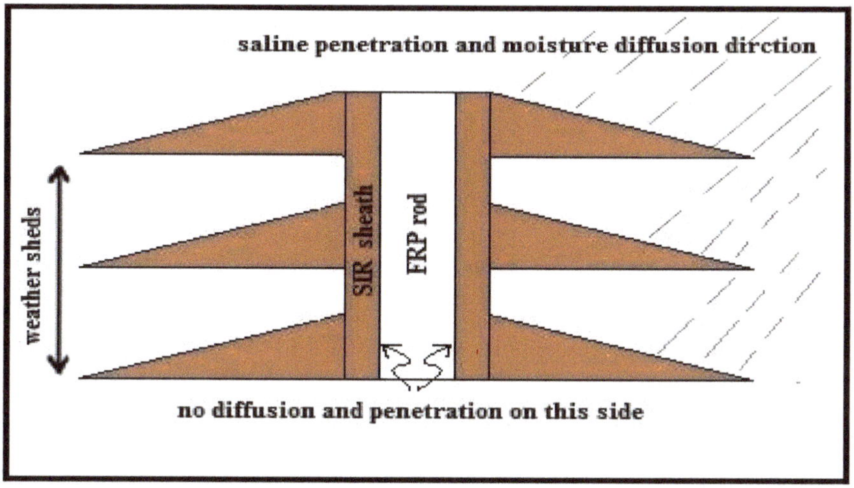

Figure 3. Two-dimensional (2-D) geometry used in simulation.

Table 1. Parameters of 2-D insulator model.

Insulator Type	Leakage Distance (cm)	Shed Spacing (cm)	Shed Diameter
Silicone rubber	40	3	12

The model provides the estimation of the local saline penetration and moisture concentration along the insulator pollution thickness. Figure 4 shows a simulation from the beginning of the penetration and diffusion process to the end of it. The curves of penetration and diffusion into the insulator sample were a two-stage growth process such as that for wetting and drying. The saline penetration in the water film increases non-linearly with the development of moisture concentration. As this process unfolds, the saline dissolved in water steadily gathers at the top of the water film. It can be clearly seen that, at the beginning of penetration and diffusion, the initial rapid growth is followed by a slowdown and ultimately drifts towards saturation. Thus, it can be concluded that the transfer of moisture significantly accelerates the penetration of saline ions on SiR materials, this method can also be used on other types of insulator models such as porcelain and glass. The negative sign with error function (*erf*) in Equation (8) states that increasing the accumulate moisture in wetting condition has the effect of increasing diffusion and penetration of saline on the insulator surface.

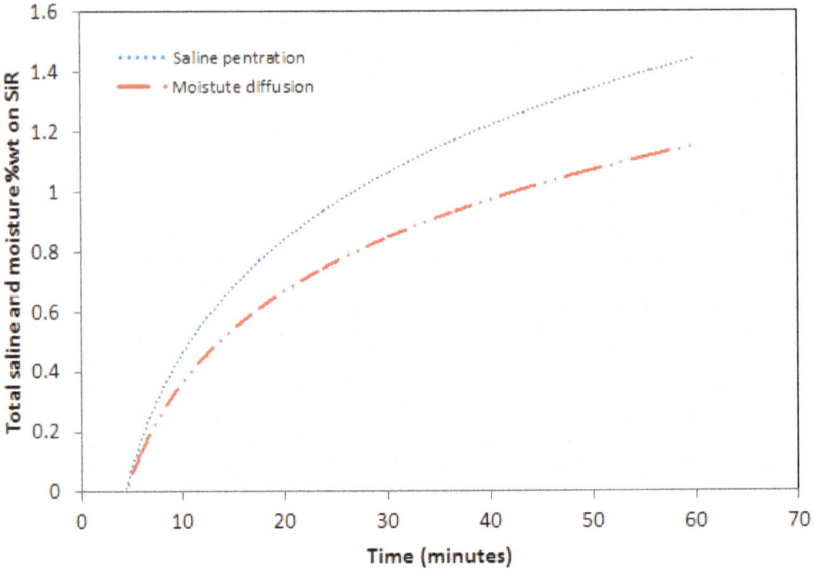

Figure 4. Characteristic curves of saline concentration and moisture diffusion.

3. Modelling of Dry Band Initiation and Formation

The development of a dry band formation on the surface of a silicone rubber model in normal cold fog can be mathematically formulated by considering the energy balance equation, which can be expressed as:

Energy in = Energy out + Energy related to change in and on insulator surface

In our case, the energy balance equation can be expressed as:

$$W_{LC} = W_{\Delta T} + W_{evap.} + W_{cond./conv.} \quad (9)$$

W_{LC} = Energy generated by leakage current
$W_{\Delta T}$ = Energy generated by change in temperature
$W_{evap.}$ = Energy loss by evaporation due to ambient temperature
$W_{cond./conv.}$ = Energy loss by convection and conduction

The model of dry band formation and initiation is shown in Figure 5. Evaporation is an important factor in cold and normal fog conditions in winter and early spring. Before the formation of the dry band, the voltage gradient along the insulator surface is uniform and its relationship to surface resistivity is:

$$E_s = \rho_s \cdot J_s \quad (10)$$

where subscript s represents the surface of insulator, J_s is the surface current density, E_s is the electric field intensity per unit area and ρ_s mass of air per unit volume on insulator surface. Equation (11) shows the relationship between current density and variable dry band length,

$$J_s = \frac{I}{\Delta L} \quad (11)$$

where ΔL is the variable length of dry band with surface current density.

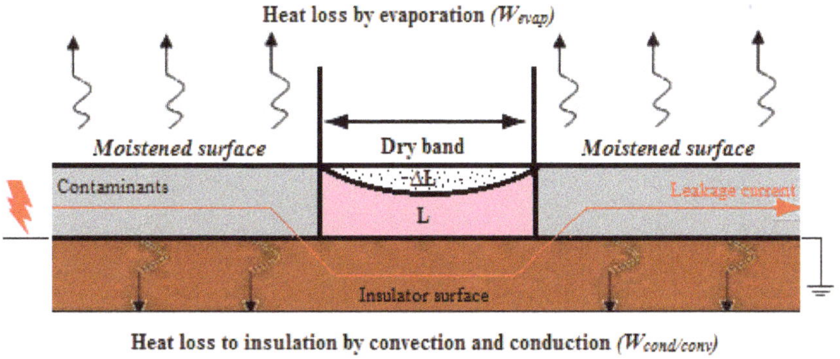

Figure 5. Modeling of energy balance on moistened insulator surface.

The power dissipated per unit area of the insulator surface is:

$$p = \rho_S \cdot J_s^2 \tag{12}$$

Due to the dissipated power the temperature of the pollution layer is rising because of heat transfer with the surroundings. Due to heat transfer, the temperature and dry-band area increased by ΔT and $\Delta s \cdot d(\Delta s)$, respectively, in a short period of time dt. Cold fog is made of condensed water droplets which are the result of a humid air mass being cooled to the dew point where it can no longer hold all of the water vapor.

Therefore, the corresponding volume $d(\Delta L) \cdot l$, and mass of cold fog is $\rho_S \cdot d \cdot (\Delta L) \cdot l$. Heat consumed by the evaporation process is given by:

$$W_{evp.} = -L_e \cdot \rho_S \cdot d(\Delta L) \cdot l \tag{13}$$

where L_e is the latent heat of fog.

The resistance of the dry band region is $l/\Delta s$. $1/\sigma$ due to a small volume of length L. Hence the heat generated by the current in AC system is:

$$W_{LC} = \frac{1}{\sigma \Delta L} \cdot l \cdot (J_S \, \Delta L)^2 \tag{14}$$

The conduction of heat is determined by the temperature difference between cold fog and insulator surface $H_{cond}(T_f - T_s)$ per unit area. The convection of heat between cold fog and the air interface on the insulator surface is $H_{conv}(T_f - T_a)$ per unit area.

The change in temperature of cold fog is in relationship with specific heat as $\Delta T \cdot mCh$. The latent heat of cold fog to moisture due to evaporation is $\Delta m \cdot Lm$. Surface resistivity changes with change of temperature along insulator surface length l in a short period of time dt so that:

$$W_{\Delta T} = C_{h.} \rho_S \cdot \Delta L \cdot l \cdot d \tag{15}$$

where ρ_s is the medium density.

If the volume of moisture is very small and does not interact with the air or insulator surface, then there is no convection or conduction. If it interacts with the insulator surface, then the area of interaction is $\Delta Am \cdot \rho \cdot l$. In a short period of time dt, the dissipated heat is given by:

$$W_{\frac{cond}{conve}} = H_c\left(T_m - T_p\right) \cdot \Delta A_m \cdot \rho \cdot l \cdot dt \tag{16}$$

By combining Equations (13)–(16), we obtain:

$$\frac{l}{\sigma\Delta L}\cdot l\cdot(J_S\ \Delta L)^2\cdot dt = -L_e\cdot\rho_s\cdot d(\Delta L)\cdot l + C_h\cdot\rho_s\cdot\Delta L\cdot l\cdot dt + H_c\ \left(T_m - T_p\right)\cdot\ \Delta A_m\ \cdot\rho\cdot l\cdot dt \tag{17}$$

If the distribution of current is uniform along the insulator surface, then Δ can be neglected. Then by dividing both side of Equation (17) by $l\cdot\frac{dt}{L}$ Equation (18) is obtained:

$$\frac{l}{\sigma}\ (J_S\ )^2 = -L_e\cdot\rho_s\cdot L\cdot\frac{dL}{dt}\cdot l + C_h\cdot\rho_s\cdot L\cdot\ l\cdot dt + L\cdot H_c\left(T_m - T_p\right)\cdot A_{m-p} \tag{18}$$

From Equation (18), it can be noted that there are five (σ, L, T, ρ_s and t) essential and critical physical parameters involved in the dry-band formation and arcing on the insulator surface. As L is the length of the dry band region, the drying rate can be calculated by using small steps dL/dt until L becomes zero. This is the point where time to dry band arcing on insulator surface will be started. The dry band is completely formed when the area L on the insulator surface becomes zero where the leakage current is intermittent.

4. Partial Arc Electric Model of Dry-Band Flashover

Several researchers have worked to make useful contributions to this subject [1,24]. However, there are several key shortcomings in the models presently available. The present model capable of handling both uniform and non-uniform distributions pollution on the surface of insulators is more relevant. There are several aspects on which arc resistance and dry-band formation depend. The models assume that if a dry band can be formed and if the arc is able to bridge the dry band will continue propagation with different contamination degree. The dry band arcing is modeled mainly in two stages: First the formation of dry band and initial arc and second the arc propagation and arc bridging.

A thorough understanding of all aspects of flashover mechanism on an insulator surface is required to explore the subject further. Such a task would necessarily include an investigation of dry band arcing under different contamination levels along the leakage distance. Propagation of the alternating current (AC) surface flashover on polluted insulators is a complex phenomenon. The length and intensity of arcs may change in milliseconds. The arc is only highly ignited in the period of peak voltage, while during reaming periods the arc ignites and reignites following the voltage. Despite the complexity of the mechanism involved in dry-band arcing, many simplifying assumptions can be made in order to obtain an acceptable mathematical modelling.

The growth of the dry band and the dry-band arcing characteristics method have been well reviewed by Jolly and Poole [25] but their model will be extended here based on fundamental mathematical equations. The model was used for the analysis of the growth of the discharge with different contamination degrees along the leakage distance of specimen. The test procedures were as follows:

CASE 1:

In this case the sample leakage distance is divided into two sections L_1 and L_2 while the corresponding surface conductivities and surface resistances are σ_1, σ_2 and r_{i1}, r_{i2}, respectively, as shown in Figure 6. L_1 is the high voltage side and L_2 the grounded side. Section L_1 was lightly polluted while L_2 was polluted with heavy pollution levels. The hydrophobicity of the high-voltage side L_1, is higher than that of the grounded side L_2. Due to this difference, the arc cannot develop along the entire surface and its length is less than L_1 due to surface strength, as shown in Equation (23). As long as the arc length is smaller than section L_1, very small parts of L_2 can hardly influence arc length X. In this case, it may assume that the probability of subsistence of surface flashover at energized condition is very low.

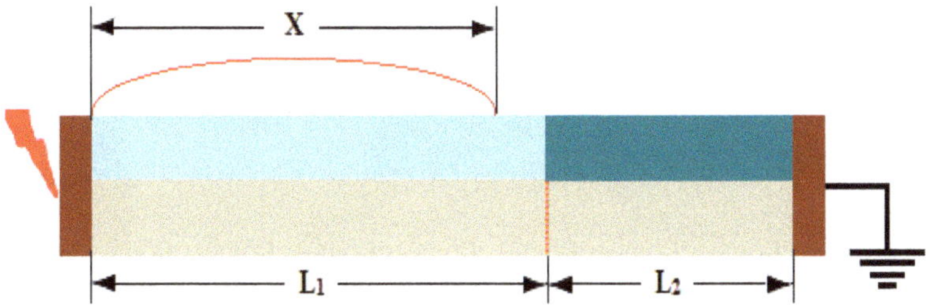

Figure 6. Polluted insulator model for dry band arcing with $X < L_1$.

If the heavily contaminated surface is located at the grounded side, the influence of the totally hydrophilic section (L_2) on arc length X can be described as follows:

$$L = L_1 + L_2 \tag{19}$$

Then the resistance of heavily polluted section L_2 that could not have connected with arc X can be written as:

$$R_i = r_{i1}(L_1 - L) + r_{i2}(L_2) \tag{20}$$

It is a well-known fact that current is necessary to sustain and ignite the arcing process on the insulator surface so that it can be derived from the Obenaus model [26] and rewritten as:

$$I = \left[\frac{nNX}{r_i(L - X)}\right]^{\frac{1}{1+n}} \tag{21}$$

It is assumed that constants n and N for arcs and insulator are comparable. From Equation (21), we can find the critical arc length Xc of the composite insulator that relates to section L_1 as follows:

$$Lim_{(X \to Xc)} = \left[\frac{Xn_1N_1}{r_{i1}(L_1 - L) + r_{i2}L_2}\right]^{\frac{1}{1+n_1}} \tag{22}$$

Equation (23) can be written in terms of sectional lengths and the ratio of surface conductivities and resistivities, and the root of an arc is close to section of L_2 of resistance per unit length. The relationship of Equation (24) shows that the L_1 leakage distance section, bridged with the arc, does not influence arc length X.

$$X_c < \left[\frac{N_1}{r_{i1}}\right]^{\frac{1}{1+n_1}} \tag{23}$$

$$X_c = \left[\frac{L_1 + kL_2}{1 + n_1}\right] \tag{24}$$

The simplified relationship of the surface conductivities and resistivities of sections L_1 and L_2 can be expressed as: $k = \frac{r_{i2}}{r_{i1}} = \frac{\sigma_1}{\sigma_2}$

where σ_1 and σ_2 are surface conductivities and r_{i1} and r_{i2} surface resistance of sections L_1 and L_2.

CASE 2:

In this case the section L_1 was heavily polluted while L_2 lightly polluted. It is assumed that the specimen is totally hydrophilic only near a high-voltage side, as presented in Figure 7. This consideration is due to long-term operating conditions. If section L_1 is hydrophilic as compared to L_2 then what influence of hydrophilic section of the creepage distance L_1 has on the arc length X.

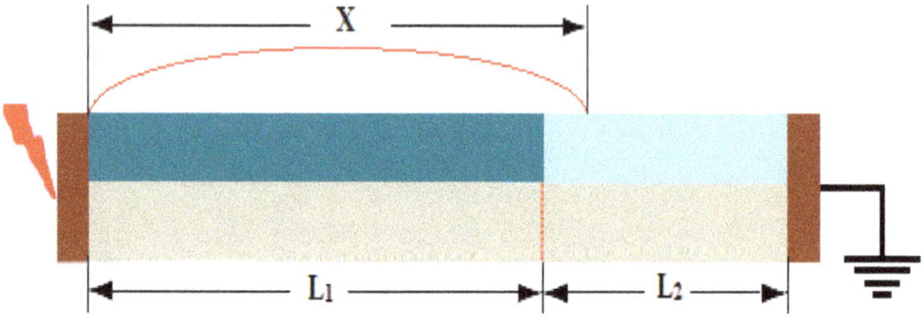

Figure 7. Polluted insulator model for dry band arcing with $X > L_1$.

When arc length X connected with section L_2 with reduced k-times is on the lightly polluted and more hydrophobic section, then the leakage distance of the sample corresponding to L_2 can be written as Equation (25).

$$L_2 = L_2 + kL_1 \tag{25}$$

$$Lim_{(X \to Xc)} = \left[\frac{Xn_2N_2}{r_{i2}(L_2 + kL_1 - L)} \right]^{\frac{1}{1+n_2}} \tag{26}$$

$$X_c > \left[\frac{N_2}{r_{i2}} \right]^{\frac{1}{1+n_2}} \tag{27}$$

$$X_c = \left[\frac{L_2 + kL_1}{1 + n_2} \right] \tag{28}$$

The relationship indicates that the L_1 section of creepage distance connected with the arc but practically L_2 does not have any influence on a critical arc length Xc. To figure out the reason, it can be assumed that discharge current and arc length X is increased resulting in a heavy pollution level and more hydrophilic surface near high voltage end, and at a certain level of contamination, as described in the division of sample creepage distance, uneven potential distribution on the specimen occurs that ignites the arc along section L_1 that causes an immediate extension for section L_2 resulting in a sudden surface flashover occurring.

5. Experimental Setup and Results

Characteristics of Dry-Band Formation

The most significant factors on the surface flashover voltage are the number, location and length of dry bands. The initiation and formation of dry band is due to many factors, such as voltage, contamination type, amount and distribution, and various environmental stresses. It is a well-known fact that, when wet and polluted insulators are energized, discharge current causes Joule heating to form dry bands. Joule heating drives the low resistance layer and evaporates wetting. Drying is more intense where the current density is high. As a result of more evaporation, more small dry bands are formed on the surface. The discharge current is associated with dry band length and if the length is sufficiently long the discharge current will decrease and the arc will extinguish, resulting in a surface flashover. Thus, a comparison is made between the dry band initiation and formation under uniform, non-uniform and discontinuous non-uniform contamination distribution on specimens. The investigation is based on contamination layer parameters, such as conductivity, layer thickness and length, and insulator surface dielectric strength. The reference insulator with no sheds consisted of two main parts. One part was a silicone rubber plate sample of rectangular shape (12 cm × 4 cm × 8 cm), mechanically connected with electrodes, one of them connected to a high voltage alternating current (HVAC) power source with AC voltage of 0–100 kV at 50 Hz, and the other one grounded as shown in Figure 8. The electrodes were made of 0.9 mm thickness copper. The contamination along the leakage distance was achieved by solid layer method has been given in IEC-60507, 1991, by brush applications of different conductivities. For experimental work the 69 kV AC test voltage was produced by a 10 kVA, 100 kV, and 50 Hz transformer. The supplied voltage can be increased manually or automatically at a rate of 1 kV/s. In the experimental setup, the analysis and processing units comprised a high frequency current transformer, a protection circuit, a LeCroy digital storage oscilloscope and a personal computer.

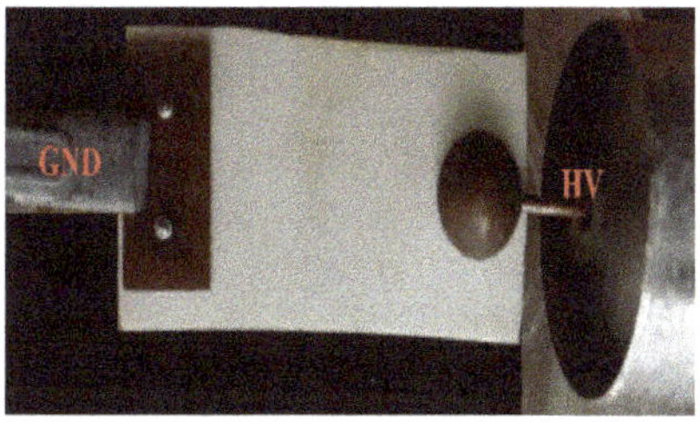

Figure 8. Experimental set up.

For sample contamination, the ratio of equivalent salt deposit density (ESDD) to non-soluble deposit density (NSDD) was 1:4. The values of ESDD and NSDD with contamination values are shown in Table 2.

Table 2. Contamination values.

Contamination Type	ESDD (mg/cm^2)	NSDD (mg/cm^2)
Uniform	0.200	0.850
Non-uniform	0.080	0.350

Initially, the distribution of voltage and layer resistivity was uniform and the sample surface infrared pictures were uniformly heated, as shown in Figure 9. This confirmed that contamination layer conductance was also uniform. As the surface becomes wet, resistivity decreases and discharge current increases. This condition is not resilient due to slightly higher resistance in some segments of the sample surface with exceeding voltage gradient in these sections. The dissipation of heat is higher at these locations such that they become dry more rapidly than the remaining surface, forming dry bands. It is clearly shown that only small dry bands are formed near the high voltage electrode as well as near areas with higher electric field strength. However, as opposed to non-uniform contamination distribution, the dry bands do not tend to elongate towards the ground electrode.

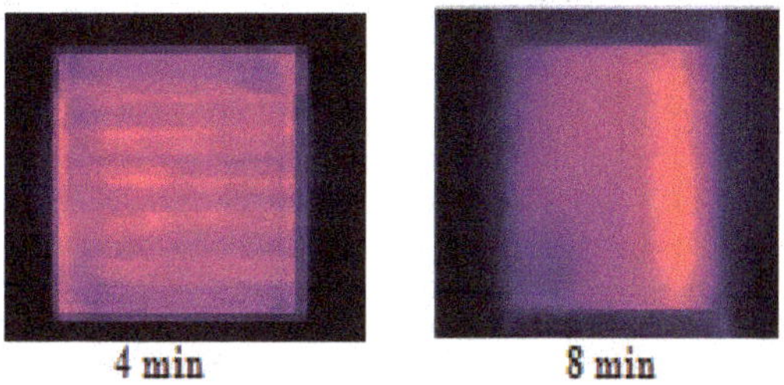

Figure 9. Dry-band development with uniform contamination record by infrared camera.

Under non-uniform contamination distribution, however, the leakage current is increased, soon leading to reduced surface conductance, as shown in Figure 10a. This strongly indicates the drying part of the non-uniform contamination distribution by the heated area of the conducting layer, with the expected formation of a complete dry band. Initially the dry band formed at the high voltage electrode is followed by a series of small discharges that gradually scatter on other part of specimen. These discharges move towards near the ground electrode region and take the form of an arc-like discharge.

On the other hand, Figure 10b shows the infrared images of dry band discharges with discontinuous non-uniform contamination distribution. It is clearly shown that the extinction of the discharges is frequently followed by reignition, temporarily bridging the other dry bands. At this state, leakage current is at its maximum value just before the start of the discharge phenomenon. It is also observed that under discontinuous non-uniform contamination the insulating surface achieves its highest dielectric strength when the specimen conductive surface carries a lot of wider dry bands located at different locations of the specimen. Multiple dry bands created at same time on an insulating surface have the ability to weaken the field strength of each other. Therefore, with the existence of multiple dry bands the field strength of each one is normally less than when it is individually created. If a number of dry bands are formed, then after a short period only one will remain and due to its higher resistance almost all the voltage will be dropped across this dry band. The dry bands formed under a discontinuous non-uniform contamination layer may be considered as a potential barrier which may efficiently weaken the dielectric strength and develop a large voltage drop on outer dry bands.

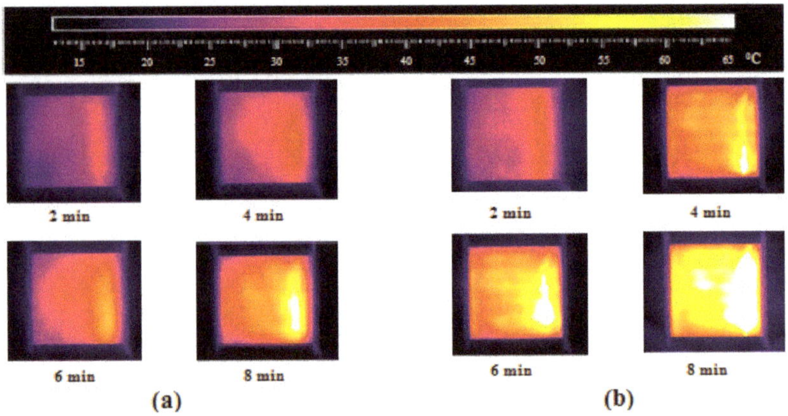

Figure 10. Dry-band development recorded by infrared camera: (**a**) under non-uniform contamination; (**b**) under non-uniform discontinuous contamination.

Results obtained indicate that with discontinuous nonuniform contamination distribution, dry-band elongation on the full leakage distance took 6 minutes, and that after 8 minutes there was a sharp rise in dry band elongation with multiple dry bands. However, with a uniform contamination layer, dry-band elongation took place only up to the 40% of the total length in 6 minutes. After that there was no more elongation of dry-band length.

6. Comparison of Models with Experimental Results

6.1. Inspection of Dry-Band Formation

This section describes the difference between the mathematical model and the test results. It is noted that in the model there is a smooth increase of resistance at the initiation and development of dry bands, while in the test the dry bands outset suddenly as shown in Figures 9 and 10. In the model (Section 3), it is anticipated that the dry band is initiated when the moist layer becomes wet. When the ΔL is very close to zero, it is shown that the dry band width is increased. However, this period is difficult to measure in the experiment. Therefore, the dry bands developed before the moisture is totally vaporized. The critical phenomena of moist film before the dry band forms are illustrate here. The equilibrium of forces between moist contact interfaces on insulator surface can be derived by the Young–Dupre equation:

$$\lambda_m Cos\,\theta = R_s\,(\lambda_s - \lambda_{ms}) \tag{29}$$

In Equation (29), the R_s is the surface roughness coefficient which depend upon the material condition, θ is the contact angle in degrees between moisture and material surface, λ_m is the surface tension of moisture, λ_s surface tension of material surface and λ_{ms} is the interfacial between the moisture and material surface.

This section explains the phenomena that observed during the experiment. For a moisture layer on the material surface, we anticipated the equilibrium state is stretched first. When the surface was hydrophilic, the length of dry bands varied from ΔL to L. However, as the moisture layer is being evaporated, the moisture layer becomes thinner and thinner, and therefore the interfacial force λ_{ms} decreases. At this point the discharge current is cut off, which caused an increase of the local electric field. Therefore, the dry band's initiation and development are started. Once the dry-band arcing is started, the arc temperature would dry the remaining part of moisture layer. At this stage, the surface is not fully dry, the dry-band arcing is weak and causes a number of multiple dry bands, which is observed in tests as shown in Figure 11.

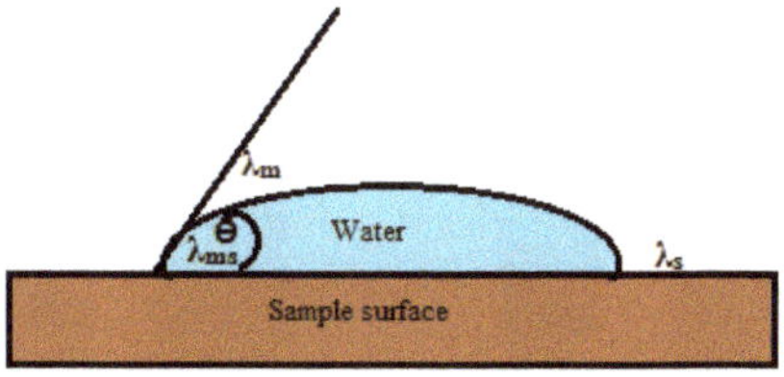

Figure 11. Interfacial equilibrium between water and sample surface.

6.2. Onset of Dry-Band Arcing

To validate the model, the experimental results of dry band arcing, arc extension and surface flashover were compared with those of the proposed mathematical model. For the experimental analysis two scenarios were configured as those on the mathematical model: (1) high-voltage side with heavy pollution, whilst the grounded side was lightly polluted, and (2) high-voltage side with light pollution and grounded side with heavy pollution. These pollution scenarios are commonly seen in the field under the operating conditions to which are exposed insulators located near the shoreline and at sites with dominant winds speed [27–30].

In the first scenario (Figure 12), when voltage is applied to a sample, local arcs are first initiated on the heavily contaminated side/part, which is essentially due to higher discharge current. The discharge current causes ohmic heating to form multiple dry bands. The voltage across the dry bands which were usually the low conductive surface parts caused air break down. This caused the dry bands to be moved towards the lightly contaminated part which became electrical in series with the heavily contaminated part of the surface. Multiple dry bands spread out onto the sample surface and nearby the electrodes, and then some of the dry bands were bridged by local arcs. Local arcs ignited and reignited many times, and then gradually developed over the surface to connect to the other arcs, thus increasing the total length of the arcs.

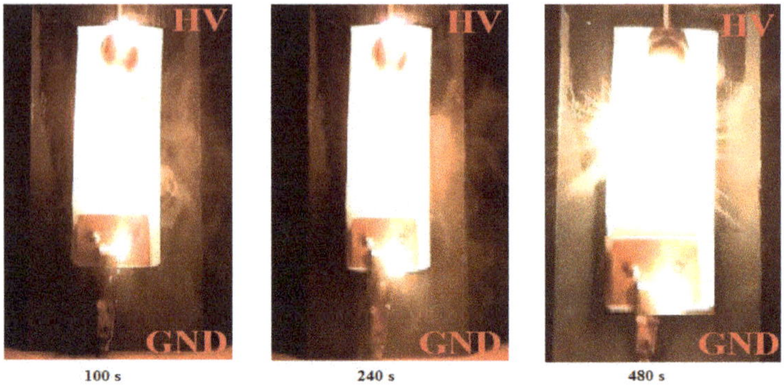

Figure 12. Image of dry-band arcing activity progression.

On the other hand, in the case of the second scenario (Figure 13), the dry-band arcing did not initiate suddenly, as the section with light contamination levels, which is more hydrophobic than the other section, can strongly limit the development of the discharge current. It was found that if the length of the arc is equal to the length of the considered leakage section (such that the considered section is completely bridged) then the arc will propagate over to the next section and that the contamination level of the remaining part of the sample will have little effect on dry-band arcing. The experimental results obtained show that the surface strength and arc length mainly depend on the leakage distance and contamination distribution.

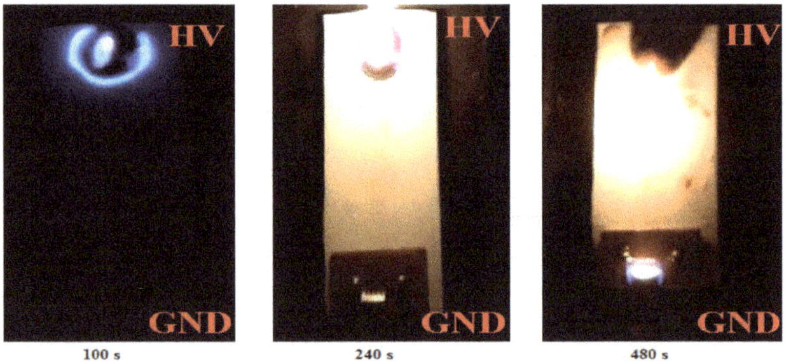

Figure 13. Image of discharge activity progression.

From the above results, it can be concluded that due to voltage drop along an arc channel the arc root transfers the potential of the electrodes next to the lightly contaminated section of the leakage distance. If the potential is sufficiently high, then after attainting the edge of the lightly contaminated section of the leakage distance, arc extension surface flashover will occur on the specimen. Surface flashover instantly occurs and is not connected with the partial arc's development. It was also observed that on the lightly contaminated section, the drop of tangential component of electric field does not exit. However, it can be observed mainly along the heavily contaminated section of the specimen.

6.3. *Impact of Wind Velocity on Surface Flashover Characteristics*

To obtain the effects of the wind velocity on the surface flashover characteristics of composite insulator. The AC surface flashover process of polluted insulator was observed at 0 m/s, 2 m/s, 4 m/s, 8 m/s 10 m/s and 12 m/s wind velocities, respectively. The tests were carried out under two conditions. One was wind but no contamination deposition condition and other was wind with contamination deposition condition. In order to wet the surface of insulator and contamination distribution uniform, $\theta = 90°$ was selected between wind direction and insulator axis. During the experiment, the relative humidity was sustained in between (80%–90%) in the environmental chamber, which helped to bond the saline mixtures on the insulator surface, and the temperature was maintained between 0 °C and 2 °C.

During tests it was observed that, when the wind velocity bellowed 8 m/s or less as shown in Figure 14, the fog drifts slowly, and approached the surface of the insulator all around it. Simultaneously, the influence of contamination on surface flashover voltage was continuous due to the result of the wetting and drying process on insulator surface. Although at some moment the effect of moistening was larger and the surface flashover voltage was higher; while other, the effect of drying was larger and the surface flashover voltage was lower. Therefore, the value of surface flashover voltage changed by a big margin at 8 m/s and for lower wind speed values. Thus, the lowest value was selected as the surface flashover voltage value. However, when the wind velocity is higher than 8 m/s, the mist drifts very fast. The fog cannot completely approach with the insulator surface, so the drying effect was higher than the wetting effect. At this situation, the flashover voltage was relatively low and stable as compared to lower wind velocities, thus the average value was preferred as the surface flashover voltage value. From Figure 14, it can be clearly seen that when the wind velocity bellows up to 8 m/s, the surface flashover voltage diversifies by a big margin, the lowest higher than the normal down 32.9%, and when the wind velocity higher than 8 m/s, the surface flashover voltage gradually decreases steadily and at 12 m/s surface flashover voltage is lower than the normal flashover voltage 6.2%.

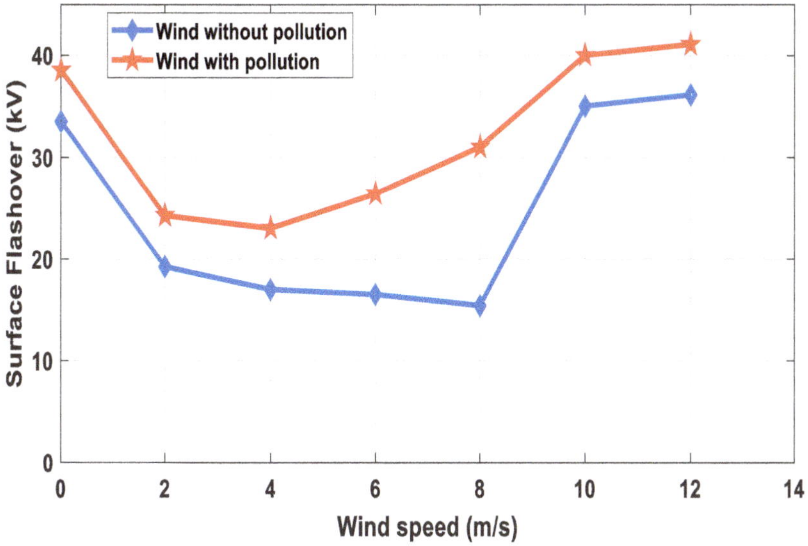

Figure 14. Influence of wind on surface flashover characteristic.

6.4. *Effect of Conductivity and Pollution Layer Length of Surface Flashover*

The dimensioning of insulators with respect to contamination is always done based on the performance characteristic of real insulators under uniform artificial contamination. However, due to the shape and in-service position, climate conditions and following the action of electrical stresses, the outdoor high-voltage insulators are actually contaminated in a non-uniform manner. The surface flashover of a uniformly contaminated insulator can be defined by a one dry band arc connected in series with a resistance of the contamination layer. In contrast, for insulators with a non-uniform contamination layer, several dry-band arcs can ignite simultaneously over their contaminated surface and may develop to a full surface flashover. This section is made to investigate the effect of uniform, non-uniform and discontinuous non-uniform contamination layer parameters, such as the conductivity and the length of the contamination layer on the material surface. Surface flashover tests are carried out following the same procedure as presented in [30].

The results of the present work and the other investigations [31] are approximately close. However, the surface flashover voltage of the present work and the work carried out by [32] are slightly lower than the surface flashover voltage with natural contamination. The main reason for this is that ions solubility of marine specification salts in artificial contamination almost slightly varies in fog, but it is not in the case of natural contamination deposition such as desert and ash, which contains weaker electrolytes. Also, the other most significant factor on the flashover strength is the length of the pollution layer. These results agree with that mentioned in [31] and would explain the observations that artificially contaminated insulators do allow surface flashover voltage lower than those naturally contaminated for the same conduction and climate conditions. It is clearly shown in Figures 15 and 16 that the length of contamination layer significantly affects the surface flashover voltage. The dielectric strength decreases with increasing of the conductivity layer. Finally, there is another strong effect observed on surface flashover due to the contamination class. It perceived that surface flashover strength is higher with uniform as compared to non-uniform and non-uniform discontinuous contamination distribution. Figure 16 shows the variation of surface flashover voltage with different pollution layer length under different types of contamination deposition. We can observe that surface flashover voltage decreases when the length of the pollution layer increases. It elapses from 36.2 kV with a uniform distribution to 33.1 kV for a surface entirely contaminated with a non-uniform discontinuous.

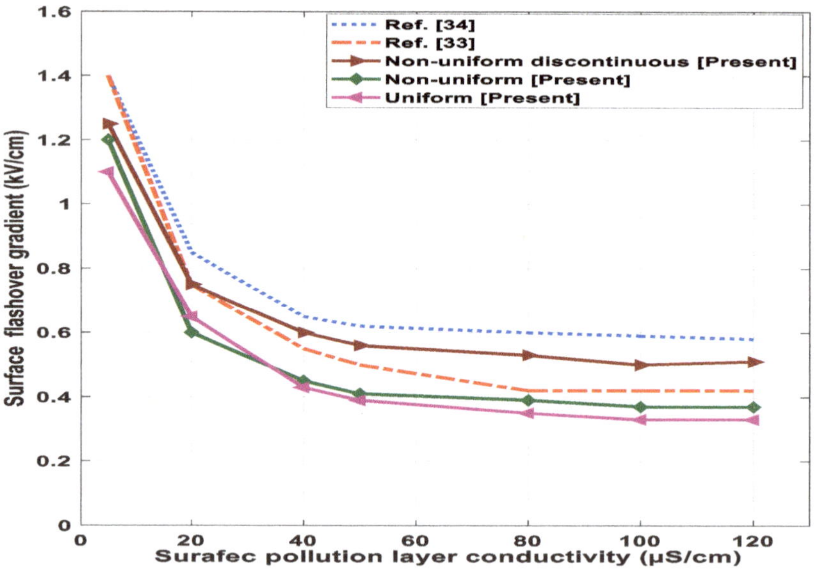

Figure 15. Surface flashover voltage versus the pollution layer conductivity with various contamination class.

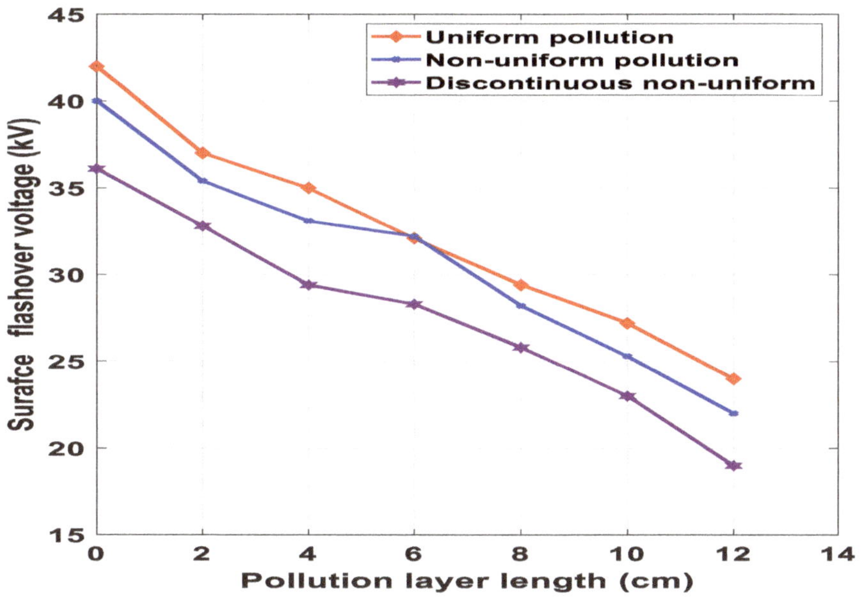

Figure 16. Surface flashover voltage versus the pollution layer length for different contamination class.

7. Conclusions

In this paper, a novel mathematical model is presented to understand the sea salt migration and deposition on high-voltage insulators near shoreline. The model introduced a new phenomenon based on saline transportation and deposition. Based on this, two more mathematical models were presented to understand the phenomenon of dry-band initiation and formation, as well as the behaviour of dry-band arcing with two different conduction states of the contamination layer on the insulator surface. Based on these, the following conclusions could be drawn:

1. From simulation and experimental work it was found that the pollution accumulation rate on the insulator surface increases with the increase of wind velocity and decreases with the increase of distance from shoreline to inland. However, when wind speed was higher than 12 m/s, the contamination density tended to be constant.
2. It was also observed that the transfer of moisture significantly accelerates the saline ions' penetration on the insulator surface.
3. The dry-band initiation and formation model presented is based on the energy balance equation. The integration of this equation results in a variation of dry-band length under moisture evaporation on the insulator surface. With the reduction of dry-band length, the surface resistance increases, which increases the discharge current on the insulator surface. The conduction and convection processes increase the surface resistance. It was observed that under this process single and multiple dry bands appear on the insulator surface.
4. Discontinuous non-uniform contamination distribution of the insulator surface leads to multiple dry bands and lower surface flashover voltage as compared to uniform and non-uniform contamination distribution.
5. A dry band arcing model was developed. It showed that the insulator surface strength and arc length mainly depend on the distribution of the pollution layer and leakage distance. The results obtained show that the configured scenarios are trustworthy to detect the dry band arcing on the insulator surface with different contamination distribution. The obtained dry-band arcing initiation and elongation rate confirm the efficiency of the proposed model.

Author Contributions: Conceptualization, M.M.H.; methodology, M.M.H., M.A.C. and A.R.; analysis, M.M.H., M.A.C. and A.R.; investigation, M.M.H.; writing—review and editing, M.M.H., M.A.C. and A.R.; visualization, M.M.H. and M.A.C.; supervision, M.M.H.; project administration, M.M.H., M.A.C. and A.R.

Funding: This research received no external funding.

Conflicts of Interest: The authors declare no conflict of interest.

References

1. Venkataraman, S.; Gorur, R.S. Extending the applicability of insulator flashover models by regression analysis. *IEEE Trans. Dielectr. Electr. Insul.* **2007**, *14*, 368–374. [CrossRef]
2. Baker, A.C.; Farzaneh, M.; Gorur, R.S.; Gubanski, S.M.; Hill, R.J.; Schneider, H.M. Insulator selection for overhead AC lines with respect to contamination. *IEEE Trans. Power Deliv.* **2009**, *24*, 1633–1641. [CrossRef]
3. El-Amine Slama, M.; Hadi, H.; Flazi, S. Investigation on influence of salts mixture on the determination of flashover discharge constant Part I: A Preliminary Study. In Proceedings of the IEEE Conference on Electrical Insulation and Dielectric Phenomena, Quebec, QC, Canada, 26–29 October 2008; pp. 674–677.
4. Sima, W.; Yuan, T.; Yang, Q.; Xu, K.; Sun, C. Effect of nonuniform pollution on the withstand characteristics of extra high voltage suspension ceramic insulator string. *IET Gen. Trans. Distrib.* **2009**, *4*, 445–455. [CrossRef]
5. Majid Hussain, M.; Farokhi, S.; McMeekin, S.G.; Farzaneh, M. Dry band formation on HV insulators polluted with different salt mixtures. In Proceedings of the 2015 IEEE Conference on Electrical Insulation and Dielectric Phenomena (CEIDP), Ann Arbor, MI, USA, 18–21 October 2015; pp. 201–204.
6. Naito, K.; Morita, K.; Hasegawa, Y.; Imakoma, T. Improvement of the dc voltage insulation efficiency of suspension insulators under contaminated conditions. *IEEE Trans. Dielectr. Electr. Insul.* **1988**, *23*, 1025–1032. [CrossRef]
7. Seta, T.; Arai, N.; Udo, T. Natural pollution test of insulators energized with HVDC. *IEEE Trans. Power Appar. Syst.* **1974**, *PAS-93*, 878–883. [CrossRef]
8. Kimoto, I.; Fujimura, T.; Naito, K. Performance of insulators for direct current transmission line under polluted condition. *IEEE Trans. Power Appar. Syst.* **1973**, *PAS-92*, 943–949. [CrossRef]
9. Houlgate, R.G.; Swift, D.A.; Cimador, A.; Pourbaix, F.; Marrone, G.; Nicolini, P. Field experience and laboratory research on composite insulators for overhead lines. *CIGRE Pap.* **1986**, *15*, 12.
10. Schneider, H.M.; Guidi, W.W.; Burnham, J.T.; Gorur, R.S.; Hall, J.F. Accelerated aging and flashover tests on 138 kV nonceramic line post insulators. *IEEE Trans. Power Deliv.* **1993**, *8*, 325–336. [CrossRef]
11. Kim, S.H.; Cherney, E.A.; Hackam, R. The loss and recovery of hydrophobicity of RTV silicone rubber insulator coatings. *IEEE Trans. Power Deliv.* **1990**, *5*, 1491–1500. [CrossRef]
12. Starr, W.T. Polymeric outdoor insulation. *IEEE Trans. Electr. Insul.* **1990**, *25*, 125–136. [CrossRef]
13. Simmons, S.; Shah, M.; Mackevich, J.; Chang, R.J. Polymer outdoor insulating materials Part 3-silicone elastomer considerations. *IEEE Electr. Insul. Mag.* **1997**, *13*, 25–32. [CrossRef]
14. Baker, A.C.; Zaffanella, L.E.; Anaivino, L.D.; Schneider, H.M.; Moran, J.H. Contamination performance of HVDC station post insulators. *IEEE Trans. Power Deliv.* **1988**, *3*, 1968–1975. [CrossRef]
15. Wilkins, R. Flashover voltage of high-voltage insulators with uniform surface pollution films. *Proc. IEE* **1969**, *116*, 457–465. [CrossRef]
16. Rizk, F.A.M. Mathematical models for pollution flashover. *IEEE Trans. Dielectr. Electr. Insul.* **1981**, *78*, 71–103.
17. Kim, S.; Cherney, E.A.; Hackam, R.; Rutherford, K.G. Chemical changes at the surface of RTV silicone rubber coating on insulators during dry-band arcing. *IEEE Trans. Dielectr. Electr. Insul.* **1994**, *1*, 106–123.
18. Morcillo, M.; Chico, B.; Mariaca, L.; Otero, E. Salinity in marine atmospheric corrosion: Its dependence on the wind regime existing in the site. *Corros. Sci.* **2000**, *42*, 91–104. [CrossRef]
19. Spiel, D.E.; Leeuw, G.D. Formation and production of sea spray aerosols. *J. Aerosol Sci.* **1996**, *27*, S65–S66. [CrossRef]
20. Feliu, S.; Morcillo, M.; Chico, B. Effect of distance from sea on atmospheric corrosion rate. *Corrosion* **1999**, *55*, 883–889. [CrossRef]
21. Gustafsson, M.E.R.; Franzen, L.G. Dry deposition and concentration of marine aerosols in a coastal area. *Atmos. Environ.* **1996**, *30*, 977–989. [CrossRef]
22. Meira, R.; Andrade, C.; Alonso, C.; Padaradtz, I.J.; Borba, J.C. Salinity of marine aerosols in a Brazilian coastal area–influence of wind regime. *Atmos. Environ.* **2007**, *41*, 8431–8441. [CrossRef]

23. Hamada, S.; Hino, S.; Kanyuki, K. Salt measurement in the coastal region. *Fac. Eng. Yamaguchi Univ.* **1986**, *01*, 255.
24. Karady, G.; Amrah, F. Dynamic modeling of AC insulator flashover characteristics. *IEE Symp. High Volt. Eng.* **1999**, *467*, 107–110.
25. Jolly, D.C.; Poole, C.D. Flashover of contaminated insulators with cylindrical symmetry under DC conditions. *IEEE Trans. Dielectr. Electr. Insul.* **1979**, *EI-14*, 77–84. [CrossRef]
26. Obenaus, F. Contamination flashover and creepage path length. *Dtsch. Elektrotechnik* **1958**, *12*, 135–136.
27. Majid Hussain, M.; Farokhi, S.; McMeekin, S.G.; Farzaneh, M. Risk assessment of failure of outdoor high voltage polluted insulators under combined stresses near shoreline. *Energies* **2017**, *10*, 1661. [CrossRef]
28. Mizuno, Y.; Kusada, H.; Naito, K. Effect of climatic conditions on contamination flashover voltage of insulators. *IEEE Trans. Dielectr. Electr. Insul.* **1997**, *4*, 286–289. [CrossRef]
29. Majid Hussain, M.; Farokhi, S.; McMeekin, S.G.; Farzaneh, M. Contamination performance of high voltage outdoor insulators in harsh marine pollution environment. In Proceedings of the IEEE 21st International Conference on Pulsed Power, Brighton, UK, 18–22 June 2017.
30. Majid Hussain, M.; Farokhi, S.; McMeekin, S.G.; Farzaneh, M. Mechanism of saline deposition and surface flashover on outdoor insulators near coastal areas Part II: Impact of various environmental stresses. *IEEE Trans. Dielectr. Electr. Insul.* **2017**, *24*, 1068–1076. [CrossRef]
31. Terrab, H.; Bayad, A. Experimental study using design of experiment of pollution layer effect on insulator performance taking into account the presence of dry bands. *IEEE Trans. Dielectr. Electr. Insul.* **2014**, *21*, 2486–2495. [CrossRef]
32. Zohng, H.P.; Dong, X.C. *Tests and Investigations on Naturally Polluted Insulators and Their Application to Insulation Design for Polluted Areas*; CIGRE Report 33–07; CIGRE: Paris, France, 1982.

energies

MDPI

Article

DC Flashover Dynamic Model of Post Insulator under Non-Uniform Pollution between Windward and Leeward Sides

Zhijin Zhang [1,*], Shenghuan Yang [1], Xingliang Jiang [1], Xinhan Qiao [1], Yingzhu Xiang [1] and Dongdong Zhang [2]

1 State Key Laboratory of Power Transmission Equipment & System Security and New Technology, Chongqing University, Chongqing 400044, China; yangshenghuan@cqu.edu.cn (S.Y.); xljiang@cqu.edu.cn (X.J.); qiaoxinhan@cqu.edu.cn (X.Q.); xiangyingzhu@cqu.edu.cn (Y.X.)
2 Nanjing Institute of Technology, Nanjing 210000, China; zddpig@163.com
* Correspondence: zhangzhijing@cqu.edu.cn; Tel.: +86-138-8320-7915

Received: 24 May 2019; Accepted: 16 June 2019; Published: 19 June 2019

Abstract: Experience shows that under unidirectional wind or certain terrain, the surface of post insulators is non-uniformly polluted between windward and leeward sides, which affects the flashover characteristics. In this paper, a formulation of residual pollution layer resistance was proposed under this non-uniformity and a typical post insulator was taken as an example to analyze and calculate its residual resistance. The theoretical resistance was verified by numerical simulations using COMSOL Multiphysics. The proposed resistance formulation was then implemented in a DC flashover dynamic model to determine the flashover voltage (U_{cal}), which was validated by artificial flashover tests. Then the factors affecting DC flashover voltage were analyzed. Research results indicate that: the residual resistance formulation agrees well with simulation results, especially when the arc length exceeds 70% of the leakage distance. The good concordance between theoretical and experimental flashover voltages with most relative error within ±10%, validates the flashover model and its residual resistance formulation. U_{cal} gets impaired under this non-uniformity. The degree of reduction is related to salt deposit density ratio (m) of windward to leeward side and leeward side area proportion (k).

Keywords: post insulator; non-uniform pollution between windward and leeward sides; residual resistance formulation; flashover dynamic model; artificial flashover tests; flashover characteristics

1. Introduction

Post insulators, which are widely used in substations and converter stations, play an important role in electrical insulation and mechanical support in AC and DC transmission systems. Natural contamination characteristics indicate that there exists non-uniform pollution distribution on the insulator surface, which behaves in three types: non-uniformity between top and bottom surfaces, along the insulator length, and the transverse direction [1]. After long-term operation, the contamination on the leeward side of insulators is more serious than the windward side, due to wind direction and rainfall angle on the site, resulting in an easily-identified boundary between them [2–4], as shown in Figure 1. The contamination at the insulator surface is wetted partially or totally in the conditions of light rain, dew, or fog, which lowers the electrical characteristics of insulators [5,6].

Figure 1. Non-uniform pollution between windward and leeward sides on insulator surface.

At present, the main research on contaminated insulators includes contamination characteristics and flashover model. Plenty of studies have been conducted through pollution tests, simulations and mathematical models. For example, literature [7,8] present two different techniques, a multi model partitioning filter (MMPF) and an artificial neural network (ANN), and use the real contamination data for MMPF modeling and the ANN training. Research results indicate that both techniques can predict accurately the ESDD (equivalent salt deposit density) of suspension insulators in different conditions of wind velocity, ambient temperature, rainfall, and so on.

In the aspect of flashover model which this paper focuses on, much research has been conducted [9–13]. In literature [9], the insulator was partitioned into triangular elements and the finite element method was adopted to determine potential distribution, pollution layer resistance, and flashover voltage. Literature [10] presented a refinement of residual resistance formulation applied to insulator open model taking into consideration the non-uniformity of current density, where the correction factor was determined by numerical simulations. The flashover dynamic model based on the corrected formulation shows good accuracy compared with experimental results. In literature [11], a 2D model of the insulator surface was established and the residual resistance and the leakage current were obtained with the finite element software. Then the resistance and current were applied in a numerical model to predict flashover voltage. In literature [12], The residual resistance was evaluated under different radii and positions of arc root by building 3D model in COMSOL Multiphysics. The simulation results demonstrate that the relationship between the residual resistance and arc length is nonlinear. Literature [13] proposed a flashover model where the pollution layer was equivalent to a rectangle. The arc was modeled by its root which was considered as an equipotential surface. However, the above research mainly focuses on the residual pollution layer resistance and flashover models under uniform contamination, and there are few studies on non-uniform pollution between windward and leeward sides.

The existing research on non-uniformity between windward and leeward sides is principally aimed to study the flashover characteristics by artificial flashover tests and test samples are mostly line suspension insulators. The related research results [14–17] reveal that, under this non-uniformity, the flashover voltage of suspension insulators decreases compared with uniform pollution. In literature [14], DC flashover tests using 7-unit suspension insulator string were carried out under this non-uniformity. Research results indicate that a reduction in the ratio W/L from 1/1 to 1/15 gave a median 35% ± 4% decrease in flashover strength, where W/L is ratio of the salt deposit density (SDD) on the windward side to that on the leeward side. Similar conclusions can be found in reference [15]. AC flashover tests were presented in literature [1,16] under this non-uniformity and what makes a difference is that there is a slighter decrease of flashover voltage than DC. Literature [17] studied flashover characteristics when insulators were polluted non-uniformly along circumference (similar to non-uniform pollution between windward and leeward sides), and preliminarily analyzed the relationship between flashover voltage and area of heavy contaminated area. However, post insulator is distinguished from suspension insulator by its geometrical structure. At present, few studies on flashover model of post insulator under non-uniform pollution between windward and leeward sides have been carried out and its flashover characteristics are still unclear in this case.

In this paper, taking a typical post insulator as the sample, the influence of this non-uniformity on DC pollution flashover characteristics was presented systematically through calculating the residual pollution layer resistance, resistance simulation validation, calculating flashover voltage by the DC flashover dynamic model and carrying out artificial flashover validation tests. Based on previous work [10,12,14,15], the contribution of this paper is to propose a non-uniform residual resistance formulation by calculating in different regions, which takes into account the arc root radius, salt deposit density ratio (m) of windward to leeward side and leeward side area proportion (k). The DC flashover dynamic model, which considers time-varying arc root radius and leakage current, is validated by experimental results under different m and k. Finally, the DC flashover characteristics of post insulator are analyzed. The research results provide experimental data and theoretical support for further revealing the electrical characteristics of post insulator and improving the external insulation selection.

2. Analysis of Residual Pollution Layer Resistance for Pollution Flashover Dynamic Model

For insulators of any shape, the discharge process under uniform pollution can be analyzed using a series model of partial arc and residual pollution layer according to the Obenaus circuit model [12,18], as shown in Figure 2.

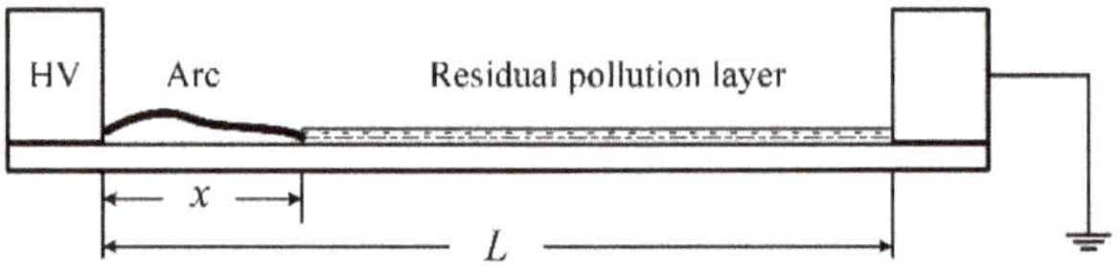

Figure 2. Obenaus circuit model.

When the partial arc is generated, with the electrode voltage drop ignored, the circuit model of Figure 2 is described in DC by the following equation:

$$U = U_{arc} + U_p = AxI^{-n} + I{\cdot}R(x) \tag{1}$$

where U is the applied voltage (V); I is the leakage current (A); x is the arc length (cm); $R(x)$ is the residual pollution layer resistance (Ω); A and n are arc constant.

As $R(x)$ is related to the arc root radius which is determined by the leakage current I, besides, x and I influence each other at any time, therefore, the discharge process is dynamic. Based on Obenaus circuit model, flashover dynamic model can be established to obtain flashover voltage and the key of dynamic model is to find analytic expression of the $R(x)$. So, the formulation of residual pollution layer resistance will be discussed first before establishing the model in the following.

2.1. Residual Pollution Layer Resistance under Uniform Pollution

With respect to the studies on residual resistance, Obenaus [18] explained that the residual resistance in series with pollution layer decreases as the arc propagates; Neumarker and Alston [12,18] assumed that the resistance per unit length of the pollution layer is constant; Open model, proposed by Rumeli [10,11], is to spread the insulator surface to a 2D equivalent surface. However, it is applied only to a uniform current distribution; Wilkins [19] proposed a new formulation where the pollution layer is equivalent to a rectangle and the constriction of current lines at the arc root is taken into account, which improves the expression of residual resistance. However, according to the study in literature [13,20], Wilkins formula cannot be applied to insulators with complex geometric shapes or non-uniform pollution.

The geometry of post insulator is complex, which cannot be simply expanded into a rectangle or equivalent to a cylinder to calculate its resistance. Considering current density distribution at surface is not uniform due to current lines constriction at the arc root, when partial arc is established, rectangular resistance model cannot reflect the non-uniformity. Therefore, it is assumed that a single dominant arc contacts in series with a residual pollution layer through the circular arc root. With thermal effects of partial arc and pollution layer ignored, the residual pollution layer resistance can be expressed as the resistance between two circular electrodes with different radii at the insulator surface, as shown in Figure 3. The two circular electrodes represent the arc root and the middle rod of post insulator, where r_0 is the arc root radius; r_1 is the rod radius, and the conductivity of the medium between two electrodes is pollution layer conductivity.

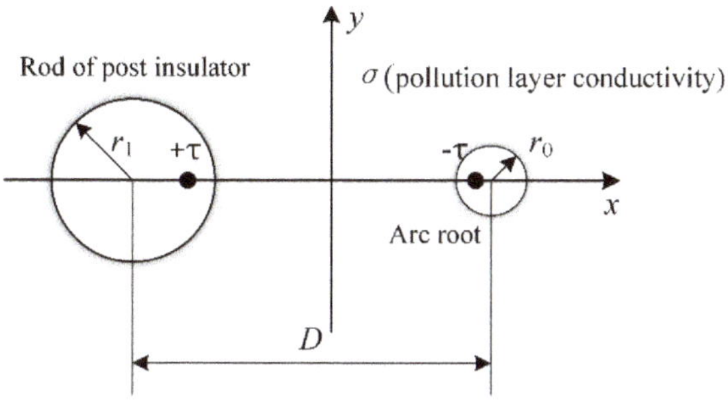

Figure 3. Schematic diagram of residual pollution layer resistance.

The resistance between two circular electrodes on an infinite conductive plane can be expressed as follow [21,22]:

$$R = \frac{1}{2\pi\sigma_e} \cosh^{-1}\left|\frac{D^2 - r_1^2 - r_0^2}{2r_1 r_0}\right| \tag{2}$$

where σ_e is the surface pollution layer conductivity at the critical flashover moment, called effective surface conductivity, S; $D = L - x + r_1$, L is insulator leakage distance, cm.

σ_s is the surface pollution layer conductivity before a partial arc is generated. In previous studies, the relationship between σ_e and σ_s can be expressed as [22]

$$\sigma_e \approx 1.25\sigma_s \tag{3}$$

The arc root radius is established to be [19]:

$$r_0 = \sqrt{\frac{I}{1.45\pi}} \tag{4}$$

where I is leakage current, A.

Meanwhile, the actual arc root radius is much smaller than that of post insulator rod, i.e., $r_0 \ll r_1$, therefore, $D^2 - {r_1}^2 - {r_0}^2 \approx D^2 - {r_1}^2$, and $D^2 - {r_1}^2 = (L - x + r_1)^2 - {r_1}^2 > 0$.

Basing on Equations (3) and (4), Equation (2) can be simplified to:

$$R = \frac{1}{2.5\pi\sigma_s} \cosh^{-1}\left(\frac{(L - x + r_1)^2 - r_1^2}{2r_1\sqrt{I/1.45\pi}}\right) \tag{5}$$

Then the identical transformation of $\cosh^{-1}\theta = \ln\left(\theta + \sqrt{\theta^2 - 1}\right)$ is applied to Equation (5) and the residual pollution layer resistance can be expressed as follow:

$$R(x, I) = \frac{1}{2.5\pi\sigma_s} \ln\left(\frac{(L - x + r_1)^2 - r_1^2}{2r_1 r_0} + \sqrt{\left(\frac{(L - x + r_1)^2 - r_1^2}{2r_1 r_0}\right)^2 - 1}\right) \tag{6}$$

where surface conductivity of pollution layer σ_s is given in literature [23] by following relation:

$$\sigma_s = (369.05 \times \text{SDD} + 0.42) \times 10^{-6}\ \text{S} \tag{7}$$

SDD is the salt deposit density of the whole insulator surface, mg/cm^2.

2.2. Residual Pollution Layer Resistance under Non-Uniform Pollution between Windward and Leeward Sides

In the case of non-uniform pollution between windward and leeward sides, the actual contamination at insulator surface can be approximated to two fan-shaped areas with different pollution degree, as shown in Figure 4. Assuming that the salt deposit densities of the windward and leeward sides are SDD_W and SDD_L respectively and the average salt deposit density is SDD, the relation is as follows

$$\begin{cases} SDD = \frac{SDD_W \cdot S_W + SDD_L \cdot S_L}{S_W + S_L} \\ k = \frac{S_L}{S_W + S_L}, m = \frac{SDD_W}{SDD_L} \end{cases} \quad (8)$$

where S_W and S_L are the surface area of windward and leeward sides, respectively, m is salt deposit density ratio of windward to leeward side which is less than 1 and k is the ratio of leeward side area to the whole area.

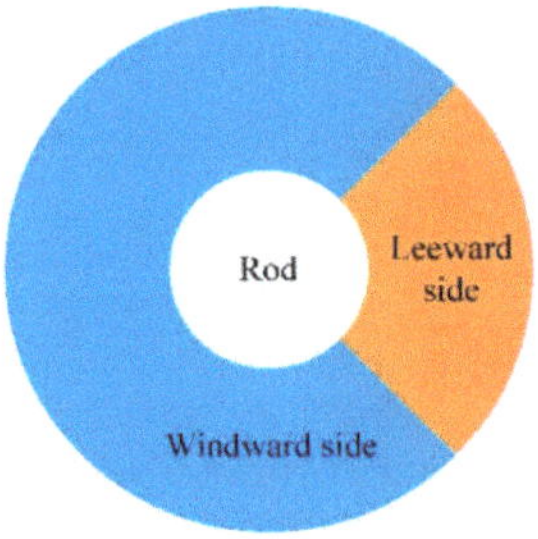

Figure 4. Schematic diagram of staining.

According to the statistical data of the natural contamination site, it was found that m of post insulator is between 0.17 and 0.76 and k is between 20% and 50%. Therefore, in order to study quantitatively the influence of different non-uniform pollution distribution on DC flashover characteristics, in the following model calculation and experimental validation, the value of m is taken as 1/8, 1/5, 1/3, and 1/1; the value of k is set to 25%, 35%, and 45% and SDD is set to 0.05, 0.10, and 0.15 mg/cm^2. Under different combinations of SDD, m and k, flashover voltage is calculated and verified by tests.

σ_W and σ_L are the surface conductivity of windward and leeward sides, respectively. When the pollution layer is at the same temperature and saturated enough, the surface conductivity of pollution layer is proportional to the corresponding SDD [15,23]. Therefore, Equation (8) can be equivalently transformed in the following form:

$$\begin{cases} \sigma_W S_W + \sigma_L S_L = \sigma_S S \\ k = \frac{S_L}{S_W + S_L}, m = \frac{\sigma_W}{\sigma_L} \end{cases} \quad (9)$$

where σ_s is the surface pollution layer conductivity under uniform pollution; the relationship between σ_W and σ_L can be established from Equation (9):

$$\sigma_W = m\sigma_L = \frac{m\sigma}{m(1-k)+k} \quad (10)$$

Under non-uniform pollution between windward and leeward sides, partial arc is always first generated at the leeward side, where the pollution degree is heavier, and propagates along its leakage distance. The surface conductivity is proportional to the thickness of pollution layer [24], therefore, the pollution layer resistance can be regarded as an overlap below and above of two conductive pollution layers with unequal areas, as shown in Figure 5. The surface conductivity should be $(m^{-1} - 1)\sigma_W$ and σ_W, respectively.

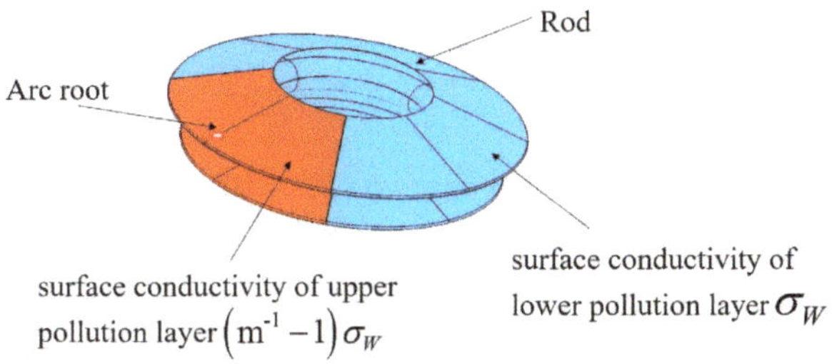

Figure 5. Conductive layers under non-uniform pollution between windward side and leeward side.

R_1 and R_2 are defined as the resistances of lower and upper conductive layer respectively, then the total residual pollution layer resistance R_{non} can be obtained by parallel calculation of R_1 and R_2:

$$R_{non} = \frac{R_1 R_2}{R_1 + R_2} \tag{11}$$

where R_1 can be calculated by Equation (6):

$$R_1(x, I) = \frac{1}{2.5\pi\sigma_w} \ln\left(\frac{(L - x + r_1)^2 - r_1^2}{2r_1 r_0} + \sqrt{\left(\frac{(L - x + r_1)^2 - r_1^2}{2r_1 r_0} \right)^2 - 1} \right) \tag{12}$$

The area of upper pollution layer, where the current line density is relatively uniform compared with the lower one, generally accounts for only 20% to 50% of the total area. Therefore, it is approximated as a rectangular conductive thin layer, as shown in Figure 6.

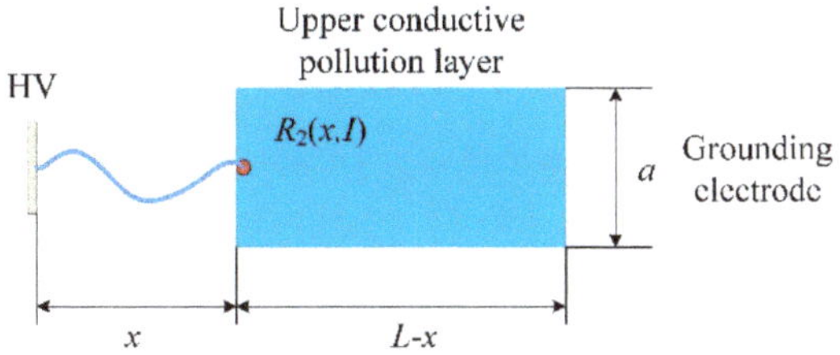

Figure 6. Schematic diagram of the upper pollution layer resistance $R_2(x, I)$.

The length of the rectangle is $L - x$, the width is $a = k{\cdot}2\pi r_1$, and the residual resistance of a narrow rectangular pollution layer can be calculated using the Wilkins formula [19]:

$$\begin{aligned} R_2(x, I) &= \frac{1}{2.5\pi(m^{-1} - 1)\sigma_W} \left[\frac{2\pi(L-x)}{T{\cdot}2\pi r_1} + \ln\frac{T{\cdot}2\pi r_1}{2\pi r_0} \right] \\ &= \frac{1}{2.5\pi(m^{-1} - 1)\sigma_W} \left[\frac{L-x}{kr_1} + \ln\frac{kr_1}{r_0} \right] \end{aligned} \tag{13}$$

Substituting Equations (12) and (13) into Equation (11), the analytic expression of $R_{non}(x, I)$ can be obtained:

$$R_{non}(x, I) = \frac{[k + m(1 - \mathrm{k})] f_1(x, I) f_2(x, I)}{2.5\pi\sigma_s{\cdot}[(1 - m) f_1(x, I) + m f_2(x, I)]} \tag{14}$$

where,

$$f_1(x,I) = \ln\left(\frac{(L-x+r_1)^2 - r_1^2}{2r_1r_0} + \sqrt{\left(\frac{(L-x+r_1)^2 - r_1^2}{2r_1r_0}\right)^2 - 1}\right) \tag{15}$$

$$f_2(x,I) = \frac{L-x}{kr_1} + \ln\frac{kr_1}{r_0} \tag{16}$$

Since the geometrical structure of each shed group is the same, one shed group with leakage distance L = 30.69 cm and rod radius r_1 = 6 cm, is taken as an example. When SDD is 0.15 mg/cm^2, $R(x, I)$ is calculated respectively under two conditions of uniform pollution (m = 1:1) and non-uniform pollution (m = 1:3, k – 25%). Residual resistance is a function not only of x but also of I. Based on previous studies [15,25] on DC leakage current under uniform pollution, we can estimate approximatively the leakage current under non-uniform pollution. In literature [25], the value of I is between 0 and 0.44 A when SDD is between 0.06 and 0.25 mg/cm^2; according to our previous flashover tests on uniform pollution of this type of post insulator, when SDD ranges from 0.05 to 0.15 mg/sm^2, I is between 0 and 0.86 A. By substituting above current range into Equation (4), r_0 is between 0 and 0.43 cm corresponding to SDD = 0.15 mg/cm^2. In order to study the influence of arc root radius on $R(x, I)$, two typical arc root radii, r_0 = 0.1 and 0.3 cm, are taken. Substituting m, k, L, r_1, and r_0 into Equations (6) and (14), the relationship between $R(x, I)$ and x is shown in Figure 7.

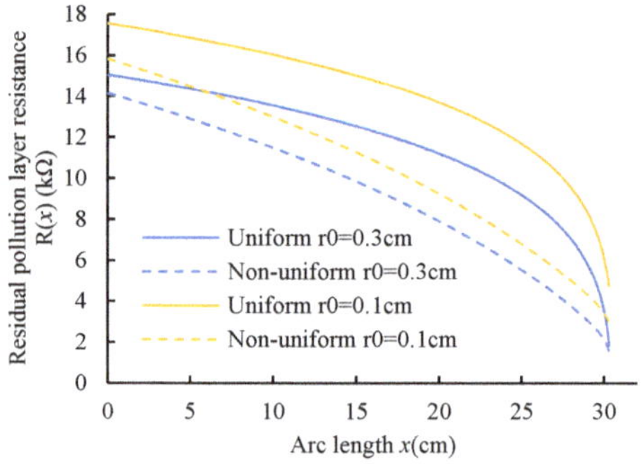

Figure 7. Relationship between residual pollution layer resistance and arc length.

It can be obtained from Figure 7 that:

(1) $R(x, I)$ decreases nonlinearly with the increase of x under both uniform and non-uniform pollution. The nonlinearity under uniform pollution is heavier than non-uniform pollution between windward and leeward sides. When x reaches the leakage distance, $R(x, I)$ drops close to zero.

(2) With the increase of x, the decreasing trend of $R(x, I)$ tends to be sharper. For example, for uniform pollution with r_0 = 0.3 cm, when x exceeds about 70% of the leakage distance (about 22 cm), $R(x, I)$ significantly decreases, which is consistent with the phenomenon that critical flashover generally occurs when arc length reaches 60%–80% of the leakage distance [15].

(3) Under certain x and r_0, $R(x, I)$ of uniformly polluted surface is larger than that of non-uniform pollution. The decrease of residual pollution layer resistance may result in the difference of flashover voltage when windward and leeward surfaces are non-uniformly polluted.

(4) Under certain pollution condition and x, when r_0 is lower, $R(x, I)$ becomes larger. The increasing amplitude of $R(x, I)$ decreases with the increase of x.

2.3. Comparative Results of Residual Pollution Layer Resistance

In order to verify the resistance formulation (14), the finite element model (FEM) is adopted for numerical simulation, considering that it is difficult to measure residual pollution layer resistance at the insulator surface during the development of arc on existing technical conditions. FEM techniques are widely applied to the modeling insulator to analyze surface potential distribution and pollution layer resistance [9–13]. In literature [10–12], 2D or 3D FEM models are established using COMSOL Multiphysics, a commercial finite element software, to simulate and calculate the residual pollution layer resistance. The results are in good agreement with the theoretical values.

Based on the actual geometry size of the post insulator, this paper applies COMSOL Multiphysics, taking a group of large and small sheds as an example, to establish 3D model of surface pollution layer. Assuming that the shape of arc root is circular [10,13], two typical arc root radii are adopted to simulate and calculate the residual resistance for different arc root positions along the leakage distance. Then the numerical values are compared with the theoretical values.

The parameters and structural diagram of the sample are shown in Table 1, where L_1 is the extended length of the large shed; L_2 is the extended length of the small shed; H_1 is the vertical distance between two large sheds; H_2 is the vertical distance between the large shed and small shed; D is the diameter of the insulator; H is the insulation height.

Table 1. Parameters of post insulator.

Shed Parameters (mm)	Leakage Distance (mm)	Structure Diagram
$L_1 = 75$ $L_2 = 58$ $H_1 = 69$ $H_2 = 39$ $D = 270$ $H = 1050$	3528	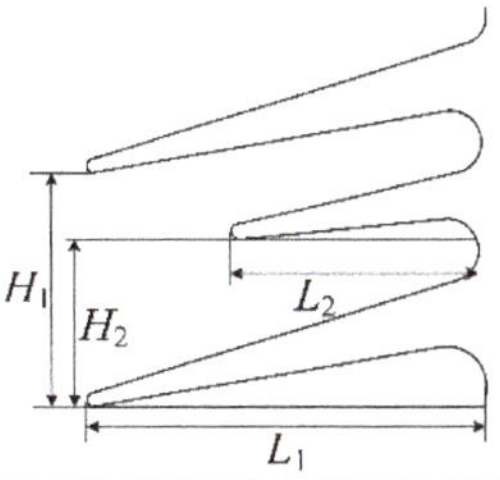

Figure 8 is a 2D cross-section diagram of the pollution layer. The thickness d of pollution layer is mainly related to NSDD (non-soluble deposit density) [26]. By gathering the wetted enough pollution layer and measuring its weight, the thickness d of the pollution layer can be obtained approximately. Our experimental results show d has an approximate relationship with SDD and NSDD in the following form:

$$d = [1.33 \times (\text{SDD} + \text{NSDD}) + 4.19] \times 10^{-3}\text{cm} \quad (17)$$

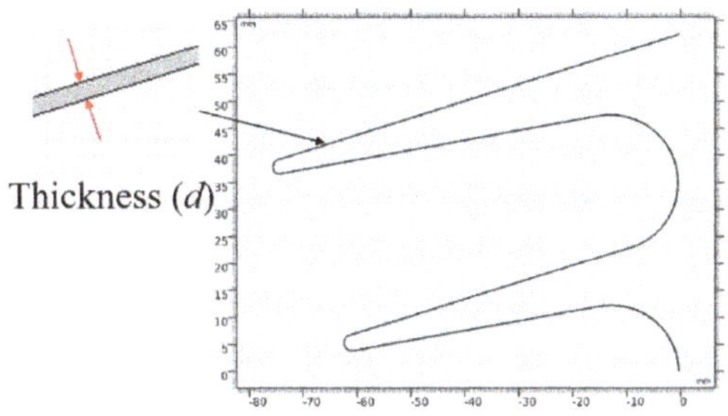

Figure 8. Two-dimensional cross-section of the pollution layer.

In the 3D pollution layer model with thickness d, the volume conductivity σ_V can be expressed as follow [10]:

$$\sigma_V = \sigma_S / d \quad (18)$$

For example, when insulator surface is uniformly polluted, SDD = 0.15 mg/cm^2 and NSDD = 0.9 mg/cm^2, d = 5.6 × 10^{-2} mm. $\sigma_S = 55.78 \times 10^{-6}$S and $\sigma_V = 0.996$ S/m, which can be calculated by Equations (7) and (18). Similarly, basing on Equations (7), (8), (17) and (18), σ_V of windward and leeward surfaces can be calculated under different non-uniform pollution distributions.

The increase of arc length is achieved by changing the position of the arc root along the leakage distance, so as to obtain the residual pollution layer resistance of different arc lengths. Using the current physical field of COMSOL Multiphysics, the potential boundary condition is adopted by setting arc root as an equipotential surface, applying a certain DC voltage (the value of the applied voltage has no effect on the resistance calculation) at the arc root, and grounding the upper end of pollution layer at the rod. The governing equations of the constant electric field are as follows:

$$J = \sigma_V E \quad E = -\nabla V \quad \nabla J = 0 \tag{19}$$

where J is volume current density, A/m^2; E is electric field intensity, V/m; V is potential, V.

The conductive layers are discretized in finite elements. According to the simulation results with different mesh sizes, the influence of FEM mesh size is not significant when the mesh quality is good because the numerical results become stable gradually. Figure 9 shows an example of the numerical results of current density and potential distribution when SDD = 0.15 mg/cm^2, NSDD = 0.9 mg/cm^2, m = 1:3, k = 25%, and r_0 = 0.3 cm. The color difference in the figure indicates the voltage drop, and the red arrow indicates the current density. It can be obtained from Figure 9 that after arc is generated at insulator surface, the current density distribution is quite non-uniform due to the constriction of the current lines around the arc root. The current density decreases rapidly in the diffusion process to the windward surface. Similarly, the potential distribution is also uneven, and the voltage drop mainly occurs at the leeward side.

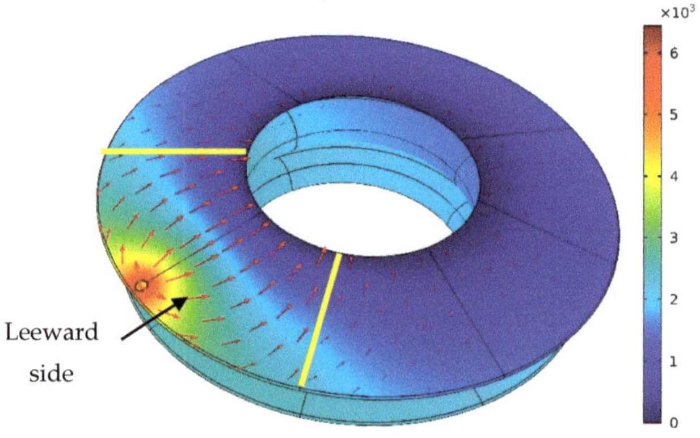

Figure 9. Distributions of the potential and current density lines.

Then the leakage current I is calculated by integrating the current density at the grounding electrode. The applied DC voltage is known, therefore, the resistance between the arc root and the grounding electrode, that is, the residual pollution layer resistance, can be obtained. Figure 10 shows the comparison between the numerical resistance values and analytical calculation by Equation (14) under the above condition.

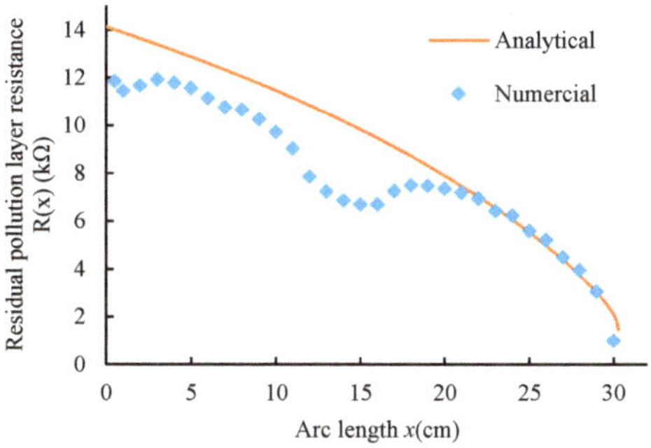

Figure 10. Comparison of $R(x)$ between numerical values and analytical calculation.

As can be obtained from Figure 10:

(1) When the arc length is between 21 cm and 30 cm, that is, the arc root is located on the upper surface of the large shed, the analytical calculation of $R(x)$ gives results in very good agreement with numerical values by COMSOL with a discrepancy lower than 6%.

(2) The discrepancy increases when the arc length ranges from 13 cm to 16 cm, that is, the arc root is located near the rod, which may be caused by the complex geometric shape near the rod of the post insulator.

(3) The analytical results obtained by the resistance formulation (14) are very close to the numerical values when the arc length exceeds about 70% of the leakage distance, where the critical flashover generally occurs [15]. The good accordance of this distance indicates that the proposed resistance formulation works well and ensures the accurate calculation of flashover voltage.

Therefore, adopting the proposed residual resistance formulation is reliable. Based on these, the pollution flashover voltage of the post insulator can be derived.

3. DC Flashover Dynamic Model and Experimental Validation

3.1. DC Pollution Flashover Dynamic Model

The flashover dynamic model of the insulator can reflect the flashover phenomenon more exactly than the static model due to taking into account instantaneous variation of the discharge parameters. Based on Obenaus model, a DC flashover dynamic mathematical model is developed in this paper. Since it is very complicated to include all these discharge parameters in this model, several simplifying assumptions are made: a single dominant arc is propagated along the leakage distance of leeward side and the temperature and humidity effects on pollution layer and arc are ignored.

The flashover model is established based on Figure 2, and its circuit equation can be expressed as (1). As x and I are time-varying and influence each other at any time, to accurately calculate flashover voltage (U_{cal}), it is necessary to recalculate I by Equation (1) at each new position of arc root [10,12,13].

The circuit Equation (1) can be rewritten as:

$$I = \frac{U - AxI^{-n}}{R(x)} \tag{20}$$

Since (20) is a non-linear equation about current I, it is difficult to obtain its analytical solution. So it can be solved numerically by the secant method [10], and Equation (20) can be rewritten to be a function of I:

$$f(I) = I - \frac{U - AxI^{-n}}{R_{non}(x)} \tag{21}$$

Its iteration format can be expressed as:

$$I_{i+1} = I_i - \frac{I_i - I_{i-1}}{f(I_i) - f(I_{i-1})} f(I_i) \tag{22}$$

By setting two initial values I_0 and I_1, which are close to the guessed analytical solution, as well as the error limit, the iteration can be started to obtain the numerical solution of I meeting the given error limit.

Hampton criterion [27] points out that the propagation of the arc should require:

$$E_p > E_{arc} \tag{23}$$

where E_p is the voltage gradient of the residual pollution layer; E_{arc} is the arc voltage gradient. At each iteration step, E_p and E_{arc} can be calculated as follows:

$$\begin{cases} E_p = \frac{U_p}{L-x} = \frac{I \cdot R(x,I)}{L-x} \\ E_{arc} = \frac{U_{arc}}{x} = AI^{-n} \end{cases} \tag{24}$$

Under a certain applied voltage, if the criterion (23) is met, the arc is propagated along insulator leakage distance, otherwise, the arc extinguishes, or the applied voltage needs to be increased

In the literature [18], the arc propagation velocity is defined as:

$$v = \mu E_{arc} \tag{25}$$

where the mobility $\mu = 25\,\mathrm{cm}^2/(\mathrm{V{\cdot}s})$ [10]. At the place of pollution flashover tests, n value is 0.52 and A is 129 for a negative DC arc [28].

The flow chart of the DC pollution flashover dynamic model is illustrated in Figure 11. Firstly, the insulator geometrical parameters (L, r_1), arc constants (n, A), non-uniform pollution degree (m, k) and values for iteration are input. At time t = 0, the initial applied voltage $U_{\min}$ is set to 2 kV and the initial arc length $x_{\min}$ is set to 1% of the total leakage distance L [10,18]. Two approximate values which start the iterative calculation of I, are $I_{\mathrm{i=0}} = 0.03$ A, $I_{\mathrm{i=1}} = 0.05$ A. The voltage increment dU = 50 V and time increment dt = 0.1 us at each step. Secondly, the arc root radius r_0 and residual pollution layer resistance R_{non} (x, I) at this time can be calculated using above initial values. Then the leakage current I is obtained by substituting r_0 and $R_{non}(x, I)$ into Equations (21) and (22). Thirdly, E_p and E_{arc} are calculated respectively using Equation (24). On the premise that the x is less than the total leakage distance L, Hampton criterion is verified by judging whether $E_p > E_{arc}$. If the propagation criterion is not satisfied, the applied voltage increases by dU. If the propagation criterion is satisfied, the arc propagates with an increased length. Finally, for the new arc length or new applied voltage, r_0, $R_{non}(x, I)$, I, E_p and E_{arc} are recalculated and the above judging steps are repeated until $x = L$. The critical flashover voltage U_{cal} is obtained at this non-uniform pollution degree.

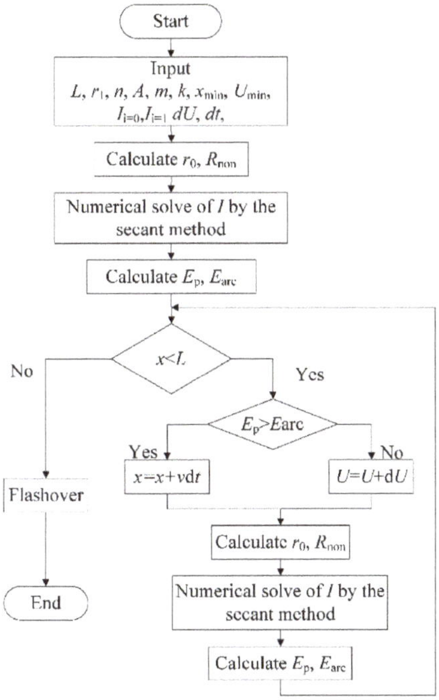

Figure 11. Flow chart of DC pollution flashover dynamic model.

3.2. Experimental Validation

The flashover tests were carried out in a multifunction artificial climate chamber, with the height of 11.6 m and the diameter of 7.8 m. The power system can provide 600 kV/0.5 A DC voltage during tests, which ensures that the voltage ripple factor is less than 3.0% when the load current is 0.5 A. The Schematic diagram of the test circuit has been shown in literature [14,15]. The test power supply satisfies the requirements recommended by [29].

The samples were polluted by solid layer method according to IEC standard [29] referring to Figure 12. Sodium chloride and kaoline were used to simulate the conductive and inert materials on the polluted insulator, respectively. In order to study the non-uniformity between the windward and leeward sides and the non-uniformity between top and bottom surfaces was not the aim of this paper, therefore, the pollution degree of the top and bottom surfaces was simplified to be the same. The ratio of SDD to NSDD is fixed at one-sixth in all the tests.

Figure 12. Schematic diagram of staining in the tests.

After applying pollution, the insulators were placed in the shade and dried naturally for 24 h. Before DC voltage was applied, the insulator surface pollution layer was wetted with steam fog. During the tests, the average voltage method [30] was adopted to obtain the average flashover voltage (U_{ave}). U_{ave} and its relative standard deviation error are established by following equation:

$$\begin{cases} U_{ave} = \sum\limits_{i-1}^{N} U_f(i)/N \\ \sigma\% = \sqrt{\frac{\sum\limits_{i-1}^{N} (U_f(i)-U_{ave})^2}{N-1}} \times \frac{100\%}{U_{ave}} \end{cases} \tag{26}$$

where $U_f(i)$ is a flashover voltage, kV; i represents the sequence number of each test; N is the number of all valid tests.

The pollution flashover tests under different non-uniformity were carried out according to above test procedure. Figure 13 shows the discharge process when SDD = 0.05 mg/cm^2, m = 1:8, and k = 45%. (Figure 13a–d represent the discharge conditions as time increases, respectively.) The area in the red box of Figure 13a is the leeside side of the insulator. It can be obtained from Figure 13 that, under non-uniform pollution between windward and leeward sides, most of the partial discharges first occur at the leeward side, and the main arc, which develops into the flashover, is also generated at the leeward side. Therefore, the assumption that the dominant arc is at the leeward side is consistent with the arc development path.

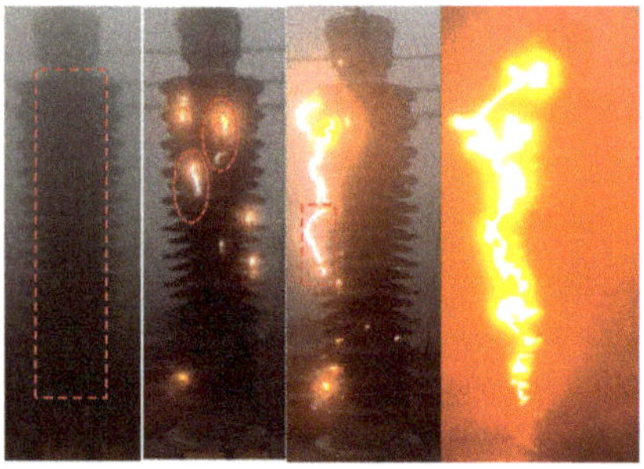

(a) (b) (c) (d)

Figure 13. Flashover process of post insulator.

In order to verify whether the assumption about the arc root radius in Section 2.2 is reasonable, the flashover test under the condition of SDD = 0.15 mg/cm^2, m = 1:3 and k = 25% was carried out and the arc root radius was measured. The arc propagation process is shown in Figure 14 where the width of the arc gradually increases. By using Image J, which is an image processing software, and taking the diameter of the post insulator (D = 270 mm) as the measuring ruler, the arc root radius of each stage was measured and the results are shown in Figure 14. The results indicate that the arc root radius is between 0 and 0.54 cm in the development process, which ensures it is reasonable to take the two arc root radii in Section 2.2, i.e., r_0 = 0.1 and 0.3 cm, to analyze its effect on the residual pollution layer resistance.

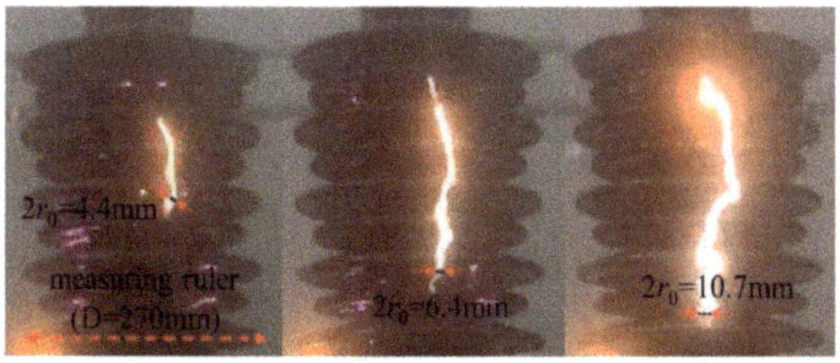

Figure 14. Measurement of arc root radius r_0.

The comparison of theoretical and experimental results is shown in Table 2. It can be obtained from the table:

Table 2. Results of validation test. SDD: salt deposit density.

Test Condition			Experimental Values U_{ave}(kV)	σ,%	Theoretical Values U_{cal} (kV)	Relative Error ΔU,%
SDD = 0.05 mg/cm^2	m = 1:1	k = 25%	83.2	3.9	90.6	8.9
	m = 1:5	k = 25%	72.0	2.1	70.1	−2.6
	m = 1:5	k = 35%	73.8	5.8	74.8	1.3
SDD = 0.10 mg/cm^2	m = 1:3	k = 35%	60.3	2.5	63.3	5.0
	m = 1:1	k = 35%	64.7	5.2	73.5	13.6
	m = 1:8	k = 45%	57.5	6.2	54.9	−4.5
SDD = 0.15 mg/cm^2	m = 1:8	k = 25%	45.0	3.4	39.2	−12.8
	m = 1:5	k = 25%	49.5	3.9	46.2	−6.7
	m = 1:8	k = 45%	50.1	4.9	44.7	−10.8

$\Delta U = (U_{cal} - U_{ave})/U_{ave} \times 100\%$, U_{ave} is experimental flashover voltage, U_{cal} is theoretical flashover voltage.

(1) The standard deviation of all test results is less than 7%, indicating that the dispersion of the test results is small.

(2) The calculated results with flashover dynamic model are in good agreement with the experimental results, most relative error within ±10%, which verifies the proposed residual resistance formulation and the DC flashover model under non-uniform pollution between windward and leeward sides in this paper. Some relative errors are higher, reaching 13.6%, possibly due to ignoring the effect of temperature changes on the residual resistance and the presence of multi-arc randomness and air-gap arc.

4. Results and Discussion

According to the flashover model above, the flashover characteristics of post insulator and the factors affecting insulator flashover voltage are analyzed. The effects of m and k on U_{cal} under different SDD are shown in the following three-dimensional diagram, Figure 15, where the three surfaces from top to bottom represent the flashover voltage distribution under SDD = 0.05, 0.10, and 0.15 mg/cm^2 respectively.

Taking the flashover voltages when SDD = 0.05 mg/cm^2 in Figure 15 as an example, the influence rule of m is illustrated as follows:

(1) The pollution flashover voltage of post insulator is related to salt deposit density ratio m. Specifically, under certain SDD and k, the more serious non-uniform pollution is (or the smaller m is), the lower the U_{cal} is. For example, the U_{cal} is 90.6, 82.6, 74.8, and 65.1 kV respectively when k = 35% and m is 1/1, 1/3, 1/5, and 1/8. Compared with the uniform pollution, the flashover voltage of m = 1/3, 1/5 and 1/8 decreases by 8.8%, 17.4%, and 28.1%, respectively.

(2) Under different k values, the downtrend of U_{cal} makes difference. The smaller k is, the sharper the trend of decrease. When k = 35% or 45%, the flashover voltages corresponding to adjacent m values only decrease by 5.7–9.7 kV, while the voltage drop of k = 25% is between 8.6 and 10.8 kV.

The decrease of U_{cal} is also related to k, and the variation trend of U_{cal} is exampled as follows when SDD = 0.10 mg/cm^2.

(3) Under certain SDD and m, U_{cal} increases slightly with the increase of k. For example, when m = 1/3 and k = 25%, 35% and 45%, U_{cal} is 60.0, 63.3 and 64.7 kV respectively, which indicates that there is an increase of 5.5% and 7.8% when k increases from 25% to 45%.

(4) The rising trend of U_{cal} is relatively milder under a higher k, resulting in relatively little effect on U_{cal}.

Based on the equation ($U = a{\cdot}\mathrm{SDD}^{-b}$), the values of U_{cal} and SDD were fitted, and its results are presented in Table 3.

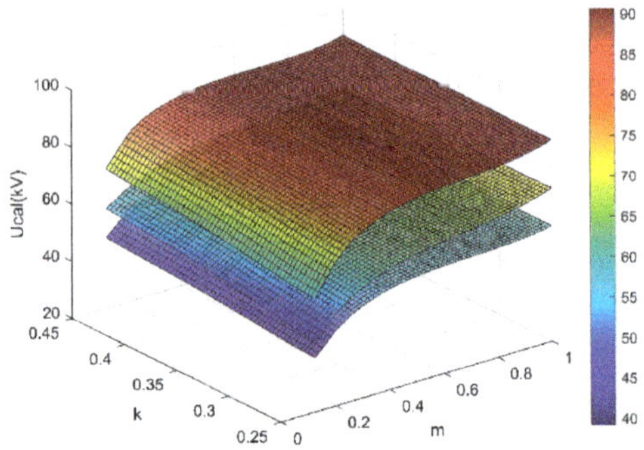

Figure 15. The effects of m and k on U_{cal} under different SDD.

Table 3. Fitting results.

m	k					
	25%		35%		45%	
	a	b	a	b	a	b
1:1	30.9	0.36	30.9	0.36	30.9	0.36
1:3	26.1	0.37	27.1	0.37	28.1	0.37
1:5	21.9	0.39	22.8	0.39	24.7	0.38
1:8	18.7	0.39	20.6	0.38	21.9	0.39

(5) All the fitting degrees are satisfied, therefore, under non-uniform pollution between windward and leeward sides, U_{cal} and SDD still satisfy the relationship of negative power function.

(6) The non-uniform pollution (m and k) has little effect on the pollution characteristic index b. b ranges from 0.36 to 0.39 and varies slightly around 0.36 (the characteristic index under uniform pollution). The effect of m and k on b is slight enough to be ignored. That is to say, the influence of SDD and the non-uniform pollution between windward and leeward sides on DC flashover voltage is independent from each other, which agrees with the experimental results in the literature [3,14,15].

(7) The coefficient a increases with the increase of m and k. For example, when k = 25%, compared with the uniform pollution, the coefficient a of m = 1/3, 1/5, and 1/8 decreases by 15.5%, 29.1%, and 39.5%, respectively. k has a slighter influence on a than m, to be specific, when m = 1/5 and k increases from 25% to 45%, and a merely increases by 12.8%.

5. Conclusions

This paper presents a residual resistance formulation under non-uniform pollution between windward and leeward sides, which is based on two circular electrodes model on a conductive plane and the narrow rectangular formula of Wilkins. The main advantage of the proposed formulation is that salt deposit density ratio, leeward side area proportion and the change of arc root radius with the leakage current can be taken into account. The analytical values of this formulation are in good agreement with the numerical results using COMSOL Multiphysics.

The proposed resistance formulation is then applied in a flashover dynamic model, which considers time-varying arc root radius and leakage current. Its results are compared with experimental results conducted in the artificial climate chamber. The good concordance validates the proposed resistance formulation and the flashover dynamic model.

The flashover characteristics of post insulator under non-uniform pollution between windward and leeward sides are analyzed using the dynamic model. The affecting factors are presented as follows:

(1) This non-uniformity lowers the DC flashover voltage U_{cal}. U_{cal} decreases markedly with the decrease of m and increases slightly with the increase of k. And the influence of m on U_{cal} is greater than that of k.

(2) Non-uniform pollution between windward and leeward sides has independent influence on U_{cal} of post insulator from SDD. The relationship between U_{cal} and SDD still fits power function.

Author Contributions: X.J. and Z.Z. conceived and designed this work; S.Y. performed the simulations and experiments; S.Y. and Z.Z. analyzed the data; X.Q., Y.X., and D.Z. assisted in the experiments and recorded the data; S.Y. wrote the paper.

Funding: This research was funded by the Jiangsu Natural Science Foundation Project (No. BK20181021).

Conflicts of Interest: The authors declare no conflict of interest.

References

1. Zhang, Z.; You, J.; Wei, D.; Jiang, X.; Zhang, D.; Bi, M. Investigations on AC pollution flashover performance of insulator string under different non-uniform pollution conditions. *IET Gener. Transm. Distrib.* **2016**, *10*, 437–443. [CrossRef]
2. Zhang, R.; Wu, G.; Yuan, T. Analysis on natural contamination characteristics of UHV AC post porcelain insulator. In Proceedings of the 2016 China International Conference on Electricity Distribution (CICED), Xi'an, China, 10–13 August 2016; pp. 1–8.
3. Zhang, Z.; Qiao, X.; Zhang, Y.; Tian, L.; Zhang, D.; Jiang, X. AC flashover performance of different shed configurations of composite insulators under fan-shaped non-uniform pollution. *High Volt.* **2018**, 3, 199–206. [CrossRef]
4. Sun, J.; Gao, G.; Wu, G.; Cao, X.; Zhu, G. Influence of pollution distribution on insulator surface on flashover characteristics. *IEEE Trans. Dielectr. Electr. Insul.* **2014**, *21*, 1637–1646. [CrossRef]
5. Karamousantas, D.C.; Chatzarakis, G.E.; Oikonomou, D.S.; Ekonomou, L.; Karampelas, P. Effective insulator maintenance scheduling using artificial neural networks. *IET Gener. Transm. Distrib.* **2010**, 479–484. [CrossRef]
6. Zhang, Z.; Qiao, X.; Yang, S.; Jiang, X. Non-Uniform Distribution of Contamination on Composite Insulators in HVDC Transmission Lines. *Appl. Sci.* **2018**, *8*, 1962. [CrossRef]
7. Vita, V.; Ekonomou, L.; Chatzarakis, G.E. Design of artificial neural network models for the estimation of distribution system voltage insulators' contamination. In Proceedings of the 12th WSEAS International Conference on Mathematical Methods, Computational Techniques and Intelligent Systems (MAMECTIS'10), Sousse, Tunisia, 3–6 May 2010; pp. 227–231.
8. Pappas, S.S.; Ekonomou, L.; Heraklion, N. Comparison of adaptive techniques for the prediction of the equivalent salt deposit density of medium voltage insulators. *WSEAS Trans. Power Syst.* **2017**, *12*, 220–224.
9. Gençoğlu, M.T.; Cebeci, M. The pollution flashover on high voltage insulators. *Electr. Power Syst. Res.* **2008**, *78*, 1914–1921. [CrossRef]
10. Bessedik, S.A.; Hadi, H.; Volat, C.; Jabbari, M. Refinement of residual resistance calculation dedicated to polluted insulator flashover models. *IEEE Trans. Dielectr. Electr. Insul.* **2014**, *21*, 1207–1215. [CrossRef]
11. Jabbari, M.; Volat, C.; Fofana, I. Application of a new dynamic numerical model to predict polluted insulator flashover voltage. In Proceedings of the 2014 IEEE Electrical Insulation Conference (EIC), Philadelphia, PA, USA, 8–11 June 2014.
12. Abimouloud, A.; Arif, S.; Korichi, D.; Ale-Emran, S.M. Prediction of DC flashover voltage of cap-andpin polluted insulator. *IET Sci. Meas. Technol.* **2019**, *13*, 279–286. [CrossRef]
13. Volat, C.; Farzaneh, M.; Mhaguen, N. Improved fem models of one- and two-arcs to predict ac critical flashover voltage of ice-covered insulators. *IEEE Trans. Dielectr. Electr. Insul.* **2011**, *18*, 393–400. [CrossRef]

14. Zhang, Z.; Zhang, D.; You, J.; Zhao, J.; Jiang, X.; Hu, J. Study on the DC flashover performance of various types of insulators with fan-shaped nonuniform pollution. *IEEE Trans. Power Deliv.* **2015**, *30*, 1871–1879. [CrossRef]
15. Zhang, D.; Zhang, Z.; Jiang, X.; Zhang, W.; Zhao, J.; Bi, M. Influence of fan-shaped non-uniform pollution on the electrical property of typical type HVDC insulator and insulation selection. *IET Gener. Transm. Distrib.* **2016**, *10*, 3555–3562. [CrossRef]
16. Zhang, Z.; You, J.; Zhang, D.; Jiang, X.; Hu, J.; Zhang, W. AC flashover performance of various types of insulators under fan-shaped non-uniform pollution. *IEEE Trans. Dielectr. Electr. Insul.* **2016**, *23*, 1760–1768. [CrossRef]
17. Boudissa, R.; Bayadi, A.; Baersch, R. Effect of pollution distribution class on insulators flashover under AC voltage. *Electr. Power Syst. Res.* **2013**, *104*, 176–182. [CrossRef]
18. Sundararajan, R.; Gorur, R.S. Dynamic arc modeling of pollution flashover of insulators under DC voltage. *IEEE Trans. Dielectr. Electr. Insul.* **1993**, *28*, 209–218. [CrossRef]
19. Wilkins, R. Flashover voltage of high-voltage insulators with uniform surface-pollution films. *Proc. Inst. Electr. Eng.* **1969**, *116*, 457–465. [CrossRef]
20. Jabbari, M.; Volat, C.; Farzaneh, M. A new single-arc AC dynamic FEM model of arc propagation on ice surfaces. In Proceedings of the 2013 IEEE Electrical Insulation Conference (EIC), Ontario, ON, Canada, 2–5 June 2013; pp. 360–364.
21. Edmonds, D.S., Jr.; Corson, M.R. The resistance between two contacts in a plane and the capacitance between paraxial cylinders. *Am. J. Phys.* **1986**, *54*, 811–816. [CrossRef]
22. Li, Y.; Yang, H.; Zhang, Q.; Yang, X.; Yu, X.; Zhou, J. Pollution flashover calculation model based on characteristics of AC partial arc on top and bottom wet-polluted dielectric surfaces. *IEEE Trans. Dielectr. Electr. Insul.* **2014**, *21*, 1735–1746. [CrossRef]
23. Topalis, F.V.; Gonos, I.F.; Stathopulos, I.A. Dielectric behaviour of polluted porcelain insulators. *IEE Proc.-Gener. Transm. Distrib.* **2001**, *148*, 269–274. [CrossRef]
24. Slama, M.; Beroual, A.; Hadi, H. Analytical computation of discharge characteristic constants and critical parameters of flashover of polluted insulators. *IEEE Trans. Dielectr. Electr. Insul.* **2010**, *17*, 1764–1771. [CrossRef]
25. Zhang, Z.; Zhao, J.; Zhang, D.; Jiang, X.; Li, Y.; Wu, B.; Wu, J. Study on the dc flashover performance of standard suspension insulator with ring-shaped non-uniform pollution. *High Volt.* **2018**, *3*, 133–139. [CrossRef]
26. Sundararajan, R.; Gorur, R.S. Role of non-soluble contaminants on the flashover voltage of porcelain insulators. *IEEE Trans. Dielectr. Electr. Insul.* **1996**, *3*, 113–118. [CrossRef]
27. Hampton, B.F. Flashover mechanism of polluted insulation. *Proc. Inst. Electr. Eng.* **2010**, *111*, 985–990. [CrossRef]
28. Zhang, Z.; Jiang, X.; Chao, Y.; Chen, L.; Sun, C.; Hu, J. Study on DC pollution flashover performance of various types of long string insulators under low atmospheric pressure conditions. *IEEE Trans. Power Deliv.* **2010**, *25*, 2132–2142. [CrossRef]
29. *Artificial Pollution Tests on High-Voltage Insulators to Be Used on dc. Systems*; IEC Tech. Rep. 1245.; IEC: Geneva, Switzerland, 1993.
30. Zhang, Z.; Jiang, X.; Sun, C.; Hu, J.; Huang, H. Study of the influence of test methods on DC pollution flashover voltage of insulator strings and its flashover process. *IEEE Trans. Dielectr. Electr. Insul.* **2010**, *17*, 1787–1795. [CrossRef]

energies

MDPI

Article

Surface Discharges and Flashover Modelling of Solid Insulators in Gases

Mohammed El Amine Slama [1,*], Abderrahmane Beroual [2] and Abderrahmane (Manu) Haddad [1]

1 Advanced High Voltage Engineering Centre, School of Engineering. Cardiff University, Queen's Buildings The Parade, Cardiff, Wales CF24 3AA, UK; haddad@cardiff.ac.uk

2 Laboratoire Ampère, University of Lyon, 36 Avenue Guy de Collongues, 69130 Ecully, France; abderrahmane.beroual@ec-lyon.fr

* Correspondence: slamame@cardiff.ac.uk

Received: 25 October 2019; Accepted: 9 December 2019; Published: 28 January 2020

Abstract: The aim of this paper is the presentation of an analytical model of insulator flashover and its application for air at atmospheric pressure and pressurized SF_6 (Sulfur Hexafluoride). After a review of the main existing models in air and compressed gases, a relationship of flashover voltage based on an electrical equivalent circuit and the thermal properties of the discharge is developed. The model includes the discharge resistance, the insulator impedance and the gas interface impedance. The application of this model to a cylindrical resin-epoxy insulator in air medium and SF_6 gas with different pressures gives results close to the experimental measurements.

Keywords: surface discharge; flashover; gas; modelling; pressure; thermal properties

1. Introduction

In order to optimize the insulation level for high-voltage components (air insulated substations (AIS), gas insulated substations (GIS) and gas insulated lines (GIL), breakers, overhead lines …), a special attention is given to creeping or surface discharges because of the thermal effects and the faults that they can produce by sparking or flashover. Then, the knowledge of the parameters characterizing this kind of discharge is essential to understand the complexity of the mechanisms involved in their development. Thus, it is fundamental to acquire such information to enable building a mathematical model that can help in optimizing the insulation efficiency.

This paper aims to carry out a review of existing models of creeping discharges and to propose an analytical approach for the calculation of flashover voltage of solid insulators in gases under lightning voltage stress.

2. Review of Surface Discharges and Flashover Models in Gases

From the insulation viewpoint, the triple junction (metal-gas-solid) constitutes the weakest point in high-voltage equipment. Indeed, when the electric field reaches a critical value, partial discharges (PDs) can be initiated in the vicinity of this region. The increase of the voltage leads these PDs to develop and to transform into surface discharges (creeping discharges) that propagate over the insulator up to flashover [1–4]. In the case of GIS and GIL, the worst case is when insulators (spacer, post-type insulator) are contaminated by metallic particles on their surfaces [5,6].

The physical mechanisms responsible for the surface discharge propagation are still not well known because of the complexity of the phenomena and the interaction of different factors, such as the interaction between the discharges, nature of gas and the proprieties of the solid insulating material, gas pressure, surface charges and pollution (metallic particle), geometrical parameters (insulator shape, electrodes form …), etc. Fundamental studies have been conducted to understand the inception and propagation of creeping discharges in various gases [7–14]. It appears from the reported results that the phenomena start with corona discharges that evolves into ramified streamers. When the streamer discharge reaches a certain length, a leader channel with streamers at its head appears.

The creeping discharge propagation dynamics in SF_6 (sulfur hexafluoride) has been investigated by many researchers [7–11]. Okubo et al. [8] reported that the creeping discharge has the same dynamics as in air (Figure 1). Tenbohlen and Schröder [9] analysed the surface discharge under lightning impulse (LI) voltage with different electrical charges deposition on the insulator surface. Figure 1 illustrates the current waveform from the inception to flashover with different electrical charges on the insulator surface. From Figure 1, some similarities with discharge current propagating in air [13] can be noted: the current increases with the leader elongation until the discharge reaches the critical length. Then, the final jump occurs causing the full flashover.

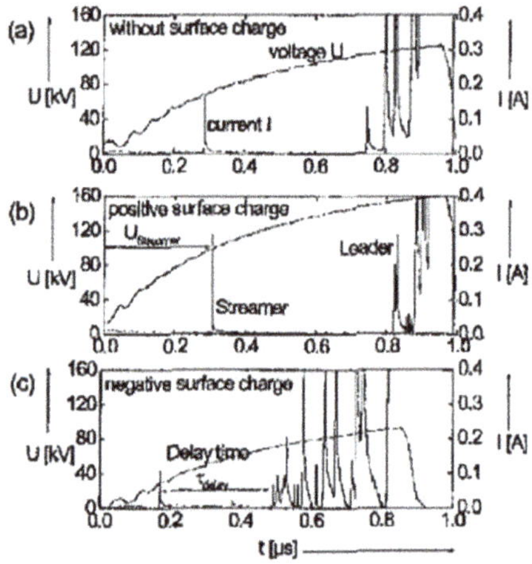

Figure 1. Instantaneous current and voltage during flashover at the surface of cylindrical epoxy insulator according to Reference [9].

Hayakawa et al. [7] analysed the mechanism of impulse creeping discharge propagation on charged PMMA (poly-methyl methacrylate) surface. Their results showed that the discharge propagation is influenced by the charged surface and can be explained by the streamer propagation and streamer-to-leader transition based on the precursor mechanism. On the other hand, according to Okubo et al. [8], Beroual [3] and Beroual et al. [10,11], the creeping discharge propagation depends on the specific capacitance of the solid insulator. The permittivity, the conductivity and the geometry of the insulator affect the propagation of the surface discharge [10–12].

Modelling and calculation of flashover voltage is not an easy task because of the interaction of different parameters, such as gas pressure and its chemical constitution, physicochemical properties of the solid insulator, nature and distribution of the surface charges, etc. Different models have been proposed in order to compute the inception voltage of creeping discharges and flashover voltage of insulator in air at atmospheric pressure [2,3,13,14]. Figure 2 depicts the different evolution steps of creeping discharge on the insulator.

According to Reference [2], the corona inception voltage depends on the equivalent capacitance of the system. It can be calculated with the following relationship [2]:

$$U_{inception} = \frac{A}{C^a} \tag{1}$$

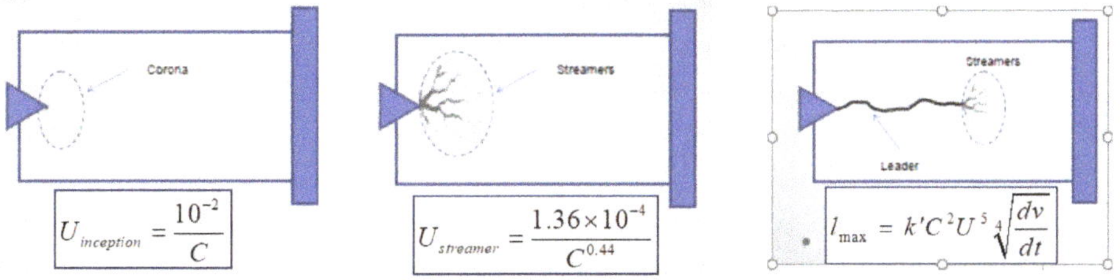

(**a**): Corona inception (**b**): Streamers (**c**): Spark (Streamers/Leader)

Figure 2. Steps of creeping discharges according to Reference [2].

The second step is the appearance of streamers (Figure 2a). The streamers voltage inception is given by [2]:

$$U_{streamer} = \frac{B}{C^b} \quad (2)$$

According to Toepler [3], the maximum (critical) length of the discharge that leads to flashover is:

$$l_{\max} = k \cdot C^2 \cdot U^5 \cdot \sqrt[4]{\frac{du}{dt}} \quad (3)$$

Then, if the voltage is increased, the discharge will be irreversible and propagates until flashover. In this case, the flashover voltage U_{fov} can be calculated as well:

$$U_{fov} = \frac{D}{C^d} \quad (4)$$

where:

- C is the equivalent capacitance,
- A, B and D are parameters that depend on the geometry and the material of insulator, the kind of the discharge and the experimental conditions (gas, pressure, temperature, humidity, electrodes shape, voltage waveform ...), respectively. Terms a, b and d are empirical parameters the values of which vary in the range 0.2–0.44.

These models are empirical and involve only the capacitance of the insulator.

In the case of SF_6, Laghari [15] proposed a relationship of flashover voltage based on the efficiency coefficient that represents the ratio of the flashover voltage for uniform electrical gradient distribution to the voltage breakdown of the same gap without an insulator with the same configuration of insulator as well:

$$V_{fov} = \frac{12.4}{\ln(V_b)} \cdot \frac{k_1}{k_2} \cdot \frac{\ln(\varepsilon_r)}{\varepsilon_r} \cdot V_b \quad (5)$$

where,

$$V_b = const + \left(\frac{E}{p}\right)_{cri} \cdot p \cdot d \quad (6)$$

V_b is the breakdown voltage calculated according to Paschen law. k_1 and k_2 are parameters that depend on the roughness and the contact nature between the insulator and the electrodes. ε_r is the permittivity of the insulator.

Hama et al. [16] proposed a semi-empirical relationship of flashover voltage based on the mechanism leader/precursor:

$$V_{fov} = \frac{X_{Leader}}{D_{pol} V_{Leader}} + V_{Leader} \quad (7)$$

where X_{Leader} and V_{Leader} are the length and the voltage of the leader discharge respectively, and D_{pol} is a coefficient that is dependent on the polarity of the applied voltage, the reduced critical electrical gradient and the shape of the electrodes, with:

$$D_{pol} = const \times \left(\frac{E}{p}\right)_{cri} \times \phi(Ry_{electrodes}) \tag{8}$$

The application of this model shows results close to the experimental measurements, but it is limited to the shape of the used insulators and the experimental conditions.

In the following, we recall the main principles of an analytical static model based on the electrical equivalent circuit and thermal discharge temperature we previously developed [1,13].

3. Principal of Circuit Model

Surface discharges are like spark (streamer/leader) discharges, i.e., a hot leader column and a streamers zone at its head [8,9,13,17]. Based on Figure 3, the voltage along the discharge can be written as follows:

$$V_d = V_l + V_s = x_l E_l + x_s E_s = x_d\, r_d\, I \tag{9}$$

where V_d, V_l and V_s are the voltages of the discharge, the leader channel and the streamers, respectively. x_l, x_s, E_l and E_s are respectively the length and the electrical gradient of the leader channel and the streamers. x_d, r_d, and I are respectively the discharge length, the discharge resistance and the current.

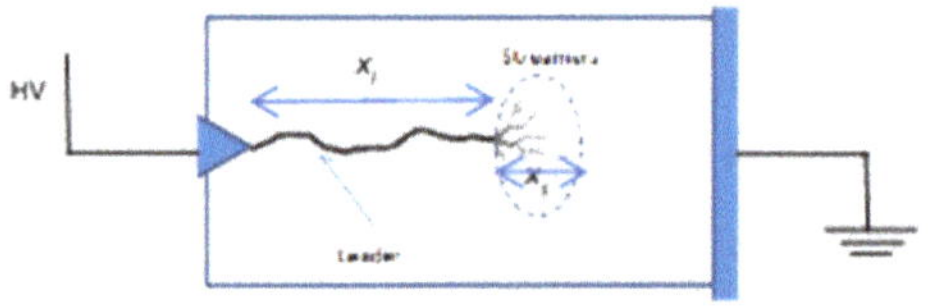

Figure 3. Illustration of leader column and streamers head of a discharge at the surface of an insulator.

The discharge resistance can be deduced from Equation (9):

$$r_d = \frac{x_l E_l + x_s E_s}{x_d I} \tag{10}$$

where $x_d = x_s + x_l$.

According to Equation (10), creeping discharge can be considered as a resistance and it can be assumed that the discharge channel is a uniform cylinder.

Many researchers published photos of surface discharges indicating that there are two regions: the main luminous discharge (leader + streamer head) and less luminous branches, as illustrated in Figures 4 and 5 [10,18]. So, the presence of those less luminous discharges can be represented as a resistor in parallel to the insulator surface. On the other hand, several research investigations demonstrate the existence of a dark current in high-pressurized gases that contribute to increase the insulator conductivity [19]. These currents contribute to the appearance of the second region (called luminous plasma), as depicted in Figures 4 and 5.

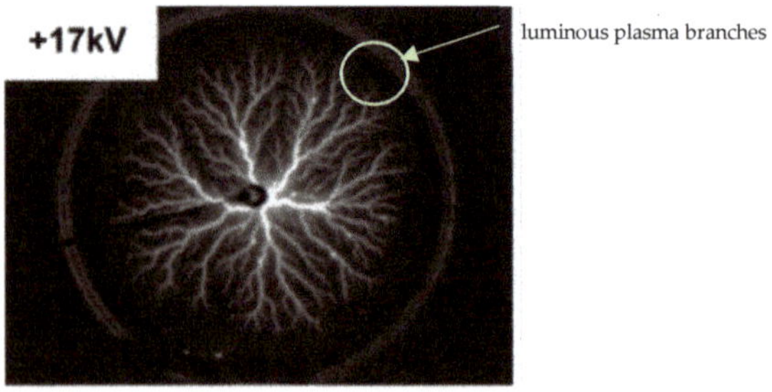

Figure 4. Surface discharge at the surface of insulator in SF6 with 3 bars under LI+ according to Reference [10].

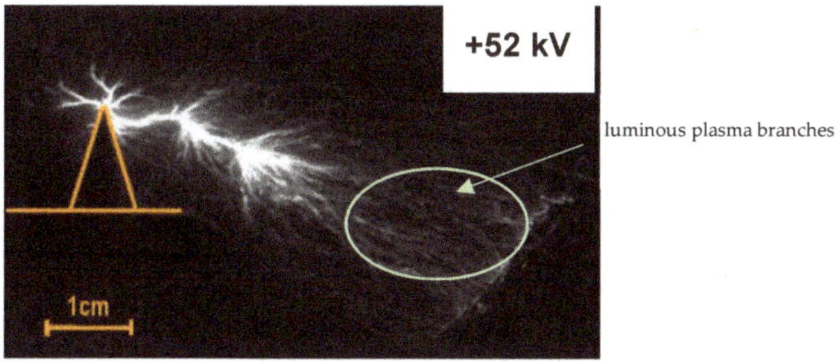

Figure 5. Surface discharge at the surface of a coated electrode in SF6 with 1 bar under LI+ according to Reference [18].

3.1. Parameters of the Circuit

The proposed model is constituted by an equivalent electrical circuit representing the electrical discharge, in series with the unbridged gap. The gap (the distance between the head of discharge and the opposite electrode) consists of a gas layer and of the solid dielectric at the interface (Figure 6). The gas layer is assumed to be equal to the diameter of the discharge channel. This model was developed elsewhere [13] in the case of air at atmospheric pressure and represents the instant when the discharge reaches a maximum length (called critical length) before the final jump [13]. In the following, the same approach [13] will be adopted with the assumption that the LI (lightning impulse) voltage waveform can be considered as a quart-cycle of sine signal with a frequency about 0.3 MHz.

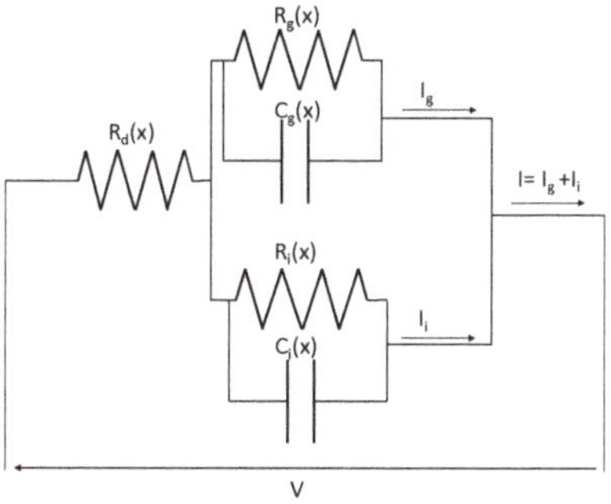

Figure 6. Insulator cylindrical model with a discharge channel and the corresponding equivalent electrical circuit.

The electrical Equation describing this circuit is:

$$V = R_d(x)I + \left[Z_g(x)//Z_i(x)\right]I \tag{11}$$

where,

$$\begin{aligned} Z_g(X) = R_g(x)//C_g(x) = \frac{R_g(x)}{1+j\omega R_g(x)C_g(x)} = \frac{r_g(L-x)}{1+\left(r_g c_g \omega\right)^2} - j\frac{r_g{}^2 c_g \omega (L-x)^2}{1+\left(r_g c_g \omega\right)^2} \\ Z_i(X) = R_i(x)//C_i(x) = \frac{R_i(x)}{1+j\omega R_i(x)C_i(x)} = \frac{r_i(L-x)}{1+(r_i c_i \omega)^2} - j\frac{r_i{}^2 c_i \omega (L-x)^2}{1+(r_i c_i \omega)^2} \end{aligned} \tag{12}$$

and

$R_d(x) = r_d x = x\frac{\rho_d}{s_d}$;

$C_i(X) = \frac{c_i}{L-x} = \varepsilon_i \frac{s_i}{L-x}$;$C_g(X) = \frac{c_g}{L-x} = \varepsilon_g \frac{s_g}{L-x}$;$R_i(x) = r_i(L-x) = \rho_i \frac{L-x}{s_i}$;$R_g(x) = r_g(L-x) = \rho_g \frac{L-x}{s_g}$

r_d is the linear resistance of the discharge channel. r_i, r_g, c_i, c_g, ε_I, ε_g, ρ_i, and ρ_g, are respectively the linear resistance, capacitance, the permittivity and the resistivity, respectively of the solid insulator and the unbridged gap. s_d is the cross-section of the discharge channel, s_i and s_g are respectively the cross-sections of the solid insulator and the layer of the unbridged gap. ω is the pulsation ($\omega = 2\pi f$, f being the frequency).

Then, the equivalent impedance of the system will be:

$$Z_{eq}(x) = r_d x + \frac{r_g r_i}{\alpha_g \alpha_i} G_1(L-x) + j\frac{r_g r_i}{\alpha_g \alpha_i} G_2(L-x) \tag{13}$$

Let us put:

$$\begin{aligned} \tau_g &= \rho_g \varepsilon_g \omega = r_g c_g \omega \\ \tau_i &= \rho_i \varepsilon_i \omega = r_i c_i \omega \end{aligned} \tag{14}$$

Product $\tau_i^2 >> 1$ and $\tau_g^2 >> 1$, then:

$$\begin{cases} \alpha_i \approx \tau_i^2 \\ \alpha_g \approx \tau_g^2 \end{cases} \tag{15}$$

The terms G_1 and G_2 are:

$$\begin{aligned} G_1 &= \frac{z_1}{z_3+z_4} \\ G_2 &= \frac{z_2}{z_3+z_4} \end{aligned} \tag{16}$$

where,

$$\begin{cases} z_1 = \left(\frac{r_g}{\tau_g^2} + \frac{r_i}{\tau_i^2}\right)(1-\tau_g\tau_i) + \left(\frac{r_g}{\tau_g} + \frac{r_i}{\tau_i}\right)(\tau_g+\tau_i) \\ z_2 = \left(\frac{r_g}{\tau_g} + \frac{r_i}{\tau_i}\right)(1-\tau_i) - \left(\frac{r_g}{\tau_g^2} - \frac{r_i}{\tau_i^2}\right)(\tau_g+\tau_i) \\ z_3 = \left(\frac{r_g}{\tau_g^2} + \frac{r_i}{\tau_i^2}\right)^2 \\ z_4 = \left(\frac{r_g}{\tau_g} + \frac{r_i}{\tau_i}\right)^2 \end{cases} \tag{17}$$

The square of the modulus of the equivalent impedance is:

$$\left|Z_{eq}\right|^2 = \gamma x^2 + 2Lx\left[r_d\left(r_d - \frac{r_g r_i}{\tau_g^2 \tau_i^2} G_1\right) - \gamma\right] \tag{18}$$

where,

$$\gamma = \left(r_d - \frac{r_g r_i}{\tau_g^2 \tau_i^2} G_1\right)^2 + \left(\frac{r_g r_i}{\tau_g^2 \tau_i^2} G_2\right)^2 \tag{19}$$

According to Reference [20], when the discharge length increases, the equivalent impedance decreases:

$$\frac{d|Z_{eq}|^2}{dx} \le 0 \tag{20}$$

By differentiating Equation (18) with respect to x, we get:

$$\frac{d|Z_{eq}|^2}{dx} = 2\gamma x + 2L\left[r_d\left(r_d - \frac{r_g r_i}{\tau_g^2 \tau_i^2} G_1\right) - \gamma\right] \le 0 \tag{21}$$

Then,

$$\frac{x}{L} - 1 \le \left[\frac{r_d}{\gamma \tau_g^2 \tau_i^2}\left(r_g r_i G_1 - \tau_g^2 \tau_i^2 r_d\right)\right] \tag{22}$$

Flashover of the solid dielectric occurs when Equation (22) is equal to zero, i.e., when the discharge length is equal to the total creeping (leakage) distance. This Equation can be considered as "the flashover condition". Therefore, the maximum (or critical) length of the discharge corresponding to flashover is:

$$x_{cri} = \frac{L}{\gamma \alpha_g \alpha_i}\left[\gamma \tau_g^2 \tau_i^2 - r_d\left(r_d - \tau_g^2 \tau_i^2 G_1\right)\right] = L \cdot n \tag{23}$$

where,

$$n = \frac{1}{\gamma\tau_g^2\tau_i^2}\left[\gamma\tau_g^2\tau_i^2 - r_d\left(r_d - r_g r_i G_1\right)\right] \quad (24)$$

where $0 < n < 1$.

The worst case can be derived from Equation (12), it corresponds to:

$$\frac{r_d}{\gamma\tau_g^2\tau_i^2}\left(r_g r_i G_1 - \tau_g^2\tau_i^2 r_d\right) \geq 0 \quad (25)$$

The term $\frac{r_d}{\gamma\tau_g^2\tau_i^2}$ is always positive, then:

$$r_g r_i G_1 \geq \tau_g^2\tau_i^2 r_d \quad (26)$$

Equation (26) can be written as:

$$\frac{\tau_g^2\tau_i^2}{G_1} \cdot \frac{r_d}{r_g r_i} = K \leq 1 \quad (27)$$

where,

$$0 < K \leq 1 \quad (28)$$

or:

$$r_d \leq K \cdot G_1 \cdot \frac{r_g r_i}{\tau_g^2\tau_i^2} \quad (29)$$

Condition (28) indicates that the discharge propagates when the ratio K is less than or equal to 1. This corresponds to the propagation criterion in which the discharge length is sufficient for causing the final jump, provoking flashover [13].

On the other hand, the power loss per unit length p_d in the discharge channel is:

$$p_d = r_d I^2 \quad (30)$$

By combining Equations (30) and (26), it yields:

$$I = \sqrt{\frac{p_d}{r_d}} \quad (31)$$

The square of the modulus of the voltage—Equation (11) is:

$$|V|^2 = |I|^2 \cdot \left|Z_{eq}\right|^2 \quad (32)$$

By substituting Equations (23), (24) and (30) in Equation (18), it yields:

$$\left|Z_{eq}\right|^2 = \beta L^2\left(\frac{r_g r_i}{\tau_g^2\tau_i^2}\right)^2 \quad (33)$$

with:

$$\beta = K^2 {G_1}^2 n^2 + \left({G_1}^2 + {G_2}^2\right)(1-n)^2 + 2K^2 G_1 n(1-n) \quad (34)$$

By substituting Equations (31) and (33) in Equation (32), the Equation of flashover voltage will be deduced as:

$$V_{FOV} = \frac{L}{\tau_g\tau_i}\sqrt{p_d \cdot \frac{r_g r_i}{r_d} \cdot \beta}. \quad (35)$$

3.2. Thermal Conductivity and Discharge Resistance

According to the solution proposed by Frank-Kamenetski [21,22], the energy dissipated by thermal conduction within the discharge channel is:

$$P_d = 16\pi\lambda_d \frac{K_B}{W_i} T^2 \quad (36)$$

By combining Equations (36) and (30), the final Equation of flashover voltage will be:

$$V_{FOV} = 4\frac{L}{\tau_g \tau_i} T \sqrt{\frac{\pi K_B}{W_i} \cdot \lambda_d \cdot \frac{r_g r_i}{r_d} \cdot \beta} \quad (37)$$

In the case of air at atmospheric pressure, the thermal conductivity is calculated according to the following Equation [23]:

$$\lambda(\theta) = \frac{\lambda_a}{1 + \frac{A_a(1-v_a)}{v_a}} \quad (38)$$

where λ_a, v_a and A_a are the thermal conductivity, volume fraction and kinetic gas coefficient for air, respectively.

Also, the discharge resistance in air at atmospheric pressure is given by [24]:

$$r_d(T) = r_{0d} \exp\left(\frac{W_i}{2K_B T}\right). \quad (39)$$

where r_{0d} is a constant in the range of operating temperatures of the discharge. W_i represents the first ionization energy of the different species constituting the discharge channel and K_B is the Boltzmann constant.

In the case of SF_6, both discharge resistance and discharge thermal conductivity are functions simultaneously of gas pressure and plasma temperature [25,26].

$$\begin{aligned} \lambda_d &= \Gamma(T,p) \\ \sigma_d &= \Sigma(T,p) \end{aligned} \quad (40)$$

According to Pinnekamp and Niemeyer [27], and Niemeyer et al. [28], the temperature of the leader discharge is between 2400 K and 2800 K. On the other hand, based on the transport parameters data of SF_6 published in the literature [25,26], the thermal conductivity was plotted as a function of gas pressure (Figure 7) and the discharge resistance against gas pressure (Figure 8) for a range of temperatures between 2500 K and 3500 K. From these figures, numerical empirical formulae of the discharge thermal conductivity and discharge resistance against pressure for a given temperature was deduced:

$$\sigma_d = A.p^{-m} \quad (41)$$

$$\lambda_d = a_0 + \ldots + a_n p^n \quad (42)$$

where p is the gas pressure and a and A are empirical parameters.

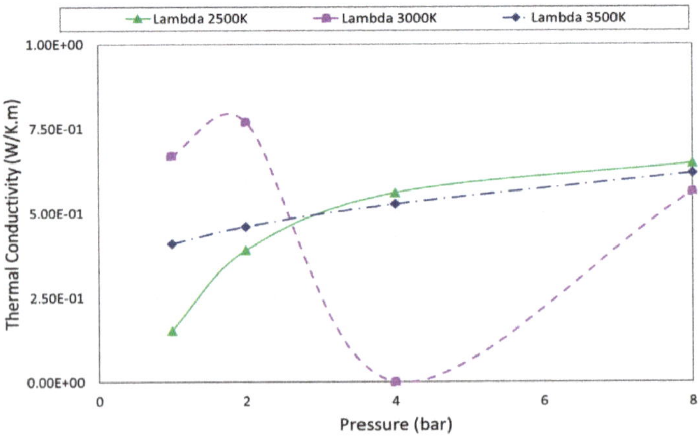

Figure 7. Discharge thermal conductance of discharge versus variation with pressure with for different temperatures.

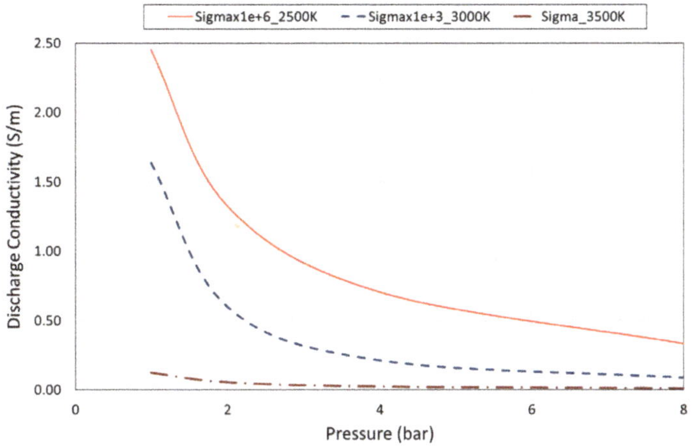

Figure 8. Discharge resistance variation versus pressure with different temperatures at 2500 and 3000 K, respectively.

4. Application

To validate the proposed model, first the model was applied for the calculation of the flashover voltage of cylindrical epoxy insulators in air at atmospheric pressure. The second application will be for the same kind of insulator in SF_6 gas medium. The computed flashover voltages are compared with the experimental data reported by other researchers, as in References [12,13,15,29]. Table 1 gives the characteristics of the used insulator in the computations.

Table 1. Characteristics of used insulators from literature used in modelling.

Insulator	Material	Diameter	Length	Reference	Gas	Pressure
1	Epoxy	25	60	[13]	Air	Atmospheric
2	Epoxy	25	60	[12]	SF_6	Variable
3	Epoxy	25	45	[29]	SF_6	Variable
4	Epoxy	30	10	[15]	SF_6	Variable

The lightning impulse voltage frequency is calculated based on the following Equation [30]:

$$f = 0.35/T_R \tag{43}$$

T_R is the rising time of the voltage front equal to 1.2 µs.

4.1. Air at Normal Atmospheric Conditions

Figure 9 illustrates the results of the application of the proposed model in air at atmospheric pressure. The model is compared with the experimental data of Reference [13], a previous model developed earlier [1] and Toepler's model. The temperature of discharge was taken between 1800 K and 2000 K, which corresponds to a leader phase on the insulator surface [13]. The resistance of air ranges from 10^{23} to 10^{25} Ω/cm, its dielectric constant being equal to 1. The effect of humidity and roughness are neglected.

By comparing flashover voltage given by Equation (36) and the other models, we can remark that the computed values are close to the measured ones and follow the same trend. According to this result, we can deduced that the impedance of the interface between the head of the discharge and the opposite electrode plays affects the result (Figure 9). It contributes to the breakdown process before the final jump of the discharge (flashover), as described in Reference [12]. The maximum deviation is 18.2% and the average deviation is less than 5%.

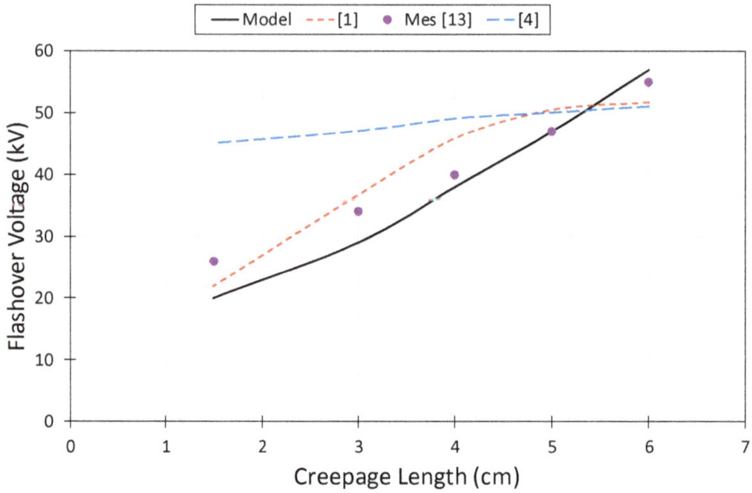

Figure 9. Comparison between calculated and measured flashover voltage versus creepage distance in air at atmospheric pressure for insulator 1.

4.2. SF_6 at Variable Pressure

In the case of SF_6, we use Equations (41) and (42), to compute the flashover voltage and its dependency on the gas pressure and temperature. The temperature of the discharge was taken between 2500 K and 3500 K.

A specific consideration for the resistance of the gas at the interface is required in the case of pressurised SF_6. In fact, experimental results concerning flashover of solid insulators on pressurised gases suggest that the discharge tends to stick to the insulator surface when the gas pressure increases [8,12]. On the other hand, according to Figures 4 and 5, the gap between the discharge's head and the ground electrode appears like an ionized cylinder. Knowing that the attachment of the pressurised gas also increases with pressure, it can be deduced that the resistance of the interface between the discharge head and the grounding electrode depends on the gas pressure as well.

Based on the data reported in the literature [25,26], the resistance of the interface can be represented as a cylindrical plasma with a temperature between 1000 K and 1500 K. In this range of temperature, the plasma resistivity increases with the gas pressure, as depicted in Figure 10. As can be observed in this figure, the assumption of a plasma with a temperature varying between 1200 K and 1400 K is a good approximation, since the resistivity is increasing with pressure for all temperatures. The dielectric constant being equal to 1 and the effects of surface charge accumulation and humidity are not considered.

Figure 11 illustrates the comparison of the calculated flashover voltage with the data of Slama et al. [12] for insulator 2 of Table 1. It can be observed that the calculated flashover voltages are close to the measured values, indicating that flashover voltage tends to be stable with the pressure increase. The maximum deviation is 8.2% and the average deviation is around 4.5%.

A comparison of the calculated flashover voltage with the data of Reference [29] obtained with insulator 3, is depicted in Figure 12. In this work, Moukengué and Feser [29] present results of flashover voltages as a function of gas pressure for different tests: one for a single impulse shot and the second for five impulse shots. It is noted that the calculated flashover voltages are close to the experimental measured values. The maximum deviation is 8.2% and the average deviation is around 6.5%.

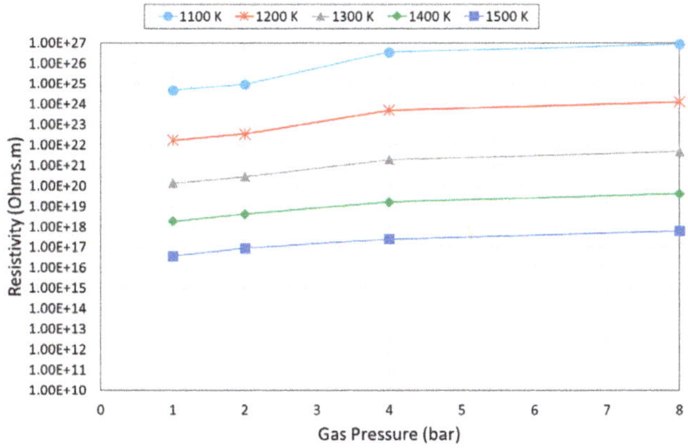

Figure 10. Resistivity of the SF_6 plasma at non-thermal regime versus gas pressure et different temperatures.

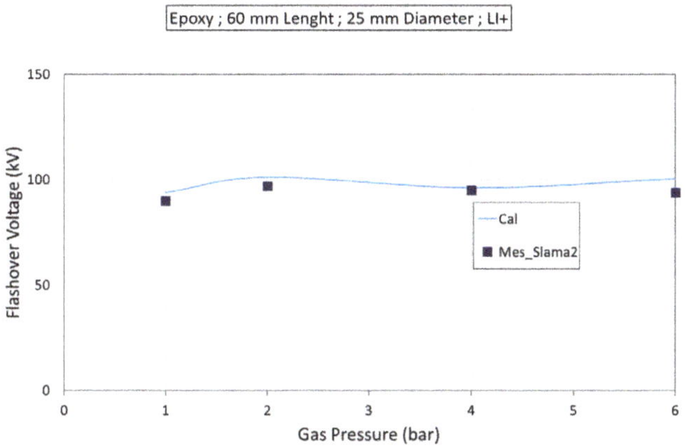

Figure 11. Comparison between calculated and measured flashover voltage versus gas pressure for insulator 2 with 60 mm length and 25 mm diameter.

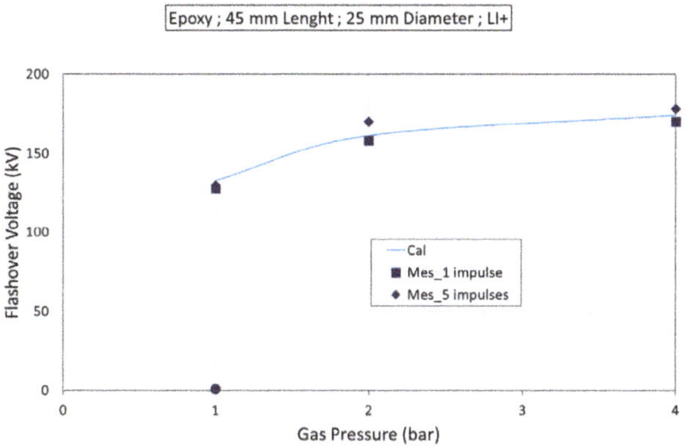

Figure 12. Comparison between calculated and measured flashover voltage versus gas pressure for a cylindrical epoxy insulator 4 with 45 mm length and 25 mm diameter.

Figure 13 shows the comparison of the results using the developed model and the data of Reference [15] with insulator 4. Again, it is observed that the calculated flashover voltages are close to the experimental ones and the maximum deviation is 10% and the average deviation is less than 4%.

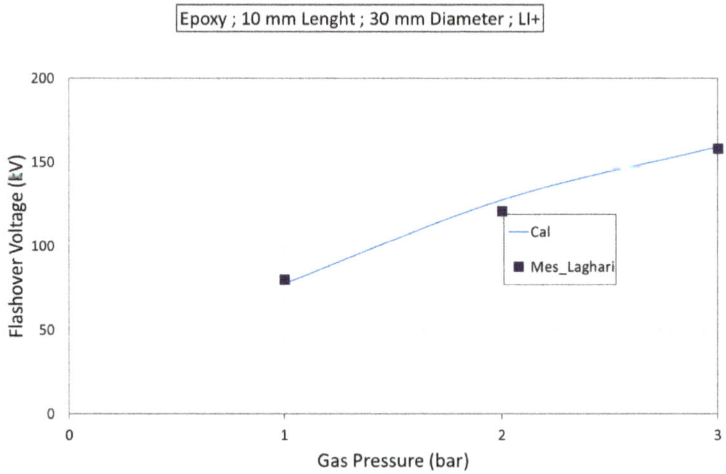

Figure 13. Comparison between calculated and measured flashover voltage versus gas pressure for a cylindrical epoxy insulator with 10 mm length and 30 mm diameter.

5. Conclusions

In this paper, a model was developed for surface discharges and flashover voltage in air at atmospheric pressure and compressed SF_6. The proposed analytical model is based on the equivalent electrical circuit representing the discharge along the insulator surface and the thermal properties of the discharge by assuming that the area between the discharge head and the ground electrode as a cylindrical plasma.

The proposed model was first applied for air at atmospheric pressure to validate it against existing models and data. It was noticed that the simulated results are very close to the experimental ones. Thus, the impedance of the interface between the head of the discharge and the opposite electrode significantly affects the result. It contributes to the breakdown process before the final jump of the final discharge that is flashover. In the case of SF_6, the application of this model to various configurations taken from the literature shows that the computed flashover voltage magnitudes are close to the measured values and exhibit similar trends.

The proposed model constitutes a first step for developing a tool for flashover prediction in ambient air and for the design of the solid insulation in GIS and GIL filled with SF_6.

Author Contributions: Conceptualization, M.E.A.S., A.B. and A.M.H.; methodology, M.E.A.S., A.B. and A.M.H.; validation, M.E.A.S., A.B. and A.M.H.; formal analysis, M.E.A.S., A.B. and A.M.H.; investigation, M.E.A.S.; writing—original draft preparation, M.E.A.S.; writing—review and editing, M.E.A.S., A.B. and A.M.H.; visualization, M.E.A.S., A.B. and A.M.H. All authors have read and agreed to the published version of the manuscript.

Funding: This research received no external funding.

Acknowledgments: The authors would like to thank Sylvain Nichele, Alain Girodet and Paul Vinson from SuperGrid Institute (France) for their help.

Conflicts of Interest: The authors declare no conflict of interest.

References

1. Slama, M.E.A.; Beroual, A.; Girodet, A.; Vinson, P. Characterization and Mathematical Modelling of Surface Solid Insulator in Air. In Proceedings of the International Symposium on High Voltage Engineering, Plzen, Czech Republic, 24–28 August 2015.
2. Al-Arainy, A.B.; Malik, N. *Experiments in High Voltage Engineering King, Saud University*; Academic Publishing and Press: Riyadh, Saudi Arabia, 2014.
3. Beroual, A. Creeping Discharges at Liquid/solid and Gas/Solid Interfaces: Analogies and Involving Mechanisms. In Proceedings of the IEEE 20th International Conference on Dielectric Liquids (ICDL), Roma, Italy, 23–27 June 2019.
4. Toepler, M. Über die physikalischen Grundgesetze der in der Isolatorentechnik auftretenden elektrischen Gleiterscheinungen. *Archiv Elektrotechnik* **1921**, *10*, 157–185. [CrossRef]
5. Cookson, A.H. Review of high-voltage gas breakdown and insulators in compressed gas. *IET* **1981**, *128*, 303–312. [CrossRef]
6. Sudarshan, T.S.; Dougal, R. Mechanisms of surface flashover along solid dielectrics in compressed gases: A Review. *IEEE Trans. Electr. Insul.* **1986**, *21*, 727–746. [CrossRef]
7. Hayakawa, N.; Ishida, Y.; Nishiguchi, H.; Hikita, M.; Okubo, H. Mechanism of impulse creepage discharge propagation on charged dielectric surface in SF6 gas. *IEEJ Trans. Power Energy* **1996**, *116-B*, 600–606. [CrossRef]
8. Okubo, H.; Kanegami, M.; Hikita, M.; Kito, Y. Creepage Discharge Propagation in Air and SF6 Gas Influenced by Surface Charge on Solid Dielectrics. *IEEE Trans. Dielectr. Electr. Insul.* **2004**, *1*, 204–304.
9. Tenbohlen, S.; Schröder, G. Discharge Development over Surfaces in SF6. *IEEE Trans. Dielectr. Electr. Insul.* **2000**, *7*, 241–246. [CrossRef]
10. Beroual, A.; Coulibaly, M.-L.; Aitken, O.; Girodet, A. Investigation on creeping discharges propagating over epoxy resin and glass insulators in presence of different gases and mixtures. *Eur. Phys. J. Appl. Phys.* **2011**, *56*, 30802–30809. [CrossRef]
11. Beroual, A.; Coulibaly, M.-L.; Aitken, O.; Girodet, A. Effect of micro-fillers in PTFE insulators on the characteristics of surface discharges in presence of SF6, CO_2 and SF6 –CO_2 mixture. *IET Gener. Transm. Distrib.* **2012**, *6*, 951–957. [CrossRef]
12. Slama, M.E.A.; Beroual, A.; Girodet, A.; Vinson, P. Barrier effect on Surface Breakdown of Epoxy Solid Dielectric in SF6 with Various Pressures. In Proceedings of the Conference on Electrical Insulation and Dielectrics Phenomena, Toronto, ON, Canada, 23 October 2016. CEIDP 2016.
13. Slama, M.E.A.; Beroual, A.; Girodet, A.; Vinson, P. Creeping Discharge and Flashover of Solid Dielectric in Air at Atmospheric Pressure: Experiment and Modelling. *IEEE Trans. Dielectr. Electr. Insul.* **2016**, *23*, 2949–2956. [CrossRef]
14. Douar, M.A.; Beroual, A.; Souche, X. Creeping discharges features propagating in air at atmospheric pressure on various materials under positive lightning impulse voltage—Part 2: Modelling and computation of discharges' parameters. *IET Gener. Transm. Distrib.* **2018**, *6*, 1429–1437. [CrossRef]
15. Laghari, R. Spacer Flashover in Compressed Gases. *IEEE Trans. Electr. Insul.* **1985**, *EI-20*, 83–92. [CrossRef]
16. Hama, H.; Inami, K.; Yoshimura, M.; Nakanishi, K. Estimation of Breakdown Voltage of Surface Flashover Initiated from Triple Junction in SF6 Gas. *IEEJ* **1996**, *116*. [CrossRef]
17. Waters, R.T.; Haddad, A.; Griffiths, H.; Harid, N.; Sarkar, P. Partial-arc and Spark Models of the Flashover of Lightly Polluted Insulators. *IEEE Trans. Dielectr. Electr. Insul.* **2010**, *17*, 417–424. [CrossRef]
18. Kessler, J.E.M. Isoliervermögen Hybrider Isoliersysteme in Gasisolierten Metallgekapselten Schaltanlagen (GIS). Ph.D. Thesis, Technische Universität München, Munich, Germany, 2011.
19. Zavattoni, L. Conduction phenomena through gas and insulating solids in HVDC GIS, and consequences on electric field distribution. Ph.D. Thesis, University of Grenoble, Grenoble, France, 2014.
20. Dhahbi-Megriche, N.; Beroual, A.; Krähenbühl, L. A New Proposal Model for Polluted Insulators Flashover. *J. Phys. D Appl. Phys.* **1997**, *30*, 889–894. [CrossRef]
21. Fridman, A.; Nester, S.; Kennedy, L.A.; Saveliev, A.; Mutaf-Yardimci, O. Gliding arc discharge. *Prog. Energy Combust. Sci.* **1999**, *25*, 211–231. [CrossRef]
22. Fridman, A. *Plasma Chemistry*; Cambridge University Press: Cambridge, UK, 2008.

23. McElhannon, W.; McLaughlim, E. Thermal Conductivity of Simple Dense Fluid Mixtures. In Proceedings of the Fourteenth International Conference on Thermal Conductivity, Storrs, CT, USA, 2–4 June 1975; Klemens, P.G., Clen, T.K., Eds.;
24. Slama, M.E.A.; Beroual, A.; Hadi, H. Analytical Computation of Discharge Characteristic Constants and Critical Parameters of Flashover of Polluted Insulators. *IEEE Trans. Dielectr. Electr. Insul.* **2010**, *17*, 1764–1771. [CrossRef]
25. Zhong, L.; Rong, M.; Wang, X.; Wu, J.; Han, G.; Lu, Y.; Yang, A.; Wu, Y. Compositions, thermodynamic properties, and transport coefficients of high temperature $C_5F_{10}O$ mixed with CO_2 and O_2 as substitutes for SF_6 to reduce global warming potential. *AIP Adv.* **2017**, *7*, 075003. [CrossRef]
26. Assael, M.J.; Koini, I.A.; Antoniadis, K.D.; Huber, M.L.; Abdulagatov, I.M.; Perkins, R.A. Reference Correlation of the Thermal Conductivity of Sulfur Hexafluoride from the Triple Point to 1000 K and up to 150 MPa. *J. Phys. Chem. Ref. Data* **2012**, *41*, 023104. [CrossRef]
27. Niemeyer, L.; Pinnekamp, F. Leader discharges in SF6. *J. Phys. D: Appl. Phys.* **1983**, *16*, 1031–1045. [CrossRef]
28. Wiegart, N.; Niemeyer, L.; Pinnekamp, F.; Boeck, W.; Kindersberger, J.; Morrow, R.; Zaengl, W.; Zwicky, I.; Gallimberti, M.; Boggs, S.A. Inhomogeneous Field Breakdown in GIS—The Prediction of Breakdown Probabilities and Voltages: Parts I, II, and III. *IEEE Trans. Power Deliv.* **1988**, *3*, 923–946. [CrossRef]
29. Moukengué lmano, A.; Feser, K. Flashover behavior of conducting particle on the spacer surface in compressed N_2, 90%N_2+10%SF_6 and SF_6 under lightning impulse stress. In Proceedings of the International Symposium on Electrical Insulation (ISEI 2000), Anaheim, CA, USA, 2–5 April 2000.
30. Ianovici et, M.; Morf, J.-J. *Compatibilité Électromagnétique*; Presses Polytechniques et Universitaires Romandes: Lausanne, Suisse, 1979.

energies

MDPI

Article

Space/Interface Charge Analysis of the Multi-Layer Oil Gap and Oil Impregnated Pressboard Under the Electrical-Thermal Combined Stress

Runhao Zou, Jian Hao * and Ruijin Liao

State Key Laboratory of Power Transmission Equipment & System Security and New Technology, Chongqing University, Chongqing 400044, China; 20161101041@cqu.edu.cn (R.Z.); rjliao@cqu.edu.cn (R.L.)
* Correspondence: haojian2016@cqu.edu.cn; Tel.: +86-18223010926

Received: 31 January 2019; Accepted: 18 March 2019; Published: 21 March 2019

Abstract: In oil-paper insulation systems, it is easy to accumulate space/interface charge under a direct current (DC) electrical field. At present, direct measurement of space/interface charge for a thick multi-layer insulation system is not possible. It is necessary to study the multi-layer oil-paper insulation system via simulation method. In this paper, the space/interface charge simulation based on the bipolar charge transport model and a simulation parameter using FEM for the multi-layer oil–paper insulation system was proposed. The influence of electrical field strength, temperature, and the combined influence of the electrical field strength and temperature on the space/interface charge behaviors were analyzed, respectively. A new method for calculating the space/interface charge density and the total charge quantity of the multi-layer oil-paper insulation under the combined action of electrical field strength and temperature was presented. Results show that the interface charge density absolute value and the total charge quantity at steady state both increases with the electrical field strength and temperature in an exponential way, respectively. Besides, temperature has a more significant influence on the charge density and the total charge quantity than the electrical field strength. The electrical field strength–temperature shifting factor $\alpha_{T'}$ was introduced for the translation of the charge density curves or the total charge quantity curves to construct the charge density main curve or the total charge quantity main curve under the combined action of electrical field strength and temperature. The equations for calculating the charge density or the total charge quantity of the multi-layer oil-paper insulation was provided, which could be used to calculate the charge density or the total charge quantity under the combined action of electrical field strength and temperature.

Keywords: space/interface charge; electrical field strength; temperature; oil-paper insulation; simulation; bipolar charge transport model

1. Introduction

Converter transformer is the key equipment of the high voltage direct current (HVDC) power grid. The outlet device of a converter transformer is mainly used to connect winding coil and bushing. The stable insulation performance of the outlet device on the valve side of the converter transformer plays a key role in the safe operation of a converter transformer [1]. The outlet device is mainly composed of a multi-layer oil-paper insulation system. Under DC voltage, the oil-paper insulation system is easy to accumulate space/interface charge. The accumulation of space/interface charges is a key factor affecting the oil-paper insulation performance of an outlet device. Space/interface charge can distort the local electrical field of the oil-paper insulation system, which will lead to insulation breakdown or cause material degradation [2–6].

At present, two types of model have been applied to describe space charge movement in insulation dielectrics, these being a unipolar charge transport model and bipolar charge transport model. Space/interface charge simulation provides an efficient way to understand the mechanism of space/interface charge migration and accumulation [7]. Guochang Li used the unipolar charge transport model to simulate the free/entrapped charge carriers' density and their effect on the electric field distribution of a single layer LDPE [8]. Shuo Jin used the bipolar charge transport model to simulate the space charge density change with the variation of applied voltage time for a single layer oil-impregnated insulation paper [9]. Kai Wu also used the bipolar charge transport model to simulate the space charge density change with the variation of temperature for a double layer oil-impregnated insulation paper [10]. S Le Roy used a bipolar charge transport model to simulate the space charge characteristics of LDPE under three different DC voltage application protocols and compared the simulation results with the experimental results attained by three different measurement methods to validate the correctness [11]. B. B. Alagoz et al. used a bipolar charge transport model to investigate the space charge's behaviors in the corona electrostatic fields and estimated the basic electrical characteristics of the system such as current draw and voltage drops [12]. B. Hamed used this model to simulate the space charge dynamic in low-density polyethylene under high DC voltage and found the appearance of a negative packet-like space charge [13].

Now the direct measurement of space/interface charge for the thick multi-layer insulation system is not possible due to the fact that the signal will decay with the increase of sample thickness [14]. A current study on space charge simulation mostly focused on a single-layer structure or a two-layer structure with the same kind of material, while the study on the multi-layer structure with an oil gap and oil-impregnated pressboard system is lack of research. Therefore, it is necessary to study the multi-layer oil impregnated pressboard and oil gap system via a simulation method.

In this paper, the bipolar charge transport model was used to simulate the space/interface charge behaviors for multi-layer oil-paper insulation using upstream finite element method (FEM). The influence of electrical field strength, temperature, and the combined action of the electrical field strength and temperature on the space/interface charge behaviors were analyzed, respectively. A new method for calculating the space/interface charge density and the total charge quantity of the multi-layer oil-paper insulation under the combined action of electrical field strength and temperature is presented.

2. Simulation Method

2.1. Charge Injection

The injection of the space charge is usually assumed to be a Schottky injection, as shown in Equations (1) and (2) [15]. Where j_h stands for the flux of holes at the anode side, j_e stands for the flux of electrons at the cathode side, A stands for the Richardson constant, 1.2×10^6 A/m^2·K^2. W_e and W_h are the injection barriers for electrons and holes. k_b is the Boltzmann constant, 1.38×10^{-23} J/K. E $(0,t)$ and E (d,t) are the electric field intensity at the anode and cathode, respectively. ε_0 is the permittivity of vacuum and ε_r is the relative dielectric constant of the insulating material. e stands for the charge quantity of one charge carrier which is 1.6×10^{-16} C.

$$j_h(0,t) = AT^2EXP(\frac{-eW_h + \sqrt{e^3E(0,t)/(4\pi\varepsilon_0\varepsilon_r)}}{K_bT}) \tag{1}$$

$$j_e(d,t) = AT^2EXP\left(\frac{-eW_e + \sqrt{e^3E(d,t)/(4\pi\varepsilon_0\varepsilon_r)}}{K_bT}\right) \tag{2}$$

2.2. Charge Carriers Movement

The charge movement in the insulation is governed by a set of self-consistent equations. Equation (3) is the Poisson equation describing the electric field distribution; Equation (4) is the transport equation which describes the migration of charge carriers; and Equation (5) is the convection equation which describes the variation of charge density.

$$\nabla(\varepsilon_0\varepsilon_r E) = \rho_v \tag{3}$$

$$J_a(t) = \mu_a\rho_a(t)\nabla\varphi(t) \tag{4}$$

$$\frac{\partial n(x,t)}{\partial t} + \frac{\partial j(x,t)}{\partial x} = s_{ef} + s_{hf} + s_{et} + s_{ht} \tag{5}$$

There are a large number of traps existing within the insulation materials [16]. Those traps are caused by the physical and chemical defects [17–21]. When charge carriers move in the medium, it is possible to be entrapped. Meanwhile, those entrapped charge carriers have a certain possibility to be detrapped. There are four types of charge carriers named free holes, free electrons, trapped holes, and trapped electrons [16]. Therefore, the variation of charge density can be described by the following Equations (6)–(9).

$$S_{ef} = -B_{ef,hf}\rho_{ef}\rho_{hf} - B_{ef,ht}\rho_{ef}\rho_{ht} - B_{eft}\rho_{ef} + N_{t,e}B_{etf}\rho_{et} \tag{6}$$

$$S_{hf} = -B_{ef,hf}\rho_{ef} * \rho_{hf} - B_{ef,ht}\rho_{ef}\rho_{ht} - B_{hft}\rho_{hf} + N_{t,h}B_{htf}\rho_{ht} \tag{7}$$

$$S_{et} = -B_{et,ht}\rho_{et}\rho_{ht} - B_{et,hf}\rho_{et}\rho_{hf} - B_{eft}\rho_{et} + N_{t,e}B_{eft}\rho_{ef} \tag{8}$$

$$S_{ht} = -B_{et,ht}\rho_{et}\rho_{ht} - B_{ef,ht}\rho_{ef}\rho_{ht} - N_{t,h}B_{htf}\rho_{ht} + B_{hft}\rho_{hf} \tag{9}$$

In the equations above, $B_{ef,hf}$, $B_{ef,ht}$, $B_{et,ht}$, and $B_{et,hf}$ are the recombination coefficient for free electrons/free holes, free electrons/trapped holes, trapped electrons/trapped holes, and trapped electrons/free holes, respectively. B_{eft}, B_{etf}, B_{hft}, and B_{htf} represent the movement of free electrons to be trapped, trapped electrons to be detrapped, free holes to be trapped, and trapped holes to be detrapped, respectively. $N_{t,e}$ is the trap quantity for electrons, and $N_{t,h}$ is the trap quantity for holes. S_{ef} is the density for free electrons, S_{hf} is the density for free holes, S_{et} is the density for trapped electrons, and S_{ht} is the density for trapped holes. Then the total density variation is the sum of those four kinds of charge carriers' change, as shown in Equation (10).

$$\rho_{total} = S_{ef} + S_{et} + S_{hf} + S_{ht} \tag{10}$$

2.3. Space Charge Movement at the Interface Between Oil gap and Oil Impregnated Pressboard

The space charge could accumulate at the interface between two different insulation materials. The discontinuity of permittivity and conductivity will cause space charge polarization, which is called Maxwell-Wagner (M-W) polarization [22]. The M-W polarization diagram is shown in Figure 1 and Equations (11)–(14). In the following equations, U stands for the electric potential, E_1 stands for the electrical field strength for dielectric 1; E_2 stands for the electrical field strength for dielectric 2; ε_1 is the relative dielectric constant for dielectric 1 while ε_2 is the relative dielectric constant for dielectric 2; l_1 and l_2 are the conductivities for dielectric 1 and 2 respectively; and ρ is the charge density. Part of the polarized charge at the interface will move into the layer structures on both sides of the interface. In order to move into the layer, charges have to overcome the energy barrier of the layer structure. This movement is described by Equation (15), which is called the Poole–Frenkel equation. In the equation, W_i stands for the barrier's energy level, which is set to 1.2 eV; A' is the probability of injection which is set to 0.4 here.

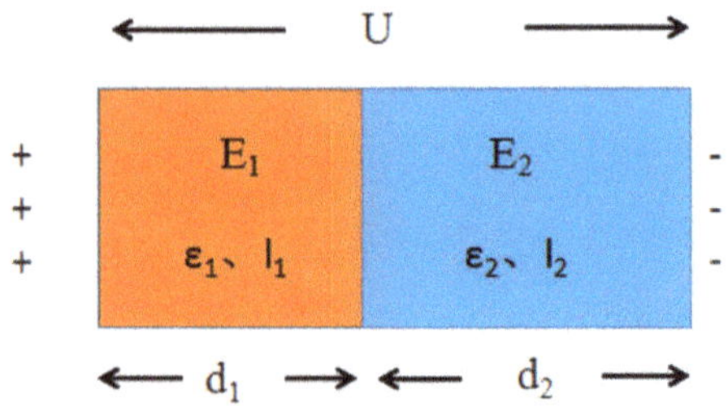

Figure 1. Diagram for Maxwell–Wagner polarization.

The model in Figure 1 illustrates that the difference of $\varepsilon_2 l_1 - \varepsilon_1 l_2$ will determine whether the polarity at the interface will be the same as the polarity on the left or right side. In this case, since the relative dielectric constant of both dielectrics is relatively close compared with the conductivity, the charges polarity at the interface is largely determined by the conductance of dielectrics. In order to solve Equations (3)–(5) with the consideration of Equations (6)–(9) and obtain the space charge density of each point at each time step, the model was meshed with a unit length ratio of 0.1009 and then the upstream finite element method was applied [23–25]. The flow chart for simulation is shown in Figure 2. The parameters for simulation were set with the reference from literature [16,26], which is shown in Table 1.

$$E_1 d_1 + E_2 d_2 = U \tag{11}$$

$$l_1 E_1 - l_2 E_2 = 0 \tag{12}$$

$$\varepsilon_2 E_2 - \varepsilon_1 E_1 = \rho \tag{13}$$

$$\rho = \frac{\varepsilon_2 l_1 - \varepsilon_1 l_2}{l_1 d_2 + l_2 d_1} U \tag{14}$$

$$j_i(d,t) = A' T^2 EXP\left(\frac{-eW_i \sqrt{e^3 E(d,t)/(4\pi\varepsilon_r)}}{K_b T}\right) \tag{15}$$

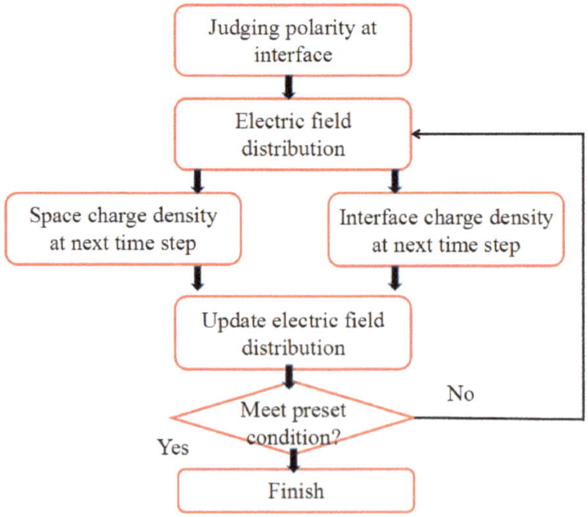

Figure 2. Flow chart for space/interface charge movement simulation.

Table 1. Parameters for simulation.

Parameters	Assigned Value for Simulation
Mobility for electrons	1×10^{-16} m^2/(V×s)
Mobility for holes	1×10^{-19} m^2/(V×s)
Trapping coefficient—electron (B_{eft})	5×10^{-3}/s
Trapping coefficient—hole (B_{hft})	5×10^{-3}/s
Detrapping coefficient—hole (B_{htf})	5×10^{-6}/s
Detrapping coefficient—electron (B_{eft})	5×10^{-6}/s
Trap concentration-e ($N_{t,e}$)	100 C/m^3
Trap concentration-hole ($N_{t,h}$)	100 C/m^3
Recombination coefficient for free electrons/free holes ($B_{ef,hf}$)	5×10^{-3} m^3/(C×s)
Recombination coefficient for free electrons/trapped holes ($B_{ef,ht}$)	1×10^{-3} m^3/(C×s)
Recombination coefficient for trapped electrons/free holes ($B_{et,hf}$)	1×10^{-3} m^3/(C×s)
Recombination coefficient ($B_{et,ht}$)	0
Barrier height for Schottky injection of electrons (W_e)	1.2 eV
Barrier height for Schottky injection of holes (W_h)	1.2 eV
Temperature (T)	293.15 K
Sample thickness for oil gap	500 μm
Sample thickness for oil impregnated pressboard	1000 μm
Relative dielectric constant for oil gap	2.2
Relative dielectric constant for oil impregnated pressboard	3.7

3. Simulation Results and Discussions

3.1. Verification of Simulation Method

The three-layer "oil impregnated pressboard (OIP) + oil gap (OG) + oil impregnated pressboard (OIP)" can be considered as the simplest structure for a thick multi-layer oil gap and oil impregnated pressboard for converter transformer's insulation. There are two interfaces in the three-layer "OIP + OG + OIP" sample. One interface has positive charge carrier accumulation while the other one has negative charge carrier accumulation, which will be helpful for analyzing both kinds of charge carriers under DC voltage. The simulation result and PEA measurement result for three-layer "OIP + OG + OIP" under DC 15 kV/mm are shown in Figure 3a,b.

The space charge injection is a homo-charge injection. The space charge carriers on both electrodes were injected into the bulk with the increase of voltage applied time, causing the space charge density in the bulk of the sample to increase, especially the charge density of the charges accumulated at the interface. The interface adjacent to the cathode has a positive charge accumulation, while the interface adjacent to the anode has negative charge accumulation. Figure 3b is the experimental result for three-layer "OIP + OG + OIP" under DC 15 kV/mm [27]. It can be seen that the simulation result shown in Figure 3a matches with the experimental result shown in Figure 3b. Figure 3c shows the charge density of the charges accumulated at both interfaces and variation with time for the three-layer "OIP + OG + OIP" sample. It can be observed that the interface charge density increases rapidly from 0 s to 600 s, the increment begins to slow down after 1200 s until it reaches saturation. This phenomenon is consistent with the charge accumulation behaviors presented in Reference [27]. The total charge amount Q of the three-layer "OIP + OG + OIP" during the DC voltage applied process was calculated based on Equation (16). Where S stands for the area of the electrode, l stands for the thickness of the sample, and $q(x)$ means the charge density at position x, $0 \leq x \leq l$. Figure 3d shows that the total charge amount Q of the three-layer "OIP + OG + OIP" during the DC voltage applied process increases quickly and then reaches a saturation value. This changing law is also the same as the result shown in Reference [27].

$$Q = S \times \int_0^l q(x) \tag{16}$$

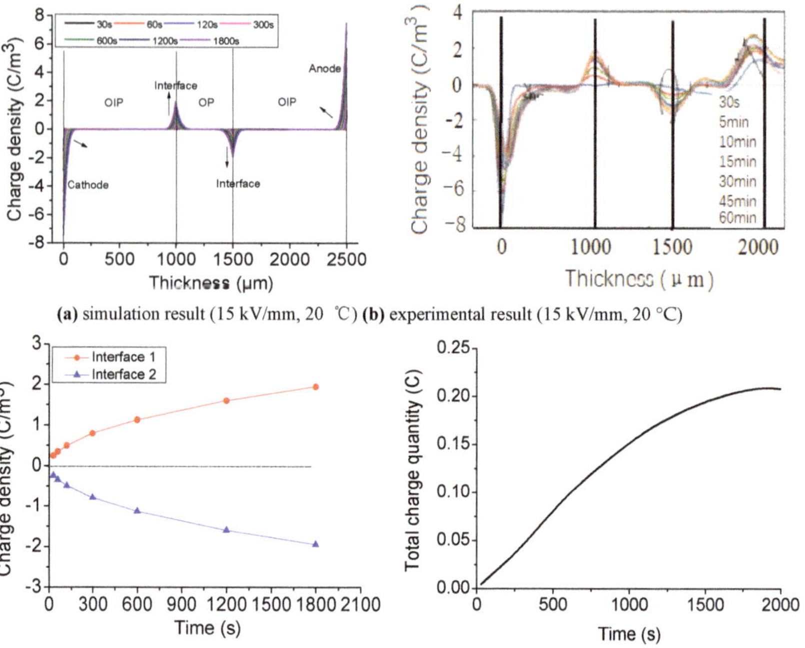

(a) simulation result (15 kV/mm, 20 ℃) **(b)** experimental result (15 kV/mm, 20 °C)

(c) charge density **(d)** total charge quantity

Figure 3. Space/interface charge simulation and PEA measurement result for three-layer "OIP+ OG + OIP" (15 kV/mm, 20 °C).

3.2. Electrical Field Strength Influence on the Space/Interface Charge Behavior

Figure 4 shows the space/interface charge simulation results for three-layer "OIP + OG + OIP" under different electrical field strengths at 20 °C. By comparing Figure 4a,b, it can be observed that the increase of electrical field strength will increase the charge density apparently. However, the polarity of the charges trapped at the interface does not change. The increase of electrical field strength from 15 kV/mm to 40 kV/mm will increase the interface charge density at a steady state from 2.4 C/m^3 to 11.5 C/m^3. Figure 4 shows that the electrical field strength has great influence on the space/interface charge density values.

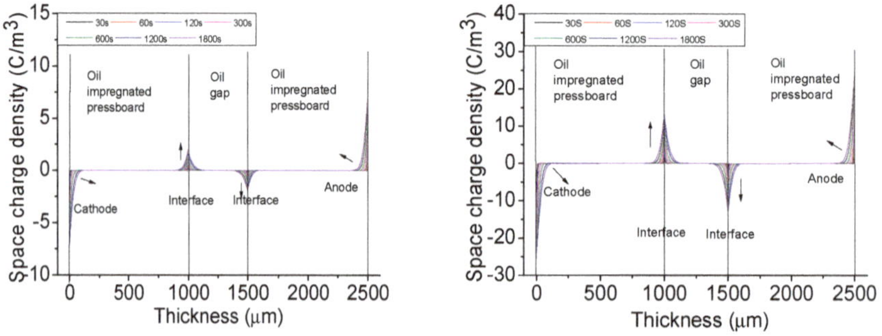

(a) 15 kV/mm **(b)** 40 kV/mm

Figure 4. *Cont.*

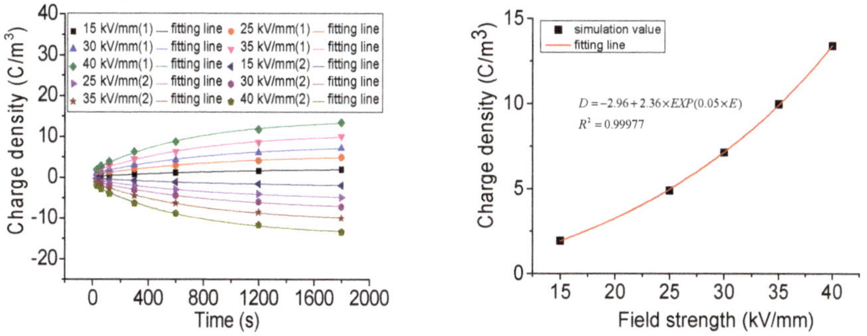

(c) charge density (d) interface charge density at steady state (1800 s)

Figure 4. Space/interface charge simulation for three-layer "OIP + OG + OIP" under different electrical field strengths (20 °C).

Figure 4c is the charge density of the positive and negative charges accumulated at both interfaces under different field strengths. It can be observed that with a lager electrical field strength applied, the increment speed of charge density before 1200 s is significantly larger. However, from 1200 s to 1800 s, the increment speed of charge density under each electrical field strength is almost identical. Figure 4d shows the charge density at steady state for the positive charges accumulated at the interface adjacent to the cathode under different electrical field strengths. The charge density at steady state also increases with the electrical field strength in an exponential way.

The oil-insulation structure of two-layer "OG + OIP", three-layer "OIP + OG + OIP", four-layer "OG + OIP + OG + OIP", five-layer "OIP + OG + OIP + OG + OIP", six-layer "OG + OIP + OG + OIP + OG + OIP", and seven-layer "OIP + OG + OIP + OG + OIP + OG + OIP" is shown in Figure 5. The oil gap thickness is 500 μm, and the oil impregnated insulation pressboard thickness is 1000 μm. The charge density absolute values of the charges accumulated at the first interface adjacent to the cathode for the oil-insulation structure of different layers were analyzed here, as shown in Figure 6. It can be found that the charge density absolute values at the steady state increase exponentially with the electrical field strength.

The fitting formula is shown in Equation (17) and Table 2. D_{steady} stands for the charge density absolute value at steady state, C/m^3. E stands for the electrical field strength, kV/mm. A_E, B_E and C_E are the fitting coefficients. At high electrical field strength, the structure's influence becomes remarkable. At 15 kV/mm, the charge density absolute value at steady state for all structures was all about 2.5 C/m^3. While at 40 kV/mm, the charge density absolute value at steady state for the two-layer structure is 19.3 C/m^3; for the three- and six-layer structures, the charge density absolute value at steady state is about 12.5 C/m^3; for the four-, five- and seven-layer structures, the charge density absolute value at steady state is about 9.5 C/m^3.

$$D_{steady} = A_E + B_E \times EXP(C_E \times E) \quad (17)$$

Table 2. Fitting coefficients for interface charge density absolute value at steady state (1800 s) under different electrical field strengths for multi-layer oil-paper insulation.

Coefficients	Two Layers	Three Layers	Four Layers	Five Layers	Six Layers	Seven Layers
A_E	−2.63	−2.96	−2.77	−3.12	−1.88	−2.91
B_E	1.56	2.36	2.31	2.6	1.55	2.43
C_E	0.066	0.048	0.043	0.041	0.057	0.043
R^2	0.99	0.99	0.99	0.99	0.99	0.99

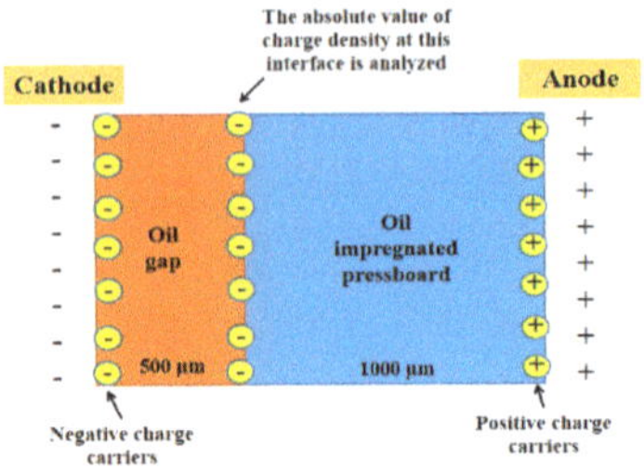

(a) two layers

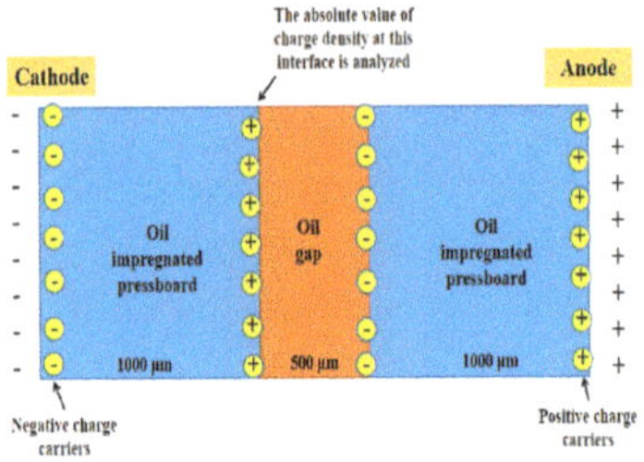

(b) three layers

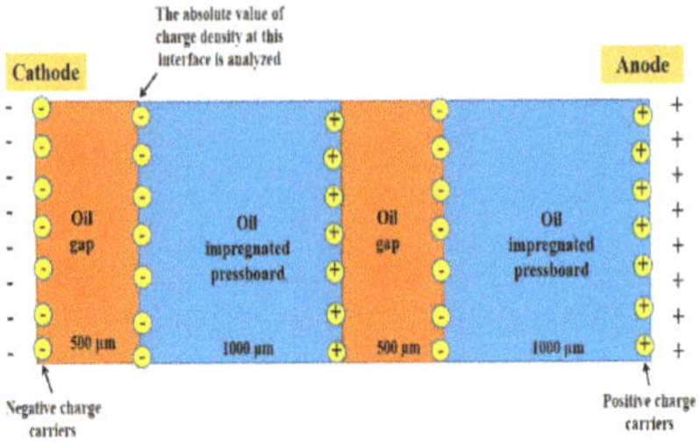

(c) four layers

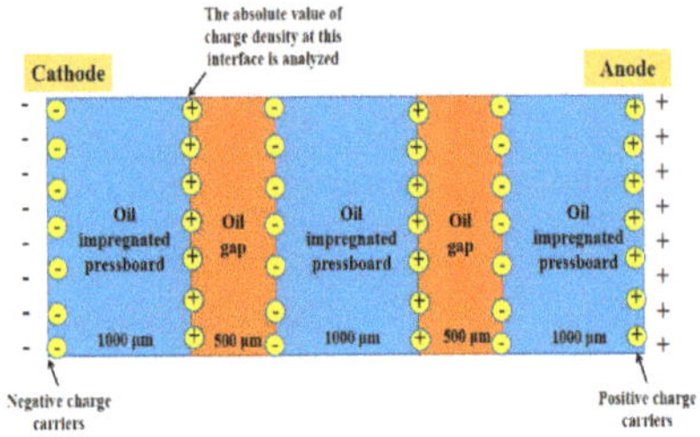

(d) five layers

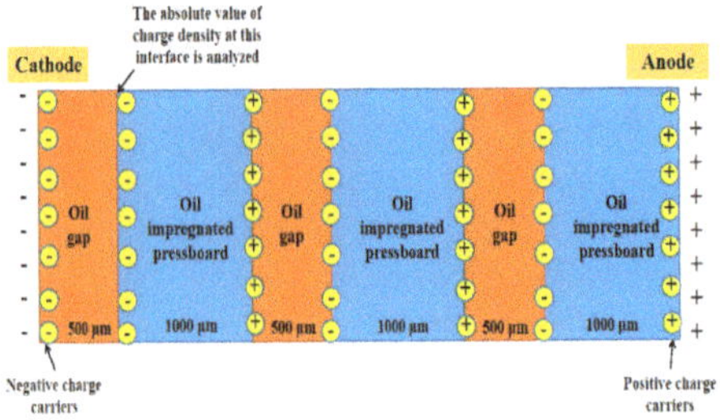

(e) six layers

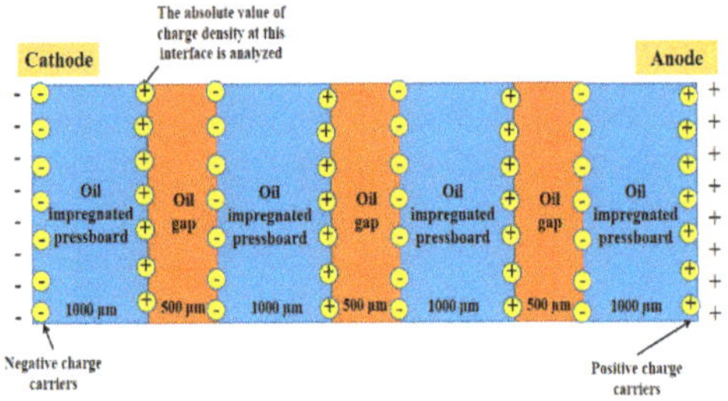

(f) seven layers

Figure 5. The oil-paper insulation structures with different layers.

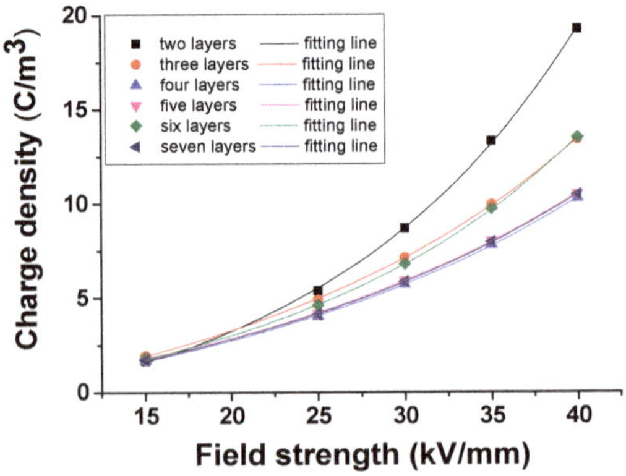

Figure 6. Interface charge density at steady state under different electrical field strengths (20 °C).

The charges accumulated at the interface are dependent on the charge injection from the electrode, the polarized charges determined by the conductivity, permittivity and thickness of dielectrics on both sides of the interface, the charge injection from the electrode, and also the charges migrated from the dielectrics and other interfaces. The interface charge migration and accumulation is illustrated in Figure 7. The accumulated charge density at the interface presents a dynamic change until the accumulated and dissipated charge tends to balance and the density value does not change. The structure of different layers contains different number of interfaces, and the distance of charge migration within the system is different, which leads to the difference of charge at the interface. In this paper, the simulation electric field strength was 15 kV/mm, 25 kV/mm, 30 kV/mm, 35 kV/mm, and 40 kV/mm, respectively. For any same electrical field strength, because of the above reasons for the generation and transfer of charges, the charge density at the interface is different for different layers of the oil-paper insulation structure, and the difference is more significant under higher electrical field strength, as shown in Figure 8.

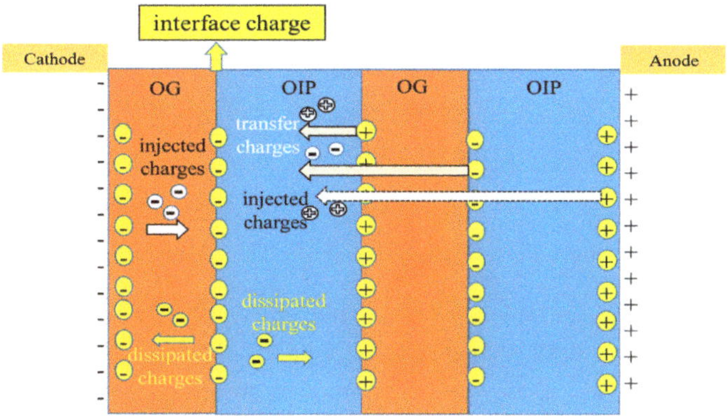

Figure 7. Schematic diagram of interfacial charge accumulation and migration.

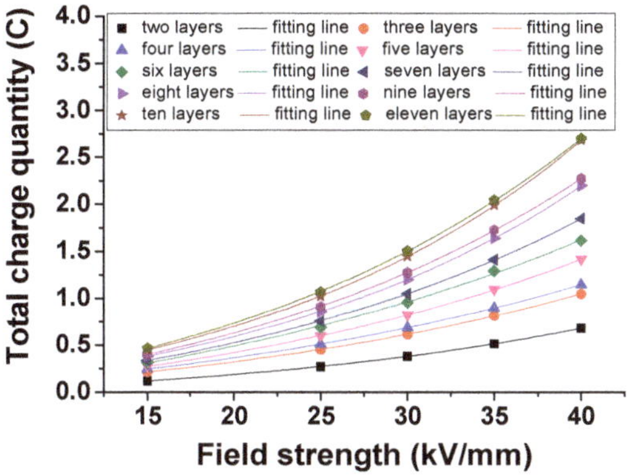

Figure 8. Total charge quantity under different electrical field strengths (20 °C) for multi-layer oil-paper insulation.

The presence of space/interface charge in a multi-layer insulation system is able to enhance locally the electrical field. Nevertheless, in real applications, at the interface, the presence of voids is always at the origin of the partial discharges phenomena, which have a very large influence on the same electrical field [28,29]. In the present model, the defects in the oil-paper insulation are characterized by trap density shown in Table 1. The trap density here is the overall characterization of defects in the oil-paper system, not local defects. In the future, it is necessary to further study the relationship between charge accumulation at the interface and partial discharge.

The relationship between the total charge quantity and electrical field strength for different layers of the oil-paper insulation system is shown in Figure 8. It can be observed, that with the increase of electrical field strength, the total charge quantity for each multi-layer oil-paper insulation system increases in an exponential way, as described in Equation (18). The fitting coefficients for the results in Figure 8 are shown in Table 3. In Equation (18), Q_e stands for the total charge quantity at steady state, and E stands for the electrical field strength, kV/mm. A_e, B_e and C_e are the fitting coefficients. From Figure 8, it can also be observed that the increase layer of the oil-impregnated pressboard will bring a bigger increment of total charge quantity than the increase layer of the oil gap.

Table 3. Fitting coefficients for total charge quantity at steady state (1800 s) under different electrical field strengths for multi-layer oil-paper insulation.

Coefficients	Two Layers	Three Layers	Four Layers	Five Layers	Six Layers	Seven Layers	Eight Layers	Nine Layers	Ten Layers	Eleven Layers
A_e	−0.16	−0.30	−0.34	−0.40	−0.66	−0.46	−0.37	−0.54	−0.44	−0.62
B_e	0.14	0.28	0.33	0.38	0.58	0.42	0.36	0.48	0.41	0.55
C_e	0.04	0.04	0.04	0.04	0.03	0.04	0.05	0.04	0.05	0.05
R^2	0.999	0.995	0.994	0.996	0.994	0.993	0.995	0.994	0.992	0.993

Since the fact that total charge quantity is the sum of the charge quantity of the whole system, the larger the system, the larger the total charge quantity. Therefore, for the simulation electric field strength at 15 kV/mm, 25 kV/mm, 30 kV/mm, 35 kV/mm and 40 kV/mm, respectively, the total charge quantity increases with the increase of insulation layers. Compared with the oil gap layer, adding an oil-impregnated pressboard layer will bring a bigger total charge quantity increase because the pressboard layer can cause more charge accumulation than the oil gap layer.

$$Q_e = A_e + B_e \times EXP(C_e \times E) \tag{18}$$

3.3. Temperature Influence on the Space/Interface Charge Behavior

Figure 9 shows the space/interface charge simulation result for the three-layer "OIP + OG + OIP" sample under DC 15 kV/mm at 40 °C and 60 °C, respectively. It can be observed that the increase of temperature significantly increased the charge density of the charges accumulated at the interfaces. High temperature brings about more charged injected into the sample. The reason for this phenomenon is mainly because the increase of temperature will give charge carriers more energy to overcome the barrier in the sample, and thus more charges will be injected into layers not only from electrodes, but also from interfaces.

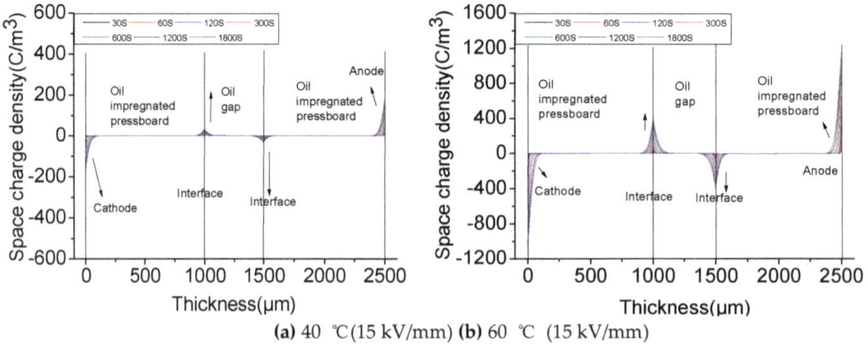

(a) 40 ℃(15 kV/mm) **(b)** 60 ℃ (15 kV/mm)

Figure 9. Space charge simulation of three-layer "OIP + OG + OIP" system under different temperatures.

The relationship between the charge density at the interface adjacent to the cathode for the oil-insulation structure with different layers at steady state and the temperature is shown in Figure 10. It can be found that the charge density absolute values at steady state increase exponentially with the temperature, as shown in Equation (19) and Table 4. In Equation (19), $D_{Tsteady}$ stands for the charge density absolute values at steady state, C/m^3. T stands for the temperature, °C. A_T, B_T and C_T are the fitting coefficients. The charge density value increases by about 200 to 400 times when the temperature increases from 20 °C to 60 °C. Due to the fact that the interface charge density values become very large at 50 °C and 60 °C, the charge density values from 20 °C to 40 °C begin to overlap with each other. At 20 °C, the charge density absolute values at steady state for all structures are about 2.5 C/m^3. At 50 °C, the difference of charge density between different layers is between 5–20 C/m^3, at 60 °C, the difference of charge density between different layers is between 30–80 C/m^3. In addition to the injection of electrode charges, this is mainly due to the more prominent behavior of charge dissipation and accumulation at the interface under high temperature. For all systems with different layer numbers, the interface charge density also increases with the increase of field strength; however, the interface charge density increases only about 5 to 10 times when the field strength increases from 15 kV/mm to 40 kV/mm. By comparing the simulation results, it can be seen that temperature has a more significant influence on the space/interface charge characteristics of the system than the electrical field strength. This trend is in accordance with the phenomenon mentioned in the literature [2,20].

$$D_{Tsteady} = A_T + B_T \times EXP(C_T \times T) \quad (19)$$

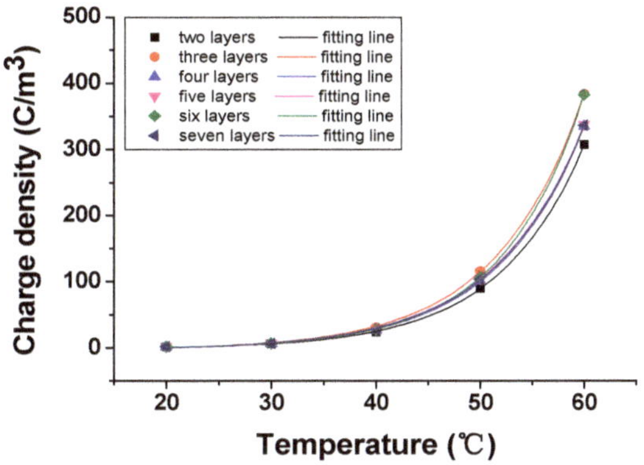

Figure 10. Absolute value of interface steady state charge density with different temperatures (15 kV/mm).

Table 4. Fitting parameters for charge density at steady state (1800 s) under different temperatures for multi-layer oil-paper insulation.

Parameters	Two Layers	Three Layers	Four Layers	Five Layers	Six Layers	Seven Layers
A_T	0.062	0.083	0.094	−0.004	0.232	0.018
B_T	0.089	0.106	0.088	0.102	0.084	0.098
C_T	0.144	0.143	0.144	0.142	0.146	0.143
R^2	0.99	0.99	0.99	0.99	0.99	0.99

The relationship between the total charge quantity and temperature for the multi-layer oil-paper insulation system is shown in Figure 11. It can be observed that with the increase of temperature, the total charge quantity for any kind of multi-layer oil-paper insulation system increases in an exponential way, which is described in Equation (20) and Table 5. In Equation (20), Q_{QT} stands for the total charge quantity at steady state with the unit C; T stands for the temperature with the unit °C. A_{QT},

B_{QT} and C_{QT} are the fitting coefficients. From Figure 11, it can be observed that at each temperature under the same DC electrical field strength 15 kV/mm, the increase of the oil-impregnated pressboard layer will bring a greater increment of total charge quantity than the increase of the oil gap layer.

$$Q_T = A_{QT} + B_{QT} \times EXP(C_{QT} \times T) \tag{20}$$

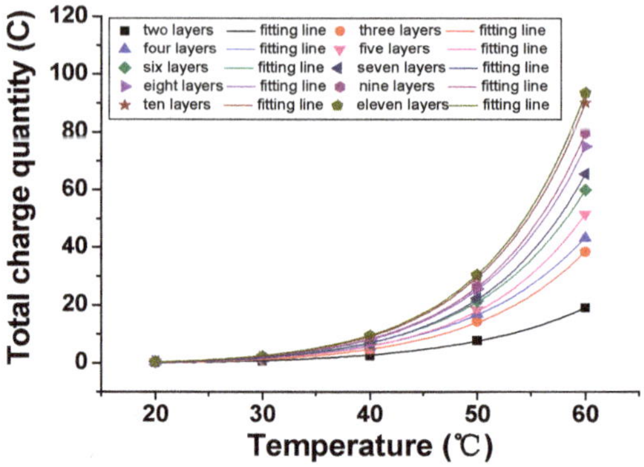

Figure 11. Total charge quantity under different temperature for multi-layer oil-paper insulation (15 kV/mm).

Table 5. Fitting parameters for the total charge quantity at steady state (1800 s) under different temperatures for multi-layer oil-paper insulation.

Coefficients	Two Layers	Three Layers	Four Layers	Five Layers	Six Layers	Seven Layers	Eight Layers	Nine Layers	Ten Layers	Eleven Layers
A_{QT}	−0.67	−0.95	−1.2	−0.98	−1.00	−1.02	−0.99	−0.88	−1.01	−0.88
B_{QT}	0.10	0.12	0.18	0.11	0.13	0.11	0.13	0.12	0.13	0.13
C_{QT}	0.09	0.10	0.09	0.10	0.10	0.11	0.11	0.11	0.11	0.11
R^2	0.99	0.99	0.99	0.99	0.99	0.99	0.99	0.99	0.99	0.99

4. Space/Interface Charge Behavior Under the Electrical-Thermal Combined Stress

4.1. Charge Density Calculation Method for the Electrical-Thermal Combined Stress

If the electrical field strength and temperature combined effect on the charge density could be quantified, it will be of great significance to understand the charge distribution of the multi-layer oil-paper insulation system used in the converter transformer. In this paper, the charge density under the combined action of electrical field strength and temperature was calculated by the translation of the charge density curves. Here, the three-layer "OIP + OG + OIP" sample was selected to illustrate the proposed method.

The charge density absolute values of the charges accumulated at the interface adjacent to the cathode for the three-layer "OIP + OG + OIP" sample at steady state (applied DC voltage for 1800 s) under the combined effect of electrical field strength and temperature is shown in Figure 12. At each temperature, the interface charge density absolute values at steady state increase with the electrical field strength in an exponential way, as described in Equation (17).

The interface charge density absolute values at steady state under 40 °C is taken as the reference temperature. The curve of the charge density changing with electrical field strength under 40 °C is called the reference curve. Then the curve of the charge density changing with the electrical field strength under 20 °C and 30 °C was moved along the x axis to the reference curve horizontally. The combination of the three curves is called the charge density main curve, as shown in Figure 12. The ratio of the electrical field strength of a point on the original curve at temperature T′ (T′ = 20 °C,

30 °C or 40 °C) before and after being moved to the main curve is defined as the electrical field strength-temperature shift factor $\alpha_{T'}$, which is defined in Equation (21). Where $E_{T'}$ is the electrical field strength of a point on the original curve at temperature T′ before being moved. $E_{ref\text{-}T}$ is the electrical field strength of that point after the curve at temperature T′ is moved to the reference temperature *T*. The $\alpha_{T'}$ value of the reference temperature T = 40 °C is defined as $\alpha_{40} = 1$. The $\alpha_{T'}$ values for temperatures 30 °C and 20 °C are defined as α_{30} and α_{20}, respectively. Based on the electrical field strength values of a point before and after being moved to the main curve, the α_{30} and α_{20} were calculated, where $\alpha_{30} = 2$ and $\alpha_{20} = 4$, as shown in Figure 12.

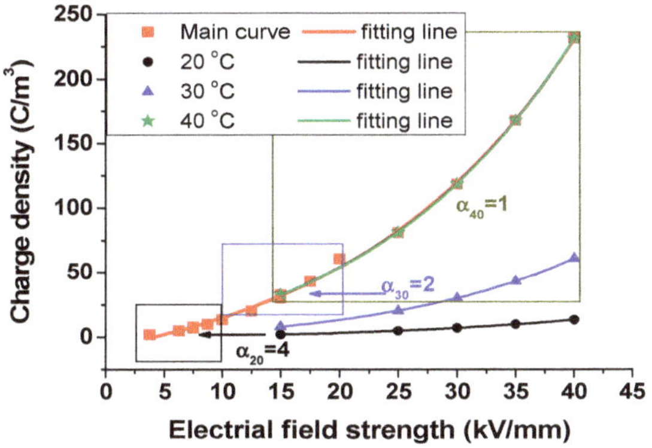

Figure 12. Interface charge density absolute values at steady state under different electrical field strengths with different temperatures.

The electrical field strength-temperature shift factor $\alpha_{T'}$ can also be expressed with an Arrhenius equation, as shown in Equation (23). Where *R* is the Boltzmann constant, 8.314J K^{-1} mol^{-1}. E_a is the activation energy, KJ/mol. *T* is the temperature before shifting, and T_{ref} is the reference temperature, K. Based on the $\alpha_{T'}$ results shown in Figure 12, the calculated activation energy of the three-layer "OIP + OG + OIP" sample is 55 kJ/mol, which is similar to the value given by Stanmm in the literature [30], which indicates that the above calculation method for the charge density under the combined action of electrical field strength and temperature is correct. Therefore, by fitting the charge density data of the main curve at the reference temperature, the formula for calculating the charge density under the combined action of electrical field strength and temperature for the three-layer "OIP + OG + OIP" sample is obtained, as shown in Equation (22). The charge density under the combined action of electrical field strength and temperature can be obtained by removing the main curve according to the translation factor $\alpha_{T'}$.

With the translation method shown above, the charge density main curves for a multi-layer oil-paper insulation system were obtained, as shown in Figure 13. The equations for calculating the steady state charge density at the interface are shown in Table 6. By using the equations in Table 6, the steady state interface charge density can be calculated under the combined action of electrical field strength and temperature for different layers of a oil-paper insulation system.

$$\alpha_{T'} = \frac{E_{T'}}{E_{\text{ref}-\text{T}}};\ \alpha_{T'} = EXP(\frac{E_a}{R}(\frac{1}{T} - \frac{1}{T_{ref}})) \tag{21}$$

$$D = -41.86 + 33.67 \times EXP(0.05 \times \frac{E}{\alpha_{T'}}) \tag{22}$$

$$\alpha_{T'} = EXP(\frac{55 \times 10^3}{8.314} \times (\frac{1}{T} - \frac{1}{313})) \tag{23}$$

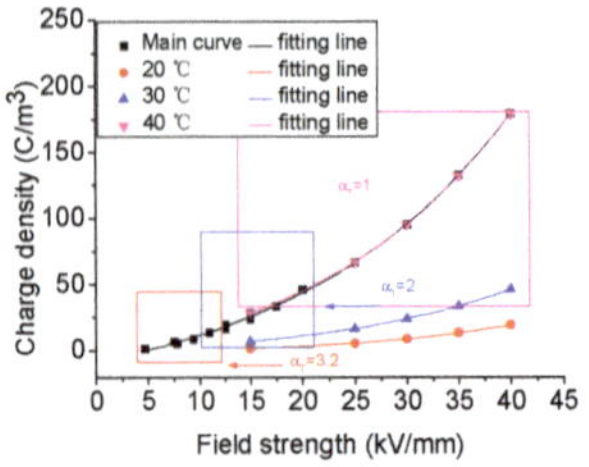

(a) two-layer structure

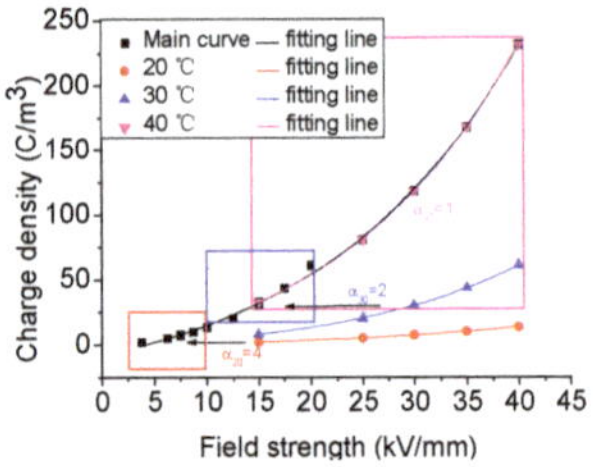

(b) three-layer structure

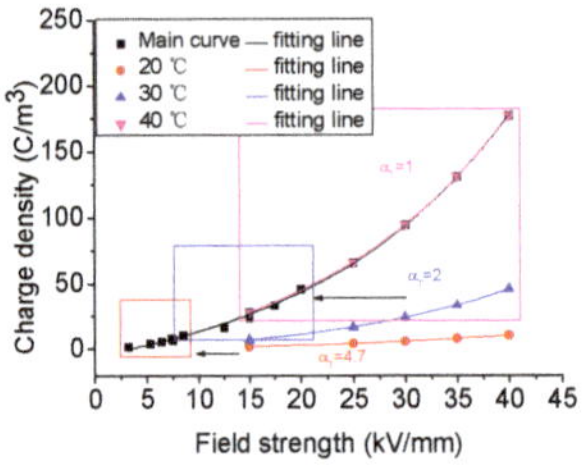

(c) four-layer structure

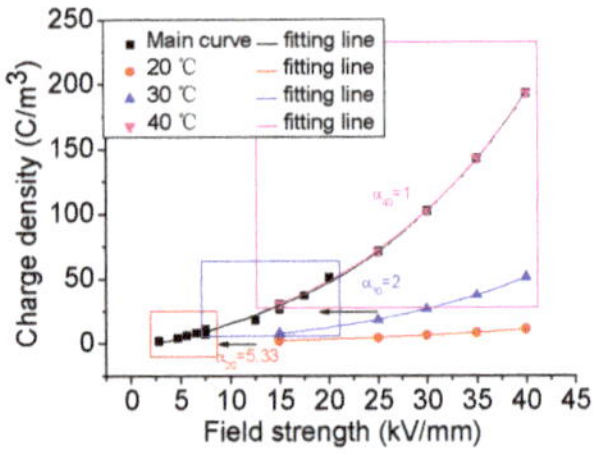

(d) five-layer structure

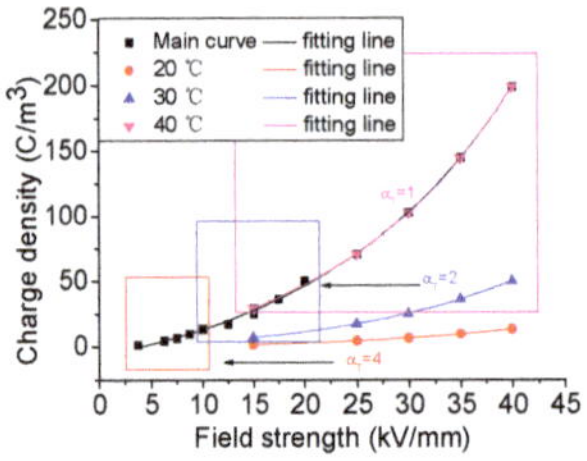

(e) six-layer structure

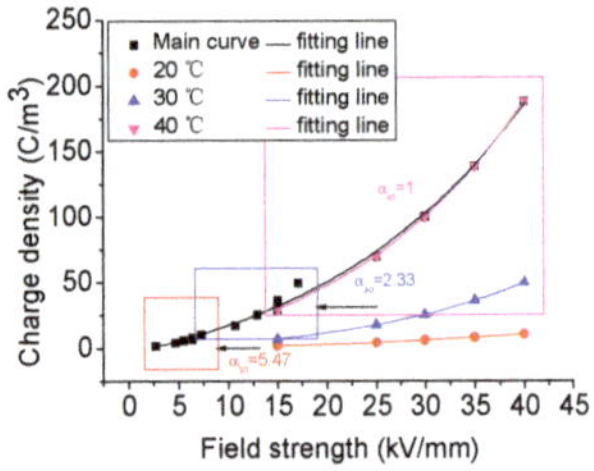

(f) seven-layer structure

Figure 13. Interface charge density absolute values at steady state under different electrical field strengths with different temperatures for multi-layer oil-paper insulation.

Table 6. Calculation equations for the charge density under the combined action of electrical field strength and temperature for multi-layer oil-paper insulation.

Sample	Equations for Charge Density
two layers	$D = -40.5 + 32.3 \times EXP(\frac{0.05 \times E}{\alpha_{T'}}); \alpha_{T'} = EXP(\frac{55 \times 10^3}{8.314}(\frac{1}{T} - \frac{1}{313}))$
three layers	$D = -41.86 + 33.67 \times EXP(\frac{0.05 \times E}{\alpha_{T'}}); \alpha_{T'} = EXP(\frac{55 \times 10^3}{8.314} \times (\frac{1}{T} - \frac{1}{313}))$
four layers	$D = -32.7 + 27.7 \times EXP(\frac{0.05 \times E}{\alpha_{T'}}); \alpha_{T'} = EXP(\frac{60 \times 10^3}{8.314}(\frac{1}{T} - \frac{1}{313}))$
five layers	$D = -32.9 + 28.6 \times EXP(\frac{0.05 \times E}{\alpha_{T'}}); \alpha_{T'} = EXP(\frac{63 \times 10^3}{8.314} \times (\frac{1}{T} - \frac{1}{313}))$
six layers	$D = -34.8 + 28.4 \times EXP(\frac{0.05 \times E}{\alpha_{T'}}); \alpha_{T'} = EXP(\frac{58 \times 10^3}{8.314}(\frac{1}{T} - \frac{1}{313}))$
seven layers	$D = -38.9 + 35.4 \times EXP(\frac{0.05 \times E}{\alpha_{T'}}); \alpha_{T'} = EXP(\frac{65 \times 10^3}{8.314} \times (\frac{1}{T} - \frac{1}{313}))$

4.2. Total Charge Quantity Calculation Method for the Electrical-Thermal Combined Stress

By using the translation method proposed above, the main curves for the total charge quantity under the combined action of electrical field strength and temperature for different layers of oil-paper insulation system can be attained, as shown in Figure 14. The equations for calculating the total charge quantity under the combined action of electrical field strength and temperature for different layers of oil-paper insulation system are shown in Table 7. By using the equations from Table 7, total charge quantity for different layers of the oil-paper insulation system can be calculated under any temperature or any electrical field strength.

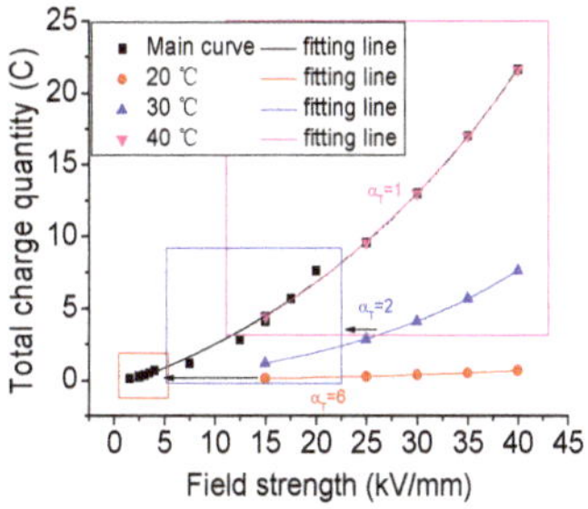

(a) two layers

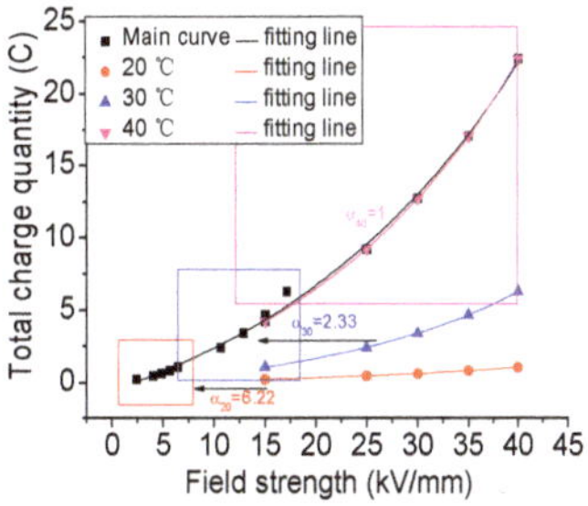

(b) three layers

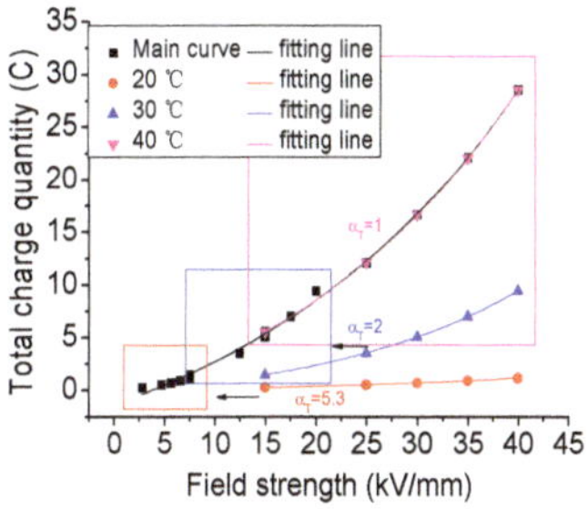

(c) four layers

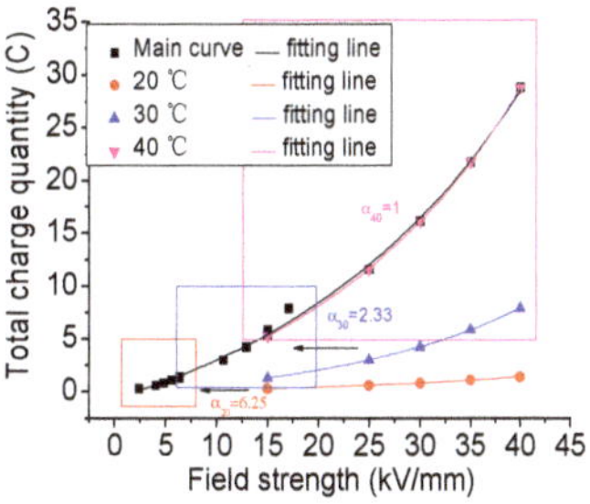

(d) five layers

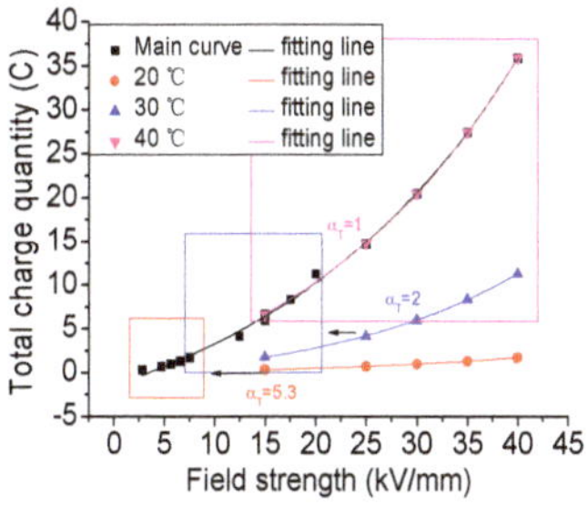

(e) six layers

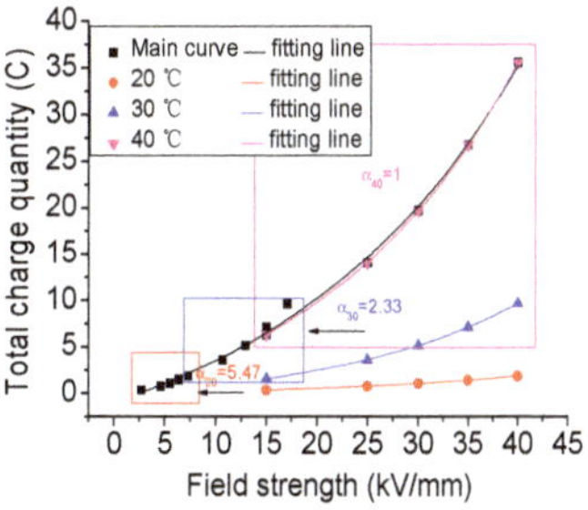

(f) seven layers

Figure 14. Total charge quantity at steady state under different electrical field strengths with different temperatures for multi-layer oil-paper insulation.

Table 7. Calculation equations for the total charge quantity under the combined action of electrical field strength and temperature for multi-layer oil-paper insulation.

Sample	Equations for Total Charge Quantity
two layers	$Q = -7.6 + 7.1 \times EXP(\frac{0.04 \times E}{\alpha_{T'}}); \alpha_{T'} = EXP(\frac{68 \times 10^3}{8.314}(\frac{1}{T} - \frac{1}{313}))$
three layers	$Q = -7.03 + 6.44 \times EXP(\frac{0.04 \times E}{\alpha_{T'}}); \alpha_{T'} = EXP(\frac{68 \times 10^3}{8.314} \times (\frac{1}{T} - \frac{1}{313}))$
four layers	$Q = -11 + 9.7 \times EXP(\frac{0.04 \times E}{\alpha_{T'}}); \alpha_{T'} = EXP(\frac{70 \times 10^3}{8.314}(\frac{1}{T} - \frac{1}{313}))$
five layers	$Q = -7.9 + 7.3 \times EXP(\frac{0.04 \times E}{\alpha_{T'}}); \alpha_{T'} = EXP(\frac{69 \times 10^3}{8.314} \times (\frac{1}{T} - \frac{1}{313}))$
six layers	$Q = -11 + 9.7 \times EXP(\frac{0.04 \times E}{\alpha_{T'}}); \alpha_{T'} = EXP(\frac{70 \times 10^3}{8.314}(\frac{1}{T} - \frac{1}{313}))$
seven layers	$Q = -9.8 + 8.9 \times EXP(\frac{0.04 \times E}{\alpha_{T'}}); \alpha_{T'} = EXP(\frac{65 \times 10^3}{8.314} \times (\frac{1}{T} - \frac{1}{313}))$

5. Conclusions

The interface charge density absolute values at steady state increase with the electrical field strength and temperature in an exponential way, respectively. Temperature has a more significant influence on the charge density than the electrical field strength. The total charge quantity of the multi-layer oil gap and oil-impregnated insulation pressboard system also increases in an exponential way with the electrical field strength and temperature increases, respectively. The temperature has a more significant influence on the total charge quantity than the electrical field strength.

A new method for calculating the space/interface charge density or the total charge quantity of the multi-layer oil-paper insulation under the combined action of electrical field strength and temperature was proposed using the translation of the charge density curves or the total charge quantity curves. The electrical field strength-temperature shifting factor $\alpha_{T'}$ was introduced in this paper. The equations for calculating the charge density or the total charge quantity of the multi-layer oil-paper insulation under the combined action of electrical field strength and temperature were provided.

Author Contributions: R.Z. analyzed the simulation model, did the simulation and wrote the paper. J.H. analyzed the data, wrote and revised the paper. R.L. contributed in the discussion.

Acknowledgments: This work was supported by the National Key R&D Program of China (2017YFB0902704), the National Natural Science Foundation of China (51707022), China Postdoctoral Science Foundation (2017M612910) and Chongqing Special Funding Project for Post-Doctoral (Xm2017040).

Conflicts of Interest: The authors declare no conflict of interest.

References

1. Lu, W.X.; Ooi, B.T. Optimal acquisition and aggregation of offshore wind power by multi terminal voltage-source HVDC. *IEEE Trans. Power Deliv.* **2003**, *18*, 201–206.
2. Tang, C.; Huang, B.; Hao, M.; Xu, Z.Q.; Hao, J.; Chen, G. Progress of space charge research on oil-paper insulation using pulsed electroacoustic techniques. *Energies* **2016**, *9*, 53. [CrossRef]
3. Du, B.X.; Zhang, J.G.; Liu, D.S. Interface charge behavior of multi-layer oil-paper insulation under DC and polarity reversal voltages. *IEEE Trans. Dielectr. Electr. Insul.* **2015**, *22*, 2628–2638. [CrossRef]
4. Huang, B.; Hao, M.; Hao, J.; Fu, J.; Wang, Q.; Chen, G. Space charge characteristics in oil and oil-impregnated pressboard and electric field distortion after polarity reversal. *IEEE Trans. Dielectr. Electr. Insul.* **2016**, *23*, 881–891. [CrossRef]
5. Ciobanu, R.; Schreiner, C.; Pfeiffer, W.; Baraboi, B. Space charge evolution in oil-paper insulation for DC cables application. In Proceedings of the 2002 IEEE 14th International Conference on Dielectric Liquids, Graz, Austria, 7–12 July 2002; pp. 321–324.
6. Hao, J.; Chen, G.; Liao, R.J. Influence of moisture on space charge dynamics in multilayer oil-paper insulation. *IEEE Trans. Dielectr. Electr. Insul.* **2012**, *19*, 1456–1464. [CrossRef]
7. Jin, S.; Ruan, J.; Du, Z.; Huang, G.; Zhu, L.; Guan, W.; Li, L.Y.; Yang, Z.F. Charge transport simulation in single-layer oil-paper insulation. *IEEE Trans. Magn.* **2016**, *52*, 1–4. [CrossRef]

8. Li, G.; Li, S.; Min, D.M.; Zhao, N.; Zhu, Y. Influence of trap depths on space charge formation and accumulation characteristics in low density polyethylene. In Proceedings of the 2013 IEEE International Conference on Solid Dielectrics (ICSD), Bologna, Italy, 30 June–4 July 2013; pp. 698–701.
9. Jin, S.; Ruan, J.; Du, Z. Charge transport in oil impregnated paper insulation under temperature gradient using transient upstream FEM. *CSEE J. Power Energy Syst.* **2015**, *1*, 3–8.
10. Zhu, Q.D.; Wu, K.; Zhu, W. Numerical simulation of space charge property in oil-paper insulation space under temperature gradient. *High Volt. Eng.* **2016**, *42*, 923–930.
11. Roy, S.L.; Teyssedre, G.; Laurent, C.; Montanari, G.C. Description of charge transport in polyethylene using a fluid model with a constant mobility: Fitting model and experiments. *J. Phys. D Appl. Phys.* **2006**, *39*, 1427–1433. [CrossRef]
12. Alagoz, B.B.; Alisoy, H.Z.; Alagoz, S.; Hansu, F. A space charge motion simulation with FDTD method and application in negative corona electrostatic field analysis. *Appl. Math. Comput.* **2012**, *218*, 9007–9017. [CrossRef]
13. Boukhari, H.; Rogti, F. Simulation of space charge dynamic in polyethylene under DC continuous electrical stress. *J. Electron. Mater.* **2016**, *45*, 5334–5340. [CrossRef]
14. Hao, M. Space Charge Behavior in Thick Oil Pressboard Insulation System for Converter Transformers. Ph.D. Thesis, Southampton University, Southampton, UK, 2015.
15. Roy, S.L.; Segur, P.; Teyssedre, G.; Laurent, C. Description of bipolar charge transport in polyethylene using a fluid model with a constant mobility: Model prediction. *J. Phys. D Appl. Phys.* **2004**, *37*, 298–305. [CrossRef]
16. Alison, J.M.; Hill, R.M. A model for bipolar charge transport, trapping and recombination in degassed cross-linked polyethene. *J. Phys. D Appl. Phys.* **1994**, *27*, 1291–1299. [CrossRef]
17. Wu, S.; Li, J.; Zhao, L.; Wang, Y.; He, Z.; Bao, L. The properties of space charge in oil-paper insulation during electrical-thermal aging. In Proceedings of the 2012 International Conference on High Voltage Engineering and Application, Shanghai, China, 17–20 September 2012; pp. 269–273.
18. Zhou, Y.; Huang, M.; Chen, W.J.; Sun, Q.H.; Wang, Y.S.; Zhang, L. Space charge characteristics of interface in oil-paper insulation under DC voltage. *High Volt. Eng.* **2011**, *37*, 2417–2423.
19. Wang, S.Q.; Zhang, G.J.; Mu, H.B.; Wang, D.; Lei, M.; Suwarno, S.; Tanaka, Y.; Takada, T. Effect of paper-aged state on space charge characteristics in oil-impregnated paper insulation. *IEEE Trans. Dielectr. Electr. Insul.* **2012**, *19*, 1871–1878. [CrossRef]
20. Tang, C. Studies on the DC Space Charge Characteristics of Oil-Paper Insulation Materials. Ph.D. Thesis, Department of Electrical Engineering, Chongqing University, Chongqing, China, 2010.
21. Hao, J.; Zou, R.H.; Liao, R.J.; Yang, L.J.; Liao, Q. New method for shallow and deep trap distribution analysis in oil impregnated insulation paper based on the space charge de-trapping. *Energies* **2018**, *11*, 271. [CrossRef]
22. Delpino, S.; Fabiani, D.; Montanari, G.C. Polymeric HVDC cable design and space charge accumulation. Part 2: Insulation interfaces. *IEEE Electr. Insul. Mag.* **2008**, *24*, 14–24. [CrossRef]
23. Takuma, T.; Ikeda, T.; Kawamoto, T. Calculation of ion flow fields of HVDC transmission lines by the finite element method. *IEEE Trans. Power Appar. Syst.* **1981**, *PAS-100*, 4802–4810. [CrossRef]
24. Du, Z.Y.; Huang, G.D.; Ruan, J.J.; Wang, G.L.; Yao, Y.; Liao, C.B.; Yuan, J.X.; Wen, W. Calculation of the ionized field around the DC voltage divider. *IEEE Trans. Magn.* **2013**, *49*, 1933–1936. [CrossRef]
25. Huang, G.D.; Ruan, J.J.; Du, Z.Y.; Liao, C.B.; Jin, S.; Wang, G.L. Improved 3-D upwind FEM for solving ionized field of HVDC transmission lines. *Proceeding CSEE* **2013**, *33*, 152–159.
26. Boufayed, F.; Teyssèdre, G.; Laurent, C.; Le Roy, S.; Dissado, L.A.; Ségur, P.; Montanari, G.C. Models of bipolar charge transport in polyethylene. *J. Appl. Phys.* **2006**, *100*, 826–856. [CrossRef]
27. Hao, J.; Huang, B.; Chen, G.; Fu, J.; Wu, G.; Wang, Q. Space charge accumulation behavior of multilayer structure oil-paper insulation and its effect on electric field distribution. *High Volt. Eng.* **2017**, *43*, 1973–1979.
28. Pompili, M.; Mazzetti, C.; Libotte, M. The effect of the definition used in measuring partial discharge inception voltages. *IEEE Trans. Electr. Insul.* **1993**, *28*, 1002–1006. [CrossRef]

29. Calcara, L.; Pompili, M.; Muzi, F. Standard evolution of partial discharge detection in dielectric liquids. *IEEE Trans. Dielectr. Electr. Insul.* **2017**, *24*, 2–6. [CrossRef]
30. Stamm, A.J. Theraml degradation of wood and cellulose. *Ind. Eng. Chem.* **1956**, *48*, 413–417. [CrossRef]

energies

MDPI

Article

Numerical Modeling of Space–Time Characteristics of Plasma Initialization in a Secondary Arc

Jinsong Li *, Hua Yu, Min Jiang, Hong Liu and Guanliang Li

Equipment State Analysis Center, State Grid Shanxi Electric Power Research Institute, Taiyuan 030001, China; yuhua16885@163.com (H.Y.); jigmi@163.com (M.J.); 18935121293@189.cn (H.L.); 15386812717@163.com (G.L.)
* Correspondence: ljskssss@sina.com

Received: 10 March 2019; Accepted: 27 May 2019; Published: 3 June 2019

Abstract: A numerical model based on the finite element simulation software COMSOL was developed to investigate the secondary arc that can limit the success of single-phase auto-reclosure solutions to the single-phase-to-ground fault. Partial differential equations accounting for variation of densities of charge particles (electrons, positive and negative ions) were coupled with Poisson's equation to consider the effects of space and surface charges on the electric field. An experiment platform was established to verify the numerical model. The brightness distribution of the experimental short-circuit arc was basically consistent with the predicted distribution of electron density, demonstrating that the simulation was effective. Furthermore, the model was used to assess the particle density distribution, electric field variation, and time dependence of ion reactions during the short-circuit discharge. Results showed that the ion concentration was higher than the initial level after the short-circuit discharge, which is an important reason for inducing the subsequent secondary arc. The intensity of the spatial electric field was obviously affected by the high-voltage electrode at the end regions, and the intermediate region was mainly affected by the particle reaction. The time correspondence between the detachment reaction and the ion source generated in the short-circuit discharge process was basically consistent, and the detachment reactions were mainly concentrated in the middle area and near the negative electrode. The research elucidates the relevant plasma process of the secondary arc and will contribute to the suppression of it.

Keywords: secondary arc; short-circuit discharge; numerical modeling; plasma discharge

1. Introduction

Because most of the faults on ultra-high voltage (UHV) and extra-high voltage (EHV) transmission lines are typically impermanent single-phase-to-ground faults, single-phase auto-reclosure (SPAR) can eliminate most of their potential effects. If a single-phase-to-ground fault occurs along the high-voltage transmission lines, a short-circuit arc discharge of large current will be incurred at the fault point. After the fault phase is switched off, the short-circuit arc will be extinguished. However, due to the electromagnetic (EM) coupling between transmission lines, a secondary arc discharge of small current will continue through the same arc path at the fault point. Therefore, the timely extinction of the secondary arc caused by the single-phase ground fault is important for the success of single-phase reclosing. To ensure the safe operation of the power transmission lines and enhance the stability of the power system, a method that enables the self-extinction of the secondary arc is urgently needed to be found [1–3].

Although physical experiments can be performed to study the secondary arc characteristics directly, such experiments are restricted by environmental conditions, require large investments, and are insufficiently flexible [4–7]. In a circuit simulation model, the fault arc is very often represented by a time-varying resistor, and it is described via nonlinear differential equations [8–10]. The arc chain model, in which arc movement is modelled with consideration of the electromagnetic force, thermal buoyancy, wind load force, and air resistance, has been adopted by many researchers to obtain the velocity equation of arc movement through the force analysis of each arc element [11–14]. However, these models do not consider the plasma produced by the discharge.

Numerical simulations are particularly suitable for analyzing and optimizing the complex plasma processes created by the air discharge, and these plasma processes can be further elucidated by comparing predictions from numerical simulations with experimental observations. Various modelling approaches have been adopted by experts, including analytical models, fluid models, non-equilibrium Boltzmann equations, Monte Carlo simulations, and particle-in-cell models [15]. Hybrid models that combine some of these models are also used [16], as further detailed in [17]. In particular, hydrodynamic fluid models have been shown to offer the advantages of efficiency, accuracy, and comprehensiveness, and have most often been employed [18,19]. However, due to extensive calculations and complex external conditions, the research has been focused on short gap discharges, such as corona discharge, dielectric barrier discharge, and other fields.

Although the secondary arc has been thoroughly studied experimentally and the arc chain model has been applied to analyze the movement of the secondary arc, few studies have been conducted on the plasma process of the secondary arc. Therefore, the aim of this study was to develop a finite element model for the secondary arc, using the simulation software COMSOL and focusing on the initial stage, and to elucidate the relevant plasma processes by comparing predictions from numerical simulations with experimental observations.

2. Model Description

2.1. Governing Equations

The most widely used formulation of a streamer propagation model in air is based on the drift-diffusion hydrodynamic approach, which considers variations in the densities of electrons and two generic types of ions (positive and negative) in space and time (see, e.g., [1–6]). This approach results in the following three partial differential convection–diffusion equations, which also account for rates of the physical processes leading to the generation and loss of charged species:

$$\frac{\partial N_e}{\partial t} + \nabla \cdot (-D_e \nabla N_e) + \beta_e \cdot \nabla N_e = f_e, \tag{1}$$

$$\frac{\partial N_p}{\partial t} + \nabla \cdot (-D_p \nabla N_p) + \beta_p \cdot \nabla N_p = f_p, \tag{2}$$

$$\frac{\partial N_n}{\partial t} + \nabla \cdot (-D_n \nabla N_n) + \beta_n \cdot \nabla N_n = f_n. \tag{3}$$

here, the subscripts e, p, and n indicate electrons and positive and negative ions, respectively; N is the density, in m^{-3}; D is the diffusion coefficient, in m^2/s; f is the net rate of the generation and loss processes, in $m^{-3}s^{-1}$; and t represents time, in s. The main processes usually considered in Equations (1) and (3) are represented by their rates: Electron impact ionization, $f_{ion} = \alpha N_e \mu_e E$; attachment of electrons to electronegative molecules (CO_2, H_2O, O_2, etc.) present in air, $f_{att} = \eta N_e \mu_e E$; detachment of electrons from negative ions, $f_{det} = k_{det} N_e N_n$; electron-ion recombination, $f_{ep} = \beta_{ep} N_e N_p$; recombination of positive and negative ions, $f_{pn} = \beta_{pn} N_p N_n$; and natural background ionization, f_0. In the expressions above, α is Townsend's ionization coefficient, in m^{-1}; μ is the mobility, in m^2/Vs; E is the electric field strength, in V/m; η is the attachment coefficient, in m^{-1}; k_{det} is the detachment coefficient, in m^3/s; and β is each respective recombination coefficient, in m^3/s. Hence, the net rates for different charged species are:

$$f_e = f_{ion} + f_{det} + f_0 - f_{att} - f_{ep}, \tag{4}$$

$$f_p = f_{ion} + f_0 - f_{pn} - f_{ep}, \tag{5}$$

$$f_n = f_{att} - f_{det} - f_{pn}. \tag{6}$$

Equations (1) and (3) must be complemented by Poisson's equation for electric potential V. The solution provides the electric field distributions affected by the space charge, which are needed to obtain the kinetic coefficients and the rates of individual processes:

$$\nabla(-\varepsilon_0\varepsilon_r\nabla V) = e(N_p - N_e - N_n), \quad (7)$$

$$-\nabla V = \mathbf{E}. \quad (8)$$

here, e is the elementary charge, ε_0 is the vacuum permittivity, and ε_r is the dielectric constant of the material (unity for air). Equations (1) and (8), with boundary and initial conditions specific to the problem, form a self-consistent model that must be solved numerically because of the strongly non-linear nature of the model.

Parameters and rate coefficients in the hydrodynamic models should be obtained from a solution of Boltzmann's equation. Local field approximations are assumed, i.e., gas properties such as drift velocities and the collisional ionization coefficient are functions only of E/N, where E is the field amplitude and N the gas number density. In this study, transport coefficients needed for simulations of discharges are obtained from [20], and the results are verified by comparing them with those obtained using the popular Boltzmann equation solver. The transport coefficients are also compared with the experimental results for air, and the achieved agreement confirms the validity of the parameters utilized [21].

The rate and kinetic coefficients used in the model are provided in Table 1. The dependencies of the ionization and attachment coefficients on the field strength are reproduced in Figure 1. The dependencies of the electron drift velocity and diffusion coefficient are approximated as $w_e = 3.2 \times 10^3 \times (E/N)^{0.8}$ m/s and $D_e = 7 \times 10^{-2} + 8 \times (E/N)^{0.8}$ m^2/s, respectively. The parameters are selected based on analysis of the literature and slightly adjusted according to the convergence of the model.

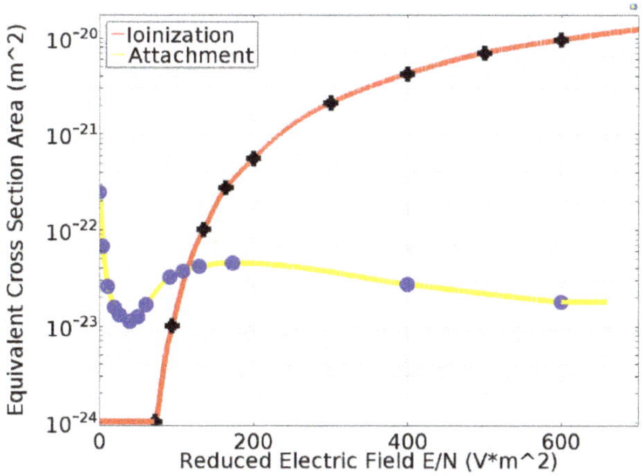

Figure 1. Ionization and electron attachment coefficients as functions of the reduced field E/N.

Table 1. Input parameters for the model.

Transport Parameters	Expression	Description
μ_p (m^2/Vs)	2.0×10^{-6}	Positive ion mobility
D_p (m^2/s)	5.05	Positive ion diffusivity
μ_n (m^2/Vs)	2.2×10^{-6}	Negative ion mobility
D_n (m^2/s)	5.56	Negative ion diffusivity
β_{ep} (m^3/s)	5.0×10^{-14}	Electron-positive ions recombination rate
β_{pn} (m^3/s)	2.07×10^{-13}	Positive-negative ions recombination rate
f_0 (1/m^3s)	1.7×10^9	Natural background ionization source item
k_{det} (m^3/s)	1×10^{-18}	Electron detachment coefficient from negative ions

2.2. Computational Domain and Meshing

The geometric model of the secondary arc simulation conducted in this study is shown in Figure 2a. Circular symmetry was exploited in constructing the geometric configuration. The whole computation domain was 1.62 m high and 0.4 m wide. The insulator string was 1 m long (i.e., the distance between the top and bottom electrode), and the radius of the center column of the insulator was 0.025 m. The ambient temperature T = 288.15 K (15 °C), and the background pressure was 1 atm (1 atmosphere = 1.01325×10^5 Pa).

The above structures were meshed by free triangles, as shown in Figure 2b. The model grid contained 10,821 triangles, with a maximum grid size of 0.06 m and a minimum grid size of 0.01 m. Near the electrodes and the ignition line, the charge density and its variation are particularly strong, which demands a very fine spatial mesh, whereas the rest of the discharge space rarely exhibits the steep gradients associated with the electrodes. As can be seen from Figure 2b, at the electrode region and ignition line a very fine resolution was employed to resolve the steep gradients, as required, but away from the axis of symmetry, a very coarse mesh was used as the charge density does not vary greatly there.

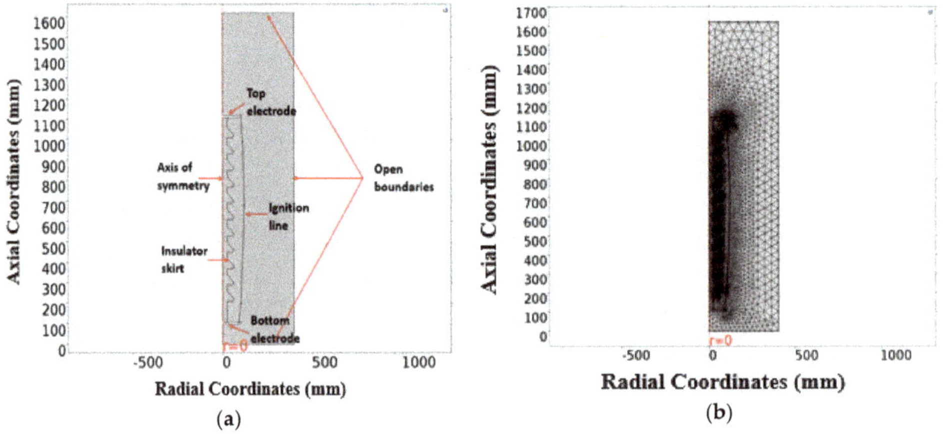

Figure 2. The geometric model and meshing result of the secondary arc simulation: (**a**) The geometric model; (**b**) meshing result.

2.3. Numerical Modelling of the Short-Circuit Arc

In the experiment study, a fuse was usually adopted to ignite the arc discharge and simulate the short circuit, and then the secondary arc was generated. For the purposes of this study, an ignition line was added to the simulation model, and the emission particle source (electron g_{e2}, positive ion g_{p2}, negative ion g_{n2}) was set on the ignition line to simulate the high-charge-density arc channel generated by the short-circuit combustion. The particle sources were established with the following Gaussian pulse functions (Figure 3) [22]:

$$\begin{cases} g_{e2} = 1 \times 10^{13} \cdot gp1(t) \\ g_{p2} = 3 \times 10^{13} \cdot gp1(t) \\ g_{n2} = 2 \times 10^{13} \cdot gp1(t) \end{cases}, \tag{9}$$

where $gp1(t)$ is the Gaussian impulse function. The width of the impulse depends on the duration of short-circuit discharge (about 0.2 s). The amplitude of the impulse has positive correlation with the short-circuit arc current (1 kA in this study). The appropriate value of the impulse at the center position was calculated to be 0.08 with a standard deviation of 0.05.

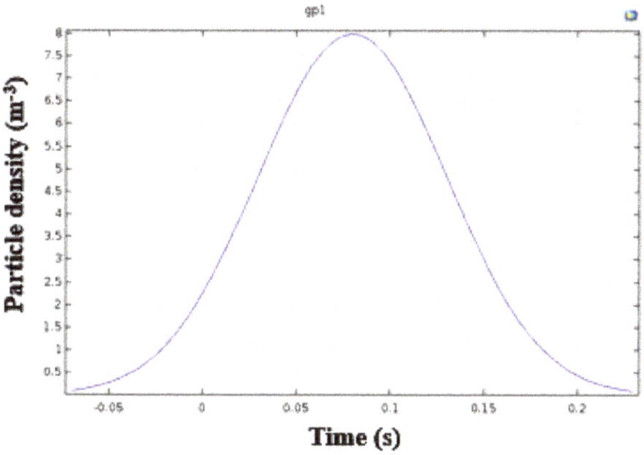

Figure 3. Gaussian impulse function.

2.4. Boundary and Initial Conditions

The boundary conditions adopted for the numerical simulation of the geometric structure shown in Figure 2a are summarized in Table 2. Because circular symmetry was adopted, the variation of three charged particles along the radial direction was zero. Considering the adsorption and recombination of negative ions by the positive electrode, the concentration of negative ions was set to zero, and the positive ion concentration was similarly set to zero on the surface of the negative electrode. On the top electrode (positive electrode), there were accumulative positive ions from ion reactions, and there were electrons and negative ions on the bottom electrode. The ignition line was loaded with the transient Gaussian pulse source according to Equation (9). These are described by the flux/source boundary condition. A zero-flux boundary was established to define the rest of the outer boundary of the calculation domain for the ion transport process [23]. The PARDISO transient solver embedded in COMSOL was utilized, and the time step of the solver was controlled by the BDF (Backward Differentiation Formulas) algorithm. The simulation time started the moment the short-circuit power was connected to 0 and ended at 0.2 s.

Table 2. Boundary conditions.

Application Location	Convection and Diffusion N_e	Convection and Diffusion N_p	Convection and Diffusion N_n
Axis of symmetry	$\frac{\partial N_e}{\partial r} = 0$	$\frac{\partial N_p}{\partial r} = 0$	$\frac{\partial N_n}{\partial r} = 0$
Top electrode	$N_e = 0$	$-\boldsymbol{n} \cdot (D_p \nabla N_p) = f_+$	$N_n = 0$
Bottom electrode	$-\boldsymbol{n} \cdot (D_e \nabla N_e) = f_-$	$N_p = 0$	$-\boldsymbol{n} \cdot (D_n \nabla N_n) = f'_-$
Ignition line	$-\boldsymbol{n} \cdot (D_e \nabla N_e) = g_{e2}$	$-\boldsymbol{n} \cdot (D_p \nabla N_p) = g_{p2}$	$-\boldsymbol{n} \cdot (D_n \nabla N_n) = g_{n2}$
Rest of the boundary	$-\boldsymbol{n} \cdot (D_e \nabla N_e) = 0$	$-\boldsymbol{n} \cdot (D_p \nabla N_p) = 0$	$-\boldsymbol{n} \cdot (D_n \nabla N_n) = 0$

Considering the sustainability of the ionic reaction, in its initial stage, three ion initial concentrations were set to $1 \times 10^{13}/m^3$, and 600 kV was loaded on the top electrode.

2.5. Experimental Platform

An equivalent single-phase experimental circuit was established according to the distributed parameter model for transmission lines, as shown in Figure 4a. Here, the inductance *L* establishes an inductive short-circuit starting current. In the experiment, this was 0.03688 H and the short-circuit current was 1 kA. The capacitance *C* represents the equivalent coupling capacitance between the faulty phase and healthy phases. This was 2.74 μF in this study. Different secondary arc currents were achieved by changing the value of the group capacitance *C* in the experiment. A voltage divider, current transformer, and oscilloscope were used to measure current and voltage waveforms in real time, and two high-speed cameras were used to record the entire discharge process at 4000 fps.

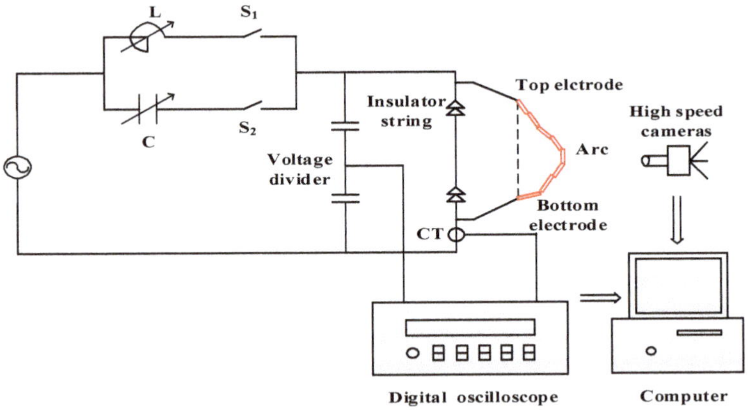

(**a**)

Figure 4. *Cont.*

(b)

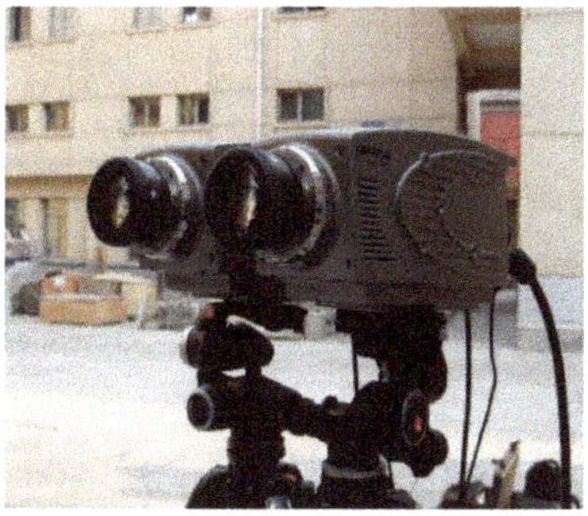

(c)

Figure 4. Experimental platform for secondary arc reproduction: (**a**) Experimental circuit; (**b**) experiment field; (**c**) high-speed cameras.

Firstly, the circuit breaker S1 was closed to simulate an inductive arc starting current. Under the action of large current, the ignition line gasified to form an arc channel. After 0.1 s, the circuit breaker S2 closed and S1 quickly opened to simulate the secondary arc. The secondary arc experiment was completed in the high-current test station of China Electric Power Research Institute, shown in Figure 4b,c.

3. Results and Discussion

3.1. Experiment Verification

Figure 5 shows images of the short-circuit arc. Under the action of large short-circuit current, the ignition line rapidly fused and strongly ionized the air around the insulator, forming a bright arc plasma channel. Over time, the arc channel continued to spread, even after the breaker cut off the power (0.1 s) due to the powerful thermal effect. It was not until 0.15 s that the attenuation phenomena such as arc passage narrowing and brightness weakening appeared obviously. Due to the strong ionization and thermal effect of the short-circuit current, there was no zero-crossing stage, which is typical in an ac arc. Due to the huge current value of the short-circuit arc and the strong ionization of the surrounding air, the movement of the arc passage was mainly radial diffusion. The force of each part of the arc passage was mainly internal electromagnetic force, and external force had little influence. Thus, the arc was warped internally with no significant upward or left–right drift.

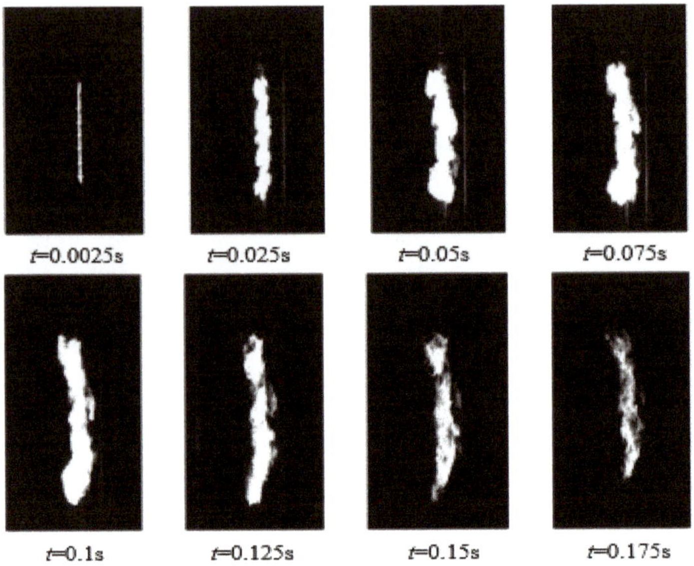

Figure 5. Short-circuit arc images at different times.

Figure 6 shows electron density distribution at different times during the short-circuit arc. At 0.0025 s, the short-circuit current started to melt the ignition line, exhibiting a luminous effect. During 0.025–0.15 s, the electrons produced by strong ionization concentrated near the ignition line, and gradually spread to the surrounding space under the action of the electric field migration and the convection diffusion of particles. After 0.15 s, the short-circuit arc decayed and the ionization region reduced. During this phase, the electron density distribution tended to return to the initial level.

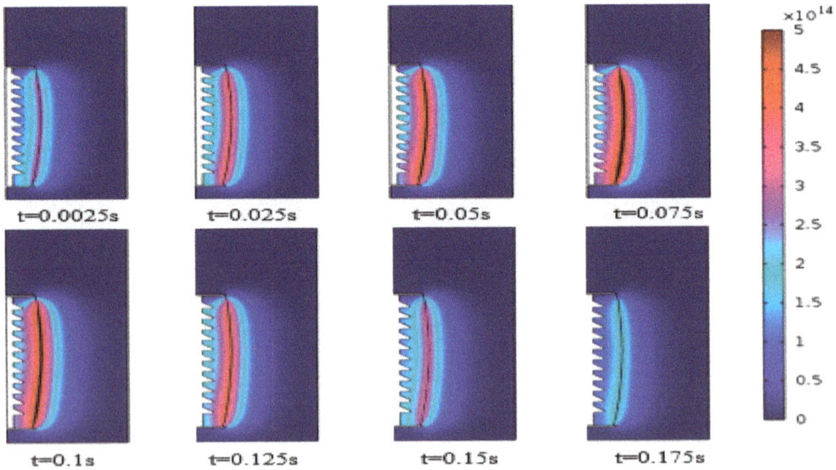

Figure 6. Electron density distribution at different times.

Luminescence is caused by the emission of photons when some equations in the plasma ionization reaction system transition from high to low energy levels. It is assumed that under the same environment, other ionization equations have the same reaction rate as the ionization reaction that emits photons. In this way, it can be considered to evaluate the ionization degree by observing the luminescence intensity with cameras, and comparing the camera images with the simulated images to verify the consistency between the experiment and simulation. Comparing Figure 6 with Figure 5, it can be seen that the brightness distribution acquired with high-speed cameras of the experimental short-circuit arc was basically consistent with the predicted distribution of electron density, demonstrating that the simulation was effective and supporting the subsequent analysis of the plasma interior.

3.2. Particle Density Distribution and Development Law

Figure 7 plots the electron concentration distribution along the ignition line at six different times. The abscissa represents the arc length between the point on the ignition line and the 0 point on the negative electrode surface. The electron density slowly increased from the cathode region to the intermediate plasma region and remained constant until close to the anode region. After that, it sharply decreased and fixed to zero on the anode surface. Notice that the electron density did not increase significantly from the cathode arc root to the intermediate plasma region, which is markedly different from general streamer discharge. This was due to the strong joule heating effect of the short-circuit arc with large current, which caused strong ionization of the surrounding air, so there was little breakdown caused by electron collision ionization. Near the anode surface, the rapid drop was caused by the absorption of electrons by the anode. During the short-circuit arc phase, the peak electron concentration reached 5.72×10^{14} m^{-3}, and the electron concentration was 1.3×10^{14} m^{-3} above what it was at the end of the simulation, which proves that the short-circuit discharge increased the concentration of space charge and provided necessary environmental conditions for the generation of a subsequent secondary arc.

Figure 8 shows the law of spatial negative ion density changing with time from the initial moment to 0.2 s on the surface of the negative pole (point 53) and the positive pole (point 58) as well as the middle point of the ignition line (point 60), which quantitatively reflects the gradual change of the concentration of transient particles in the short-circuit discharge process. It is not difficult to see that over time, the concentration of negative ions rose and then levelled off. When the ion reaction approached the end of simulation time, the ion concentration was higher than the initial level, which proves that the short-circuit discharge increased the spatial ion concentration and provided necessary environmental conditions for the subsequent secondary arc. Due to the difference of diffusion, convection, and adsorption coefficients between positive ions and negative ions, the changing curves of concentrations of positive ions and negative ions had slight differences despite showing the same trend.

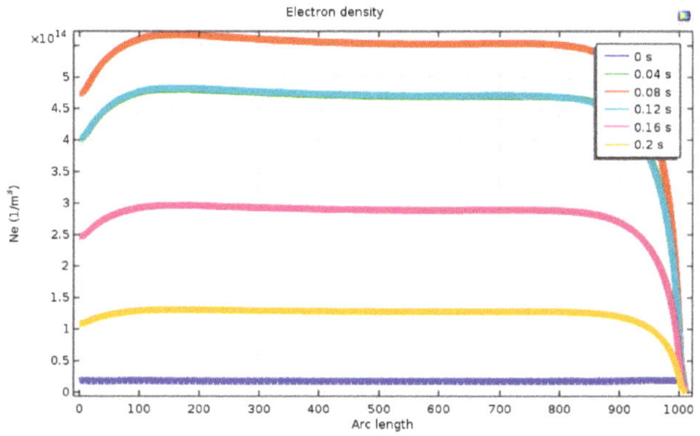

Figure 7. Electron density distribution along the ignition line at key time nodes during the discharge.

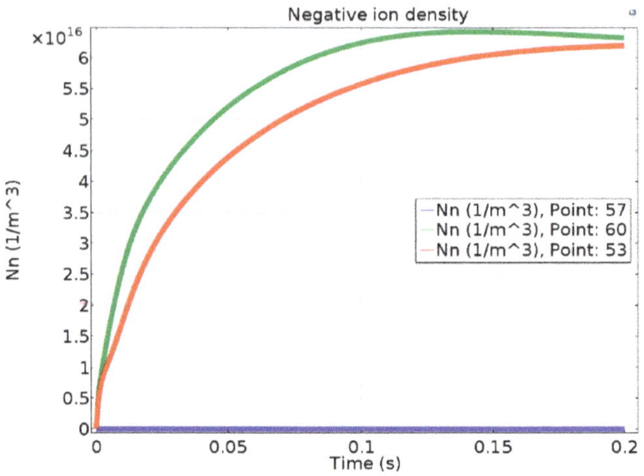

Figure 8. The negative ion density at the positive electrode (point 58), the negative electrode surface (point 53), and the central point (point 60) during the discharge.

3.3. Spatial Distribution of the Electric Field During Discharge

Figure 9 depicts the spatial electric field distribution in the initial stage of discharge, which was mainly caused by the high voltage applied by the electrode. In Figure 9, the intensity of the electric field is given by the cloud diagram, and the direction of the electric field is described by the red arrow. The electric field line started at the positive pole, crossed the whole space, and ended at the negative pole. Moreover, a large field intensity was generated at the maximum geometric curvature radius. This field intensity caused the point discharge.

Because the electric field generated by the ions was considerably different from the electric field generated by the electrodes, the field intensity effects at different times cannot be intuited from the cloud map, and are better represented by a one-dimensional graph. Figure 10 plots the electric field intensity as a function of the time in the middle region of the discharge space (r = 120 mm; Z = 680 mm, as an example). With increased discharge time, the electric field intensity at this spatial point showed an S-shaped upward trend. The essential reason for this rising trend is that a large number of ions are generated in the discharge process, and the electric field intensity generated by ions follows Gaussian electric field distribution, as described in Equations (7) and (8). Because the evaluated point was close to the short-circuit discharge area, the electric field tended to increase. When the discharge entered the later stage, the ions generated migrated under the action of the electric field and spread to other regions, causing the rate of the electric field intensity increase to attenuate gradually. The electric field intensity reaches the peak at about 0.2 s and then slowly decays. It takes 1 s usually for the extinction of the secondary arc, and if the simulation time is sufficiently long, the final electric field intensity could be predicted to return to its initial level [24].

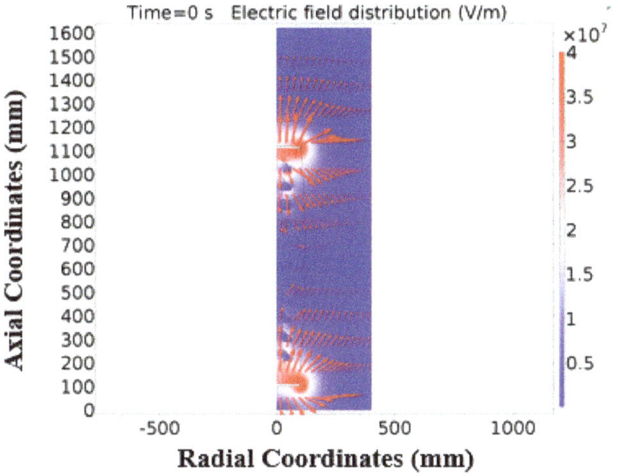

Figure 9. Spatial electric field distribution at the initial moment.

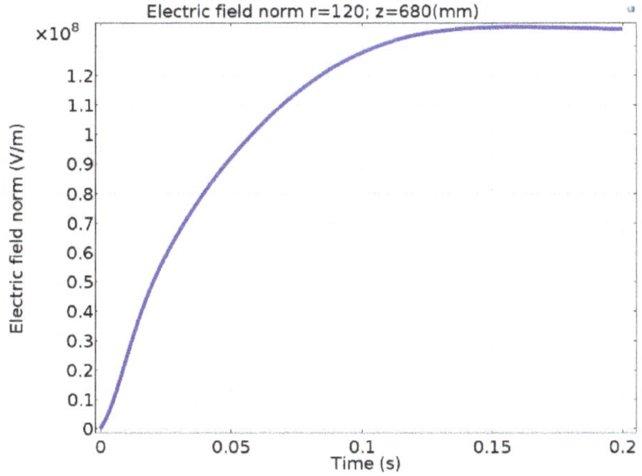

Figure 10. Electric field intensity over time at the midpoint of the discharge area.

To study the development of the electric field intensity over time, the transversal at $r = 100$ between the positive and negative electrodes was selected as an additional evaluation object. The development of the axial electric field intensity during the initial, peak, final, and intermediate stages of the discharge process was analyzed. Figure 11 shows that the electric field generated by the high voltage of the electrode was the strongest near the electrode and the lowest in the middle of the discharge region. Evidently, the contribution of ions to the spatial electric field was smaller than that of the high-voltage electrode; nonetheless, the former cannot be ignored. The contribution of ions to the spatial electric field was the largest in the intermediate discharge region, and the effect was small near the electrode.

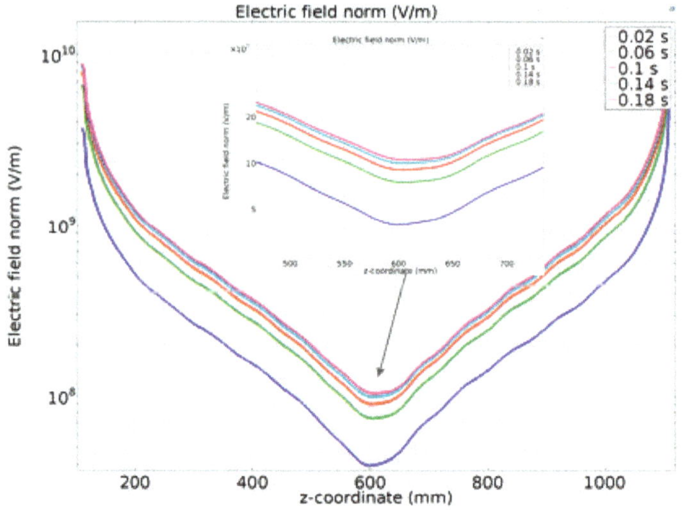

Figure 11. Development of axial electric field intensity over time at r = 100 mm.

3.4. *Particle Reactions in Discharge Process*

Figure 12 shows the time-dependent calculation results of the average detachment reaction rate in the discharge area. The time correspondences of the detachment reaction and ion source generated during the short-circuit discharge (Figure 3) is basically consistent: the detachment reaction rate increased sharply during the initial stage and peaked at the same time as the short-circuit discharge ion sources simulated by the Gaussian pulse. According to the formula $f_{\text{det}} = k_{\text{det}} N_{\text{e}} N_{\text{n}}$, the detachment of electrons from negative ions is influenced by both the concentration of electrons and the concentration of negative ions. This result occurs because numerous negative ions and electrons are rapidly generated in the short-circuit discharge. After the completion of the short-circuit discharge, the detachment reaction speed decreases gradually, unlike the rapid decay of ion sources concentration. That is because the recombination reaction process is influenced by the slow diffusion and migration of ions in space.

Considering the assumption of electrical neutrality during the initial stage of the discharge process, the detachment reaction rate in the initial stage was relatively uniform. Therefore, this section focuses on the spatial distribution of the detachment rate in the peak stages of the short-circuit discharge. On the surface plot Figure 13, one can observe the detachment reaction rate at the discharge peak was the highest near the ignition line and the lowest near the positive electrode. This is confirmed in Figure 14, which shows the detachment rate along the transversal at r = 100 between the positive and negative electrodes. As the bodies involved in the detachment reaction were negative ions and electrons, which were absorbed and neutralized near the positive electrode, the detachment reaction speed near the positive electrode was reduced. During the later stage of discharge, the negative ions and electrons generated by the short circuit tended to be uniform under the action of the electric field migration and diffusion. Therefore, the distribution of the detachment reaction rate gradually returned to the initial state.

Furthermore, the average recombination reaction (both electron-positive ions and positive-negative ions) rate was studied during the discharge. The results demonstrate that they are consistent with the trend of the detachment reaction rate, with differences only in magnitude. The influencing factors and mechanisms of each stage are also consistent with the detachment reaction, with differences caused only by the respective reaction coefficients. These reaction coefficients are affected by the ion collision cross sections and are selected for normal temperature and pressure conditions.

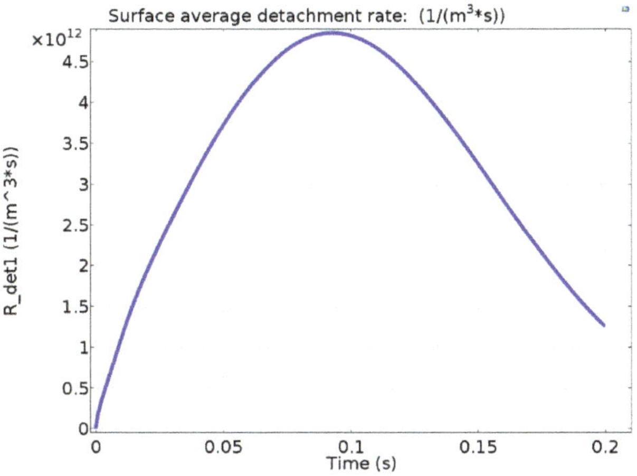

Figure 12. Time dependence of the average detachment reaction in the discharge space.

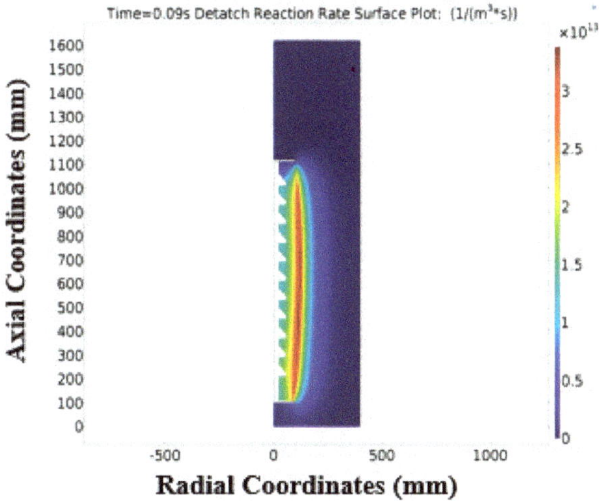

Figure 13. Detachment reaction rate surface plot at the discharge peak.

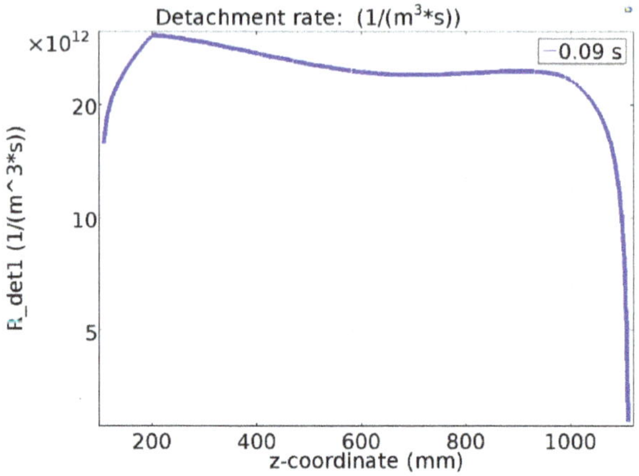

Figure 14. Detachment reaction rate at $r = 100$ mm at the peak stage of discharge.

4. Conclusions

In this study, COMSOL multi-physical field coupling analysis software was used to simulate the short-circuit discharge process during the initial stages of a secondary arc fault based on the relevant partial differential equations. The model particularly focused on the complex plasma created by the short-circuit discharge (initial stage of the secondary arc), which has been little researched in relevant studies. This could increase our fundamental understanding of the secondary arc and long gap alternating-current arc in the free air from a microscopic discharge mechanism. It could help to develop the secondary arc suppression technique and ensure the timely extinction of the secondary arc, and thus contribute to optimizing single-phase auto-reclosing (SPAR). This is of great significance to the safe operation of power transmission lines and enhances the stability of the power system. An experiment platform was established to verify the numerical model. Then the particle density distribution, electric field variation, and time dependence of ion reactions during the discharge were analyzed. The main conclusions are as follows:

(1) The brightness distribution obtained by high-speed cameras of the experimental short-circuit arc was basically consistent with the predicted distribution of electron density, demonstrating that the simulation was effective and supported the subsequent analysis of the plasma interior.
(2) With the short-circuit discharge, the electron density along the ignition line first increased and then decreased, and its distribution was quite different from the general streamer discharge. Over time, the concentration of negative ions rose and then levelled off, and due to the differences of diffusion, convection, and adsorption coefficients between positive ions and negative ions, the changing curves of concentrations of positive ions and negative ions had slight differences despite showing the same trend. Near the end of the simulation time, there was a considerably larger number of charged particles than the initial level, which provided the necessary environmental conditions for subsequent secondary arc generation.
(3) The initial stage of discharge was mainly point discharge in space. The spatial electric field intensity showed an S-shaped upward trend in the discharge process. The end regions were Ssignificantly affected by the high-voltage electrode, whereas the middle area was mainly affected by the particle reaction.
(4) The time correspondence between the detachment reaction and the ion source generated in the short-circuit discharge process was basically consistent, and the detachment reactions were mainly concentrated in the middle area and near the negative electrode. The average recombination reaction rates were consistent with the trend of the detachment reaction rate during the discharge, with differences only in magnitude.

Author Contributions: Conceived and designed the numerical model, J.L.; performed the experiments and analyzed the data, J.L., H.Y. and M.J.; wrote the main manuscript text, J.L.; gave important advice on the model, H.L. and G.L.; all authors read and approved the final manuscript.

Funding: This research received no external funding.

Conflicts of Interest: The authors declare no conflict of interest.

References

1. Esztergalyos, J.; Andrichak, J.; Colwell, D.H.; Dawson, D.C.; Jodice, J.A.; Murray, T.J.; Nail, G.R.; Politis, A.; Pope, J.W.; Rockefeller, G.D.; et al. Single-phase tripping and auto reclosing of transmission lines. *IEEE Trans. Power Deliv.* **1992**, *7*, 182–192.
2. Prikler, L.; Kizilcay, M.; Bán, G.; Handl, P. Modeling secondary arc based on identification of arc parameters from staged fault test records. *Int. J. Electr. Power* **2003**, *25*, 581–589. [CrossRef]
3. Cong, H.; Du, S.; Shu, X.; Li, Q. Research portfolio and prospect on physical characteristics and modelling methods of secondary arcs related to high-voltage power transmission lines. *Int. J. Appl. Electromagn.* **2018**, *56*, 195–209. [CrossRef]
4. Tsuboi, T.; Takami, J.; Okabe, S.; Aoki, K.; Yamagata, Y. Study on a field data of secondary arc extinction time for large-sized transmission lines. *IEEE Trans. Dielectr. Electr. Insul.* **2013**, *20*, 2277–2286. [CrossRef]
5. Tavares, M.C.; Talaisys, J.; Camara, A. Voltage harmonic content of long artificially generated electrical arc in out-door experiment at 500 kV towers. *IEEE Trans. Dielectr. Electr. Insul.* **2014**, *29*, 1005–1014. [CrossRef]
6. Li, Q.; Cong, H.; Sun, Q.; Xing, J.; Chen, Q. Characteristics of secondary ac arc column motion near power transmission line insulator atring. *IEEE Trans. Power Deliv.* **2014**, *29*, 2324–2331. [CrossRef]
7. Gu, S.; He, J.; Zeng, R.; Zhang, B.; Xu, G. Motion characteristics of long ac arcs in atmospheric air. *Appl. Phys. Lett.* **2007**, *90*, 051501. [CrossRef]
8. Terzija, V.V.; Wehrmann, S. Long arc in still air: Testing, modelling and simulation. *EEUG News* **2001**, *7*, 44–54.
9. Johns, A.T.; Aggarwal, R.K.; Song, Y.H. Improved techniques for modeling fault arcs on faulted EHV tranmission system. *IET Proc. Gener. Transmis. Distrib.* **1994**, *141*, 148–154. [CrossRef]
10. Kizilcay, M. Evaluation of existing secondary arc models. *EEUG News* **1997**, *3*, 49–60.
11. Gu, S.; He, J.; Zhang, B.; Xu, G.; Han, S. Movement simulation of long electric arc along the surface of insulator string in free air. *IEEE Trans. Magn.* **2006**, *42*, 1359–1362.
12. Horinouchi, K.; Nakayama, Y.; Hidaka, M.; Yonezawa, T.; Sasao, H. A method of simulating magnetically driven arcs. *IEEE Trans. Power Deliv.* **1997**, *12*, 213–218. [CrossRef]
13. Cong, H.; Li, Q.; Xing, J.; Siew, W.H. Modeling study of the secondary arc with stochastic initial positions caused by the primary arc. *IEEE Trans. Plasma Sci.* **2015**, *43*, 2046–2053. [CrossRef]
14. Sima, W.; Tan, W.; Yang, Q.; Luo, B.; Li, L. Long AC Arc movement model for parallel gap lightning protection device with consideration of thermal buoyancy and magnetic force. *Proc. CSEE* **2011**, *31*, 138–145.
15. Birdsall, C.K.; Langdon, A.B. *Plasma Physics via Computer Simulation*; CRC Press: Boca Raton, FL, USA, 1991.
16. Shi, J.; Kong, M. Cathode fall characteristics in a DC atmospheric pressure glow discharge. *J. Appl. Phys.* **2003**, *94*, 5504–5513. [CrossRef]
17. Bogaerts, A.; Gijbels, R. Numerical modeling of gas discharge plasmas for various applications. *Vacuum* **2003**, *69*, 37–52. [CrossRef]
18. Davies, A.J.; Davies, C.S.; Evans, C.J. Computer simulation of rapidly developing gaseous discharges. *Proc. IEE Sci. Meas. Technol.* **1971**, *118*, 816–823. [CrossRef]
19. Singh, S.; Serdyuk, Y.; Summer, R. Adaptive numerical simulation of streamer propagation in atmospheric air. In Proceedings of the COMSOL Conference, Rotterdam, The Netherlands, 23–25 October 2013.
20. Singh, S. *Computational Framework for Studying Charge Transport in High-Voltage Gas-Insulated Systems*; Chalmers University of Technology: Gothenburg, Sweden, 2017; pp. 14–17.
21. Georghiou, G.E.; Papadakis, A.P.; Morrow, R.; Metaxas, A.C. Numerical modelling of atmospheric pressure gas discharges leading to plasma production. *J. Phys. D Appl. Phys.* **2005**, *38*, R303–R328. [CrossRef]
22. Cong, H.; Li, Q.; Du, S.; Lu, Y.; Li, J. Space Plasma Distribution Effect of Short-Circuit Arc on Generation of Secondary Arc. *Energies* **2018**, *11*, 828. [CrossRef]

23. Tran, T.N.; Golosnoy, I.O.; Lewin, P.L.; Georghiou, G.E. Numerical modelling of negative discharges in air with experimental validation. *J. Phys. D Appl. Phys.* **2011**, 015203. [CrossRef]
24. Cong, H.; Li, Q.; Xing, J.; Li, J.; Chen, Q. Critical length criterion and the arc chain model for calculating the arcing time of the secondary arc related to AC transmission lines. *Plasma Sci. Technol.* **2015**, *17*, 475. [CrossRef]

energies

MDPI

Article

Calculation of Ion Flow Field of Monopolar Transmission Line in Corona Cage Including the Effect of Wind

Zhenyu Li and Xuezeng Zhao *

Department of Mechatronics Control and Automation, School of Mechatronics Engineering, Harbin Institute of Technology, Harbin 150001, China; constantlzy@163.com

* Correspondence: zhaoxz@hit.edu.cn; Tel.: +86-152-4508-7779

Received: 11 September 2019; Accepted: 14 October 2019; Published: 16 October 2019

Abstract: In this work, the ion flow field of a monopolar transmission line inside the corona cage of a square cross-section is iteratively calculated concerning the effects of wind. The electric field distribution is solved analytically using the charge simulation method (CSM). Meanwhile, the upwind finite volume method (UFVM) with 2nd order accuracy is presented for the distribution of space charge density. Additionally, a dual mesh grid is established in the calculation domain, the interlaced geometric construction of the mesh assures a quick and effective convergence rate. In the final part, a reduced-scaled experiment is designed to examine the feasibility and accuracy of this approach, electric field and ion current density on the bottom side are measured by field mills and Wilson plates. The data numerically computed fits well with that acquired by measurement.

Keywords: corona discharge; electric field analysis; ion flow field; space charge density; UFVM

1. Introduction

In operating HVDC transmission lines of a power system, the phenomenon of corona discharge is a leading cause of radiation interference (RI), noise interference (NI), and corona loss (CL) [1]. Thus, investigation on the ion flow field distributed around the conductors receives considerable attention in the design of HVDC transmission lines.

Commonly, one of the main obstacles in solving the ion flow field is the nonlinearity between the electric field and space charge density. The vast majority of solutions calculate the electric field and the space charge density iteratively and the iteration process ends once the criteria are met. In the meantime, wind flow affects the distribution of the electric field around transmission lines to certain degree as well. All the above-mentioned issues increase the difficulty of calculating the ion flow field.

In the past few decades, research on the ion flow field calculation has varied in terms of the methods utilized to calculate the electric field and the space charge density in the domain of interest.

With regard to the electric field, Janischewskyj and Gela [2] introduced finite element method (FEM) to solve the electric field numerically; afterwards, this method was frequently adopted and well-developed. In 1983, Takuma et al. [3] applied CSM to calculate the nominal electric field without space charge, while FEM was used to solve the electric field induced by space charges. Since then, this approach has been broadly applied in electric field calculation [4,5]. CSM offers satisfactory accuracy, whereas the calculation domain is restricted to an infinite field above the ground or axisymmetric structure. Simultaneously, a drawback of FEM is that the accuracy of the electric field close to the conductor surface is not as expected because of the steep gradient.

In regard to the calculation of space charge density, method of characteristics (MOC) is diffusely utilized [6–10], space charge density is calculated along electric field lines with given initial charge density on the conductor surface. This approach relies on the Deustch's assumption, which assumes the space charge affects the amplitude of an electric field rather than its direction. Xiao calculated the ion flow field around a cross-over transmission line in the 3D domain with the MOC method. However, the effectiveness of this method is unsatisfactory if the influence of wind flow is under consideration. Lu et al. [11] proposed an upwind FEM, which avoids non-physical instability of the numerical calculation. Zhou et al. [12] induced upwind weighting function to FEM for the purpose of eliminating the oscillations in simulation of charge conservation. Levin [13] established dual mesh based on the triangulation grid in calculation domain; the new mesh is called donor cell and the space charge density is hereby solved in accordance with Gauss' Law. Then, upwind the FVM method were used in several research projects [5,14–17] in which the numerical stability, effectiveness, and accuracy of the solution process was improved substantially. Yang et al. [18] proposed an upstream meshless method to solve the current continuity equation.

For the purpose of implementing an indoor experiment and control of environmental parameters, a corona cage is designed in where the phenomenon of corona discharge initializes on a relatively lower voltage level. Bian et al. [19] and Lekganyane et al. [20] investigated the ion flow field in a square cross-section cage and compared the result with that of an indoor test line; Zhou et al. [8] presented a comprehensive study on the ion flow field distribution in a cylindrical cage employing a mesh-based method and MOC. However, there is paucity of published research concerning the effect of wind flow on the corona discharge of corona cage.

However, solution of the ion flow field generally concerns the numerical stability, calculation accuracy and the impact of wind flow. The referenced articles barely meet these requirements at the same time. Therefore, it is necessary to develop a method that offers a quick, stable, and accurate solution of ion flow field.

In this paper, the calculation domain is tessellated in the form of a dual mesh. Next, CSM is utilized for a nominal electric field in the absence of space charges. Simultaneously, electric fields generated by space charges is available if the space charges density is known, by this means, the accuracy of the calculated field is guaranteed even on the conductor surface. The 2nd order upwind FVM is employed to calculate the space charge density distribution. Eventually, the calculated result is validated with that derived by experiments.

The importance and originality of this study consists of the nominal electric field in a square cross-section being solved by means of proper placement of the simulation charges, the more accurate solution of space charge density distribution involving the impact from wind flow, as well as the applicability of this approach in presence of wind flow. The calculation process provides rapid convergence rate as the analytically calculated electric field is less time-consuming compared to the traditional method using FEM.

2. Mathematical Description

2.1. Governing Equations and Simplifying Assumption

Generally, the ion flow field in the ambient of conductor is governed by Poisson's equation and the current density conservation equation [3]:

$$\begin{cases} \nabla \boldsymbol{E} = -\rho^- / \varepsilon_0 \\ \nabla \cdot \boldsymbol{J}^- = 0 \\ \boldsymbol{J}^- = \rho^- (b\boldsymbol{E} + \boldsymbol{W}) \end{cases} \tag{1}$$

where,

b is the ion mobility, 1.5×10^{-4}, m^2/V/s;
$\boldsymbol{E}$ is the electric field, V/m;
ρ is the negative space charge density, C/m^3;
$\boldsymbol{J}^-$ is the negative ion current density vector, A/m^2;

W is the wind velocity vector, m/s; and
ε_0 is the permittivity of air equals 8.854×10^{-12}, F/m.

Further, certain assumptions are proposed in advance in order to reduce the complexity of calculation and acquire satisfactory precision:

(a) The thin ionization layer close to the conductor surface is neglected;
(b) The ion mobility remains unchanged throughout the solution process;
(c) Influence exerted by ion diffusion is ignored;
(d) Kaptzov's assumption [21] which presumes the electric field on conductor surface remains constant after the applied voltage reaches the onset value is adopted.

2.2. Boundary Conditions

Before proceeding to the solution process. Boundary conditions and initial conditions of the calculation domain are listed in Table 1.

Table 1. Boundary conditions and initial conditions.

Distribution Variables	Conductor Surface	Cage Wall
Electric potential	V_{app}	0
Space-charge density	ρ_s	$\frac{\partial \rho}{\partial n}$
Electric field	E_{on}	$\frac{\partial E}{\partial n}$

where,

V_{app} is the voltage supplied on the conductor, V;
ρ_s refers to the space charge density on the conductor surface, C/m^3;
E_{on} is the onset electric field, V/m; and
E_{on} is assumed to be constant on the conductor surface according to Kaptzov's assumption, the explicit value is attained using Peek's empirical formula [22]:

$$E_{\text{on}} = 30m(1 + \sqrt{\frac{0.0906}{r}}) \tag{2}$$

where,

m is the roughness factor set to 0.65; and
r is the radius of the conductor, m.

An appropriate guess of the initial value of the charge density on conductor surface determines the accuracy of the result and diminishes the iteration process. The empirical formula introduced in [23] is referred to in this paper:

$$\rho_s = \frac{E_g}{E_c} \frac{8\varepsilon_0 V_c (V_{app} - V_c)}{r H_{con} (5 - 4V_c / V_{app})} \tag{3}$$

where,

E_g is the ground level electric field under the conductor, V/m;
E_c is the nominal electric field on the conductor, V/m;
V_c is the onset voltage of the conductor, V;
V_{con} is the conductor voltage, V; and

H_{con} is the height of the conductor, m.

Takuma [3] assumes that the initial charge density is evenly distributed on the conductor surface, yet it is not suitable for the situation where the transmission line is placed inside corona with a square cross-section cage. Because distances between the conductor surface to the cage wall are diverse, which differs from the situation above the ground or in the coaxial cage. Hence, in this work, charge density is set to be linearly dependent on E_c. To be specific, for each node on the conductor surface, E_c is calculated by CSM, neglecting space charges. Thus, the initial charge density of this node is achieved by substitute corresponding E_c into Equation (3).

3. Solution Process

The distributions of electric field and space charge density are solved iteratively, in this process, the initial charge density on the conductor surface is updated in each iteration in case that the condition of convergence is not satisfied. The detailed procedure is organized and illustrated in Figure 1.

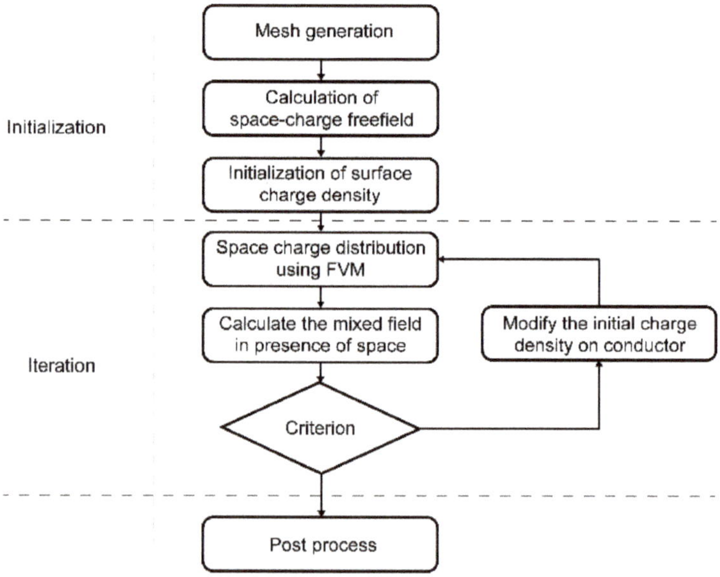

Figure 1. Flowchart of the method.

3.1. Discretization of Calculation Domain

The calculation is conducted in the 2D cross-section of the cage, and the calculation domain is divided in form of dual mesh. Specifically, a Delaunay Triangular mesh is generated in the first place; after that, polygon cells are constructed via connecting the barycenter and the midpoints of triangular cell edges, which share common vertexes [24].

The meshing of the calculation domain is demonstrated in Figure 2. Meshing of the area in the vicinity of the conductor surface and the cage wall are finer in the cause of the need of more accurate calculation results. As a result of the grid independency test, the difference of the calculated result is less than 1%, while the domain is tessellated into 1879 cells.

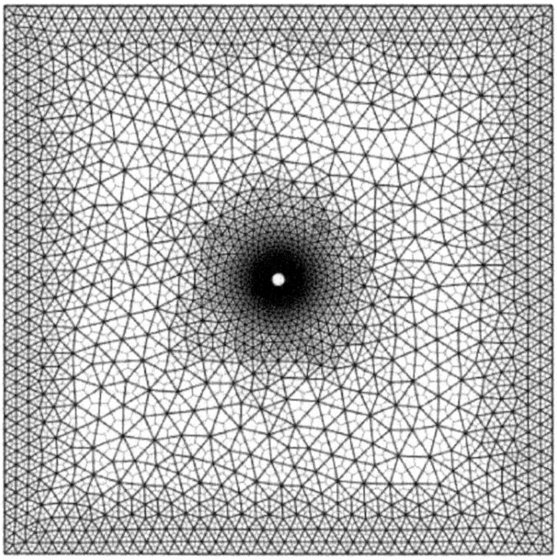

Figure 2. Tessellation of calculation domain.

Since the electric field is calculated analytically, the electric field in each single point can be calculated defectively. Regarding to the space charge density, they are stored in the nodes of the triangulation. The effectiveness of the solution process is ensured resulting from the interlaced meshing scheme.

3.2. Calculation of Electric Field

The calculation of electric field composes of two parts: the nominal field and the field induced by space charges.

Regarding the nominal field to be calculated by CSM, there are 40 simulation charges equally distributed inside the conductor.

In contrast to the circumstance of outdoor lines above the ground, extra simulation charges are placed outside the cage wall to maintain the zero potential of the grounded cage wall; as a result, as demonstrated in Figure 3, there are 160 simulation charges equidistantly arranged with interval d_c, the perpendicular distance between the cage wall and simulation charges are two times of d_c. Image charges are placed symmetrically in the opposite sides of the ground. The match points are placed right on the conductors and corona walls, respectively.

According to the principle of CSM, values of the simulation charges must ensure the potential on the conductor surface and cage wall to be V_{app} and zero. Thus, coefficient equations are listed below:

$$\begin{cases} P_{cond} \cdot Q_{cond} + P_{cage} \cdot Q_{cage} = V_{app} \\ P'_{cond} \cdot Q_{cond} + P'_{cage} \cdot Q_{cage} = 0 \end{cases} \tag{4}$$

where Q_{cond} and Q_{cage} are simulation charges of the conductor and cage, P_{cond}, P_{cage} are the potential coefficients regarding the conductor surface, and P'_{cond}, P'_{cage} are the potential coefficients for the cage wall.

The nominal electric field is therefore obtained by superposing the field caused by simulation charges. The space charges-induced field can be calculated if the space charges are known. Accordingly, the ion flow field is available as follows:

$$E_m = \sum_i \frac{Q_i}{4\pi\varepsilon_0}\left(\frac{\vec{r_i}}{r_i^2} - \frac{\vec{r'_i}}{r_i'^2}\right) + \sum_j \int_{s_j} \frac{\rho_j}{4\pi\varepsilon_0}\left(\frac{\vec{r_j}}{r_j^2} - \frac{\vec{r'_j}}{r_j'^2}\right) ds_j \tag{5}$$

where,

s_j is the area of the polygon cell, m^3;
r_i and r_j are the distances between the observation point to the source, m;
r'_i and r'_j are the distances between the observation point to the image points, m;
Q_i are the simulation charges consist of Q_{cond} and Q_{cage}, C; and
ρ_j is the charge density on triangulation nodes, C/m^3.

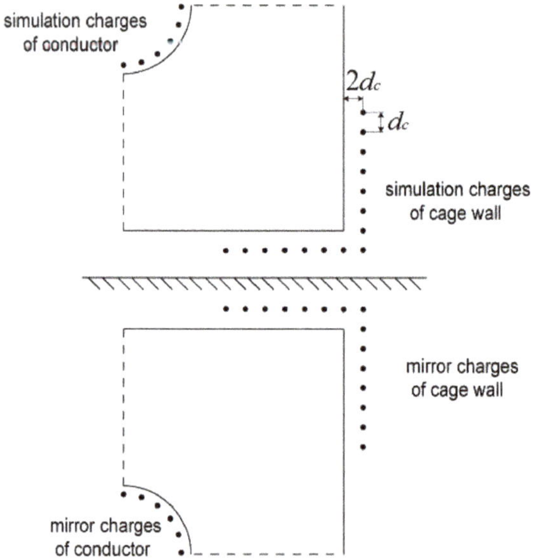

Figure 3. CSM method.

3.3. Calculation of Ion Current Density

FVM ensures the maximum principle of charge density on cell boundaries resulting in numerical stability in the calculation process. Additionally, the upwind scheme satisfies the physical fact that the migration of space charges is affected by upwind stream only [25]. Particularly, this method is feasible for calculation in the presence of wind flow.

By substituting Equation (3) into Equation (2), following equation is obtained:

$$\nabla \cdot \left[\rho^{-}(b\mathbf{E} + \mathbf{W})\right] = 0 \tag{6}$$

Afterwards, Equation (6) is converted to integral form:

$$\iint_{s} \rho(b\mathbf{E} + \mathbf{W})dl = 0 \tag{7}$$

where,

s and l are the area and boundary of the cell.

Next, Equation (7) is rewritten in the form of linear equations in the light of Figure 4:

$$\sum\nolimits_{n=1}^{n} \rho \mathbf{V}_n L_n = 0 \tag{8}$$

where,

$\boldsymbol{V}_n = b\boldsymbol{E}_n + \boldsymbol{W}_n$, $\boldsymbol{E}_n$ and $\boldsymbol{W}_n$ are the outward normal component of electric field and wind speed on cell edges, and
L_n is the length of the *i*th cell; and
n presents the serial number of the edges.

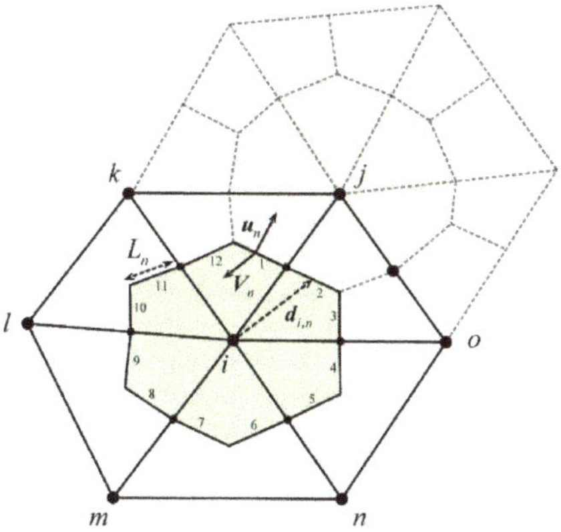

Figure 4. Control volume of UFVM.

Finally, a system of linear equations is established and the distribution of charge density is therefore achieved.

Normally, charge density is defined as the average value of adjacent nodes [15], or determined by the upwind scheme. Nevertheless, these methods are only of 1st order accuracy which is insufficient for engineering requirements.

With an aim to promote the accuracy of space charge density, a 2nd order upwind scheme is utilized. Charge densities on the edges are solved directly. Apparently, the charge densities on cell edges can be expressed according to the Taylor series expansion as Figure 4:

$$\begin{cases} \rho_{i,n} = \rho_i + \nabla\rho_{i,n} \cdot \boldsymbol{d}_{i,n} & if \;\; \boldsymbol{V}_n \cdot \boldsymbol{u}_n > 0 \\ \rho_{i,n} = \rho_n + \nabla\rho_{n,i} \cdot \boldsymbol{d}_{n,i} & if \;\; \boldsymbol{V}_n \cdot \boldsymbol{u}_n < 0 \end{cases} \tag{9}$$

where,

$\rho_{i,n}$ is the charge density on the edge, C/m^3;
$\boldsymbol{d}_{i,n}$ is the vector direct from the *i*th node to the corresponding neighboring nodes; and
$\nabla\rho_{i,n}$ and $\nabla\rho_{n,i}$ are the gradient of corresponding upwind node.

For the known charge densities on the *i*th node and its adjacent nodes, the following over determined matrix equation can be established; thus, the gradient of charge density on the *i*th node is obtained:

$$\left| \begin{matrix} \Delta x_1 & \Delta y_1 \\ \Delta x_2 & \Delta y_2 \\ \cdots & \cdots \\ \Delta x_n & \Delta y_n \end{matrix} \right|_{A_{n\times 2}} \times \left| \begin{matrix} \frac{\partial \rho_i}{\partial x} \\ \frac{\partial \rho_i}{\partial y} \end{matrix} \right|_{X_{2\times 1}} = \left| \begin{matrix} \rho_i - \rho_1 \\ \rho_i - \rho_2 \\ \cdots \\ \rho_i - \rho_n \end{matrix} \right|_{B_{n\times 1}} \tag{10}$$

3.4. *Terminal Criteria and Initial Charge Density*

The iteration procedure will terminate while the following conditions are met.

$$\frac{E_c - E_0}{E_0} < \delta_E \tag{11}$$

$$\frac{1}{N}\sum_{i=1}^{N}\frac{\left|\rho_{m,i}-\rho_{m-1,i}\right|}{\left|\rho_{m-1,i}\right|}<\delta_{\rho} \tag{12}$$

where,

δ_E and δ_ρ are relative terminal criteria;
E_c is the electric field on conductor surface, V/m; and
$\rho_{m,i}$ and $\rho_{m-1,i}$ are consecutive space charge densities of the iteration process in ith cell, C/m^3.

However, the initial charge densities on the conductor surface need to be modified in each iteration so as to maintain the electric field on conductor surface as E_0, the principle abides by the following equation:

$$\rho_{m-1}=\rho_m\left(1+\mu\cdot\frac{E_c-E_0}{E_c+E_0}\right) \tag{13}$$

where,

ρ_{m-1}, ρ_m are charge densities of two consecutive iteration on conductor surface, C/m^3; and
μ is the acceleration factor equals to two.

Distributions of the space charge density of different wind speed are indicated in Figure 5. The convergence rate of the calculation under −120 kV voltage supply and 10 m/s wind speed is shown in Figure 6; the iteration process has a good convergence and ends after about 45 times of iteration. It takes fewer than 10 min since the coefficient matrix of the electric field is prepared once and for all.

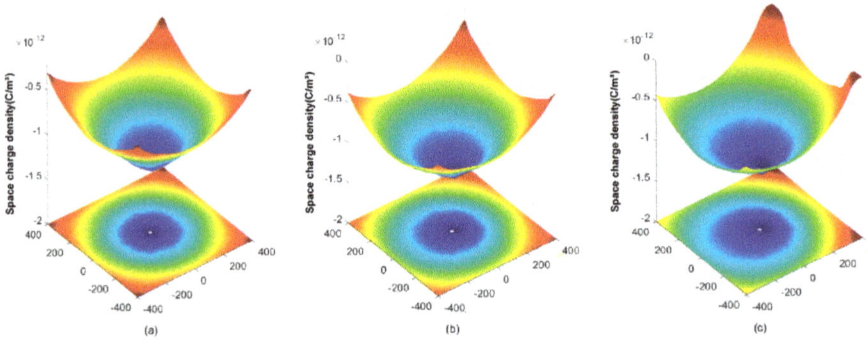

Figure 5. Space charge density under 100 kV supply voltage. (**a**) space charge density under 0 m/s wind speed; (**b**) space charge density under 5 m/s wind speed; and (**c**) space charge density under 10 m/s wind speed.

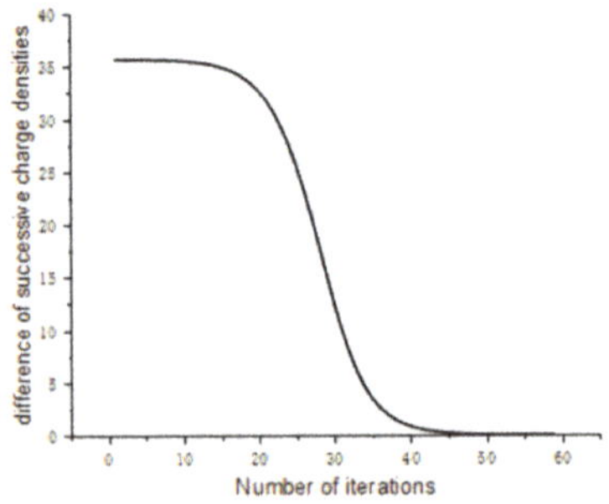

Figure 6. Convergence to terminal criteria.

4. Validation

4.1. Design of the Experiment

The corona cage is frequently used, which offers much more intensive field strength on relatively low voltage level. The reduced size of experimental facility enables indoor experiments.

The dimension of the main section of the corona cage is 1500 × 800 × 800 (mm), there are two guard sections size at 300 × 800 × 800 (mm) in order to eliminate the impact results from end effect [26]. The conductor of 10 mm radius is hung along the center of the cross-section, and the two ends of the conductor are fixed on a steel frame with two composite insulators as a connection. The schematic diagram and experimental equipment are indicated in Figures 7 and 8.

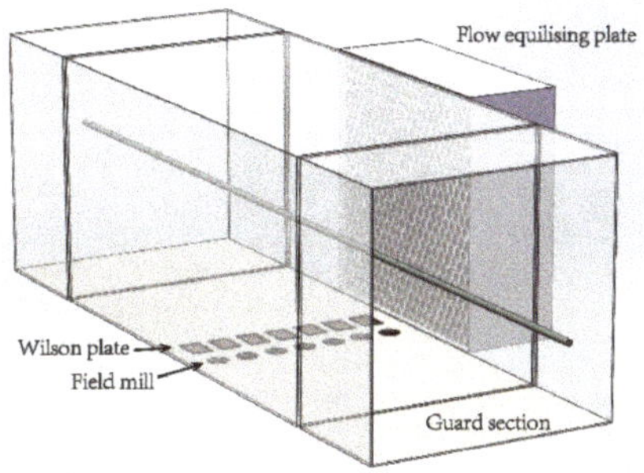

Figure 7. Schematic diagram of the experimental setting.

Figure 8. (**a**) The general view of the corona cage; (**b**) The electric field mill; (**c**) The Wilson plate; (**d**) Flow equalizing plate; (**e**) adjustable speed fan; (**f**) digital high precision anemometer; and (**g**) scope coder.

Seven Wilson plates and field mills are parallelly laid on the bottom side of the cage equidistantly for the measurement of ion current density and electric field.

Yokogawa scope coder DL 850 is used to record sampling signals of the ion flow current collected by Wilson plates, this equipment provides high sampling rate which is up to 100 Ms/s and enables continuous synchronous measurement of multi-channel.

The voltage was supported by a DC source with fluctuation less than 5%. The experiment was conducted at standard air pressure in a high voltage laboratory. The relative humidity was 20% to 30% and the temperature ranged from 10 °C to 15 °C.

An adjustable speed fan with four gears is placed next to the cage, facing one side of the cage, and it provides lateral wind speed of 0–10 m/s. The impact of flow divergence is moderated by a flow equalizing plate, the wind speed is measured by a digital high precision anemometer (AS-8336). The desired wind speed is achieved by adjusting the gear and distance between the fan and the equalizing plate. Since even distribution of the wind speed is the prerequisite of the experiment, wind speed values of 16 points are measured in the cross-section of the cage. The measurement shows that the standard deviation of wind speeds is less than 5%. The average measured result is demonstrated in Figure 9.

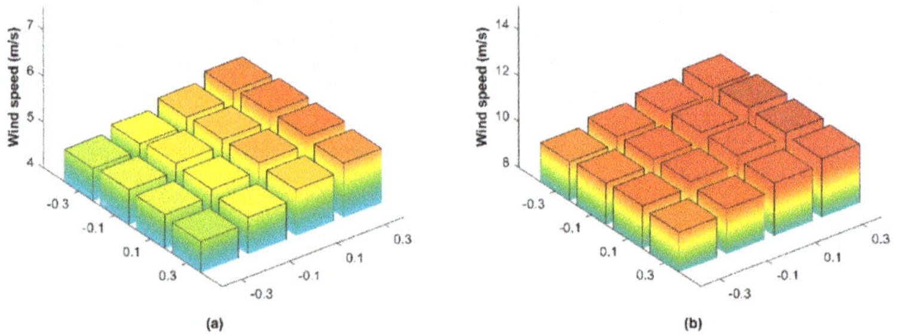

Figure 9. (**a**) measured wind speeds of 5 m/s; and (**b**) measured wind speeds of 5 m/s.

4.2. Discussion of Numerical and Measured Results

For the purpose of investigating the impact caused by varied supply voltage and wind speed, the voltage applied on the conductor is adjusted to −80 kV, −100 kV, and −120 kV, the wind speed is specified at 0 m/s, 5 m/s and 10 m/s. Both the numerically resolved and practically measured results of the electric field and the ion current density on the bottom side of the cage are demonstrated in Figure 10. It shows that the numerical result fits well with that is measured in experiment. As a result, the method used in this work is qualified to evaluate the ion flow field in a corona cage.

The electric field and ion flow current reach their peaks right under the conductor and decrease with the distance from the center point when the wind speed is zero. As the wind speed rises, the curve shifts in the same direction with the wind flow. In comparison with the electric field, the degree of the shift on the ion current density is larger because it is more affected by the movement of the space charges.

In order to illustrate in a more intuitive manner, the cross-section of interest is bisected by vertical center line which is illustrated in Figure 11, the section in where the wind flow and the electric field move in opposite directions is defined as upwind; on the contrary, the other section is downwind.

In the upwind section, electric field and ion current density weakens because the wind flow reduces the density of space charges. This effect decreases as approaching the conductor due to the increasing electric field force.

For both electric field and ion current density, absolute values in the upwind section is slightly lower than that in the downwind section. The absolute value declines along with the increasement of wind speed. In addition, the degree of the impact results from wind speed mounts as the electric field strength decreases. Additionally, the ion current density on lower voltage level is more sensitive to the wind speed owing to weaker field strength.

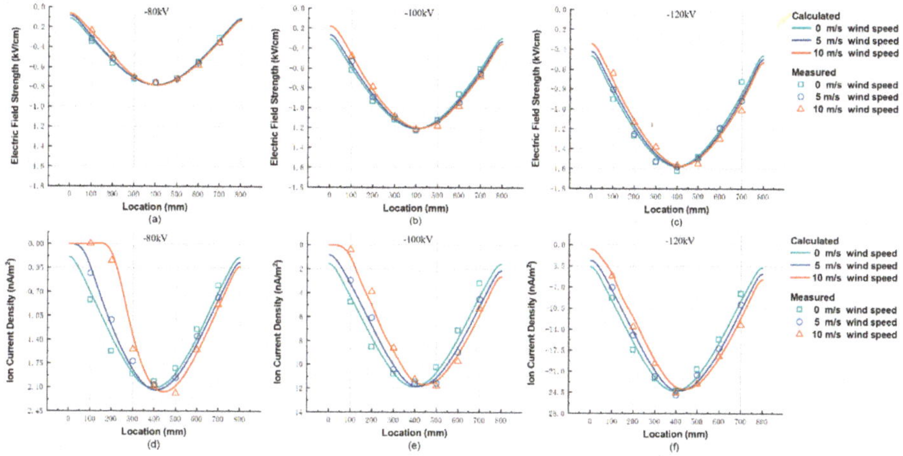

Figure 10. (**a**) Electric field on the bottom side under −80 kV supply voltage; (**b**) electric field on the bottom side under −100 kV supply voltage; (**c**) electric field on the bottom side under −120 kV supply voltage; (**d**) ion current density on the bottom side under 80 kV supply voltage; (**e**) ion current density on the bottom side under 100 kV supply voltage; (**f**) ion current density on the bottom side under 120 kV supply voltage.

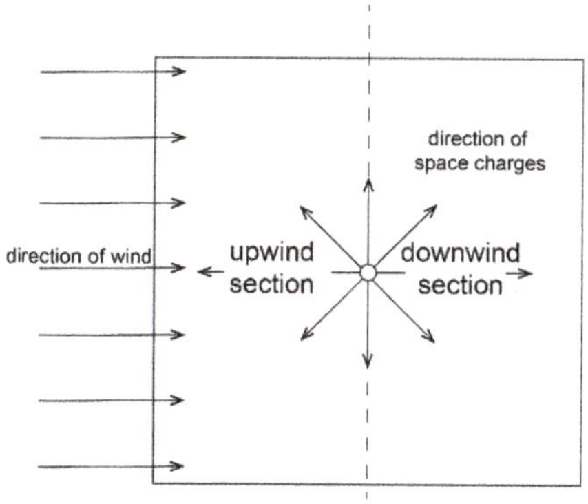

Figure 11. Schematic diagram of the cross-section.

In upwind and downwind section, the wind flow exerts contrarily influence on the electric field and the ion flow current. To be explicit, increased wind speed weakens the electric field intensity and ion current density in the upwind section. For the downwind section, the result is opposite. The explanation is that the charge density in the upwind section is larger than that of the downwind section which is rather obvious, as shown in Figure 5.

5. Conclusions

The method combining CSM and upwind FVM, which concerns the impact of wind flow, is proposed in this paper, experiment is designed to examine the numeric results. The results indicate that such a method provides a relatively accurate evaluation of the ion flow field under a corona discharge conductor. For the merits of the presented method, the analytically calculated electric field offers a more precise electric field in the vicinity of the conductor than that is solved in traditional FEM; simultaneously, it reduces calculation time and enhances numerical stability to a large extent by means of avoiding iterative calculation of two numeric methods. Moreover, 2nd-order UFVM improves the accuracy of space charge density and is competent for the conditions, including wind impact. The solution process converges effectively and stably. The dependence of the ion flow field on the wind is studied, which approves the influence of wind flow on the ion flow field.

Author Contributions: Conceptualization, Z.L. and X.Z.; Methodology, Z.L.; Software, Z.L.; Validation, Z.L.; Formal Analysis, Z.L.; Investigation, Z.L.; Resources, Z.L. and X.Z.; Data Curation, Z.L.; Writing-Original Draft Preparation, Z.L.; Writing-Review & Editing, Z.L.; Visualization, Z.L.; Supervision, X.Z.; Project Administration, X.Z.; Funding Acquisition, X.Z.

Funding: This research received no external funding.

Conflicts of Interest: The authors declare no conflict of interest.

References

1. Maruvada, P.S. *Corona Performance of High-Voltage Transmission Lines*; Research Studies Press Baldock: Herfordshire, UK, 2000.
2. Janischewskyj, W.; Cela, G. Finite Element Solution for Electric Fields of Coronating DC Transmission Lines. *IEEE Trans. Power Appar. Syst.* **1979**, *PAS-98*, 1000–1012. [CrossRef]
3. Takuma, T.; Ikeda, T.; Kawamoto, T. Calculation of ION Flow Fields of HVDC Transmission Lines By the Finite Element Method. *IEEE Trans. Power Appar. Syst.* **1981**, *PAS-100*, 4802–4810. [CrossRef]
4. Abdel-Salam, M.; Farghally, M.; Abdel-Sattar, S. Finite element solution of monopolar corona equation. *IEEE Trans. Electr. Insul.* **1983**, *EI-18*, 110–119. [CrossRef]
5. Li, X.; Ciric, I.R.; Raghuveer, M.R. Highly stable finite volume based relaxation iterative algorithm for solution of DC line ionized fields in the presence of wind. *Int. J. Numer. Model. Electron. Netw. Devices Fields* **1997**, *10*, 355–370. [CrossRef]
6. Davis, J.L.; Hoburg, J.F. HVDC transmission line computations using finite element and characteristics method. *J. Electrost.* **1986**, *18*, 1–22. [CrossRef]
7. Fortin, S.; Zhao, H.; Ma, J.; Member, S.; Dawalibi, F.P. A New Approach to Calculate the Ionized Field Transmission Lines in the Space and on the Earth Surface. In Proceedings of the 2006 International Conference on Power System Technology, Chongqing, China, 22–26 October 2006.
8. Zhou, X.; Cui, X.; Lu, T.; Fang, C.; Zhen, Y. Spatial distribution of ion current around HVDC bundle conductors. *IEEE Trans. Power Deliv.* **2012**, *27*, 380–390. [CrossRef]
9. Guillod, T.; Pfeiffer, M.; Franck, C.M. Improved coupled ion-flow field calculation method for AC/DC hybrid overhead power lines. *IEEE Trans. Power Deliv.* **2014**, *29*, 2493–2501. [CrossRef]
10. Zhang, B.; Mo, J.; He, J.; Zhuang, C. A Time-domain Approach of Ion Flow Field around AC-DC hybrid Transmission Lines Based on Method of Characteristics. *IEEE Trans. Magn.* **2015**, *52*, 7205004. [CrossRef]
11. Lu, T.; Feng, H.; Cui, X.; Zhao, Z.; Li, L. Analysis of the ionized field under HVDC transmission lines in the presence of wind based on upstream finite element method. *IEEE Trans. Magn.* **2010**, *46*, 2939–2942. [CrossRef]
12. Zhou, X.; Lu, T.; Cui, X.; Zhen, Y.; Liu, G. Simulation of ion-flow field using fully coupled upwind finite-element method. *IEEE Trans. Power Deliv.* **2012**, *27*, 1574–1582. [CrossRef]
13. Levin, P.L.; Hoburg, J.F. Donor Cell-Finite Element Descriptions of Wire-Duct Precipitator Fields, Charges, and Efficiencies. *IEEE Trans. Ind. Appl.* **1990**, *26*, 662–670. [CrossRef]
14. Zhang, B.; He, J.; Zeng, R.; Gu, S.; Cao, L. Calculation of Ion Flow Field Under HVdc Bipolar Transmission Lines by Integral Equation Method. *IEEE Trans. Magn.* **2007**, *43*, 1237–1240. [CrossRef]
15. Long, Z.; Yao, Q.; Song, Q.; Li, S. A second-order accurate finite volume method for the computation of electrical conditions inside a wire-plate electrostatic precipitator on unstructured meshes. *J. Electrost.* **2009**, *67*, 597–604. [CrossRef]
16. Yin, H.; He, J.; Zhang, B.; Zeng, R. Finite volume-based approach for the hybrid ion-flow field of UHVAC and UHVDC transmission lines in parallel. *IEEE Trans. Power Deliv.* **2011**, *26*, 2809–2820. [CrossRef]
17. Yin, H.; Zhang, B.; He, J.; Zeng, R.; Li, R. Time-domain finite volume method for ion-flow field analysis of bipolar high-voltage direct current transmission lines. *IET Gener. Transm. Distrib.* **2012**, *6*, 785–791. [CrossRef]

18. Yang, F.; Liu, Z.; Luo, H.; Liu, X.; He, W. Calculation of ionized field of HVDC transmission lines by the meshless method. *IEEE Trans. Magn.* **2014**, *50*, 7200406. [CrossRef]
19. Bian, X.; Yu, D.; Meng, X.; Macalpine, M.; Wang, L.; Guan, Z.; Yao, W.; Zhao, S. Corona-generated space charge effects on electric field distribution for an indoor corona cage and a monopolar test line. *IEEE Trans. Dielectr. Electr. Insul.* **2011**, *18*, 1767–1778. [CrossRef]
20. Lekganyane, M.J.; Ijumba, N.M.; Britten, A.C. A comparative study of space charge effects on corona current using an indoor corona cage and a monopolar test line. In Proceedings of the 2007 IEEE Power Engineering Society Conference and Exposition in Africa, Johannesburg, South Africa, 16–20 July 2007.
21. Kaptzov, N.A. *Elektrische Vorgänge in Gasen und im Vakuum*; VEB Deutscher Verlag der Wissenschaften: Berlin, Germany, 1955; pp. 488–491. ISBN 978-3-446-42771-6.
22. Peek, F.W. *Dielectric Phenomena in High Voltage Engineering*; McGraw-Hill Book Company, Inc: New York, NY, USA, 1920.
23. Abdel-salam, M.; Al-hamouz, Z. A finite-element analysis of bipolar ionized field. *IEEE Tans. Ind. Appl.* **1995**, *31*, 477–483. [CrossRef]
24. Zhou, X.; Cui, X.; Lu, T.; Zhen, Y.; Luo, Z. A time-efficient method for the simulation of ion flow field of the AC-DC hybrid transmission lines. *IEEE Trans. Magn.* **2012**, *48*, 731–734. [CrossRef]
25. Li, X. Numerical Analysis of Ionized Fields Associated with HVDC Transmission Lines Including Effect of Wind. Ph.D. Thesis, The University of Manitoba, Winnipeg, MB, USA, 1997.
26. Urban, R.G.; Reader, H.C.; Holtzhausen, J.P. Small corona cage for wideband HVac radio noise studies: Rationale and critical design. *IEEE Trans. Power Deliv.* **2008**, *23*, 1150–1157. [CrossRef]

energies

MDPI

Article

Charge-Simulation-Based Electric Field Analysis and Electrical Tree Propagation Model with Defects in 10 kV XLPE Cable Joint

Jiahong He [1,*], Kang He [1] and Longfei Cui [2]

[1] School of Electric Engineering, Southeast University, Nanjing 210096, Jiangsu, China; 220192809@seu.edu.cn
[2] NR Electric Company Limited, Nanjing 211102, China; cuilf@nrec.com
* Correspondence: hejiahong@seu.edu.cn

Received: 31 October 2019; Accepted: 26 November 2019; Published: 27 November 2019

Abstract: The most severe partial discharges and main insulation failures of 10 kV cross-linked polyethylene cables occur at the joint due to defects caused by various factors during the manufacturing and installation processes. The electric field distortion is analyzed as the indicator by the charge simulation method to identify four typical defects (air void, water film, metal debris, and metal needle). This charge simulation method is combined with random walk theory to describe the stochastic process of electrical tree growth around the defects with an analysis of the charge accumulation process. The results illustrate that the electrical trees around the metal debris and needle are more likely to approach the cable core and cause main insulation failure compared with other types of the defects because the vertical field vector to the cable core is significantly larger than the field vectors to other directions during the tree propagation process with conductive defects. The electric field was measured around the cable joint surface and compared with the simulation results to validate the calculation model and the measurement method. The air void and water film defects are difficult to detect when their sizes are less than 5 mm^3 because the field distortions caused by the air void and water film are relatively small and might be concealed by interference. The proposed electric field analysis focuses on the electric field distortion in the cable joint, which is the original cause of the insulation material breakdown. This method identifies the defect and predicts the electrical tree growth in the cable joint simultaneously. It requires no directly attached or embedded sensors to impact the cable joint structure and maintains the power transmission during the detection process.

Keywords: cable joint; charge simulation method; electrical tree; random walk theory

1. Introduction

The power cable has been extensively used in power transmission and distribution systems due to its large ampacity and ability to be installed underground [1,2]. The cables have been used for more than 40 years of service in the developed countries, and the cable failure accidents frequently occur in large cities due to the aging and manufactured defects of cables during the long-term operation [3,4]. Therefore, it is necessary to develop an effective and efficient method to detect the defects and predict the aging rate based on the electrical tree model in the cable to maintain safe power transmission.

With the improvement of high voltage (HV) and insulation technology, cable manufacturing has replaced the traditional oil-filled paper used as the insulation material with cross-linked polyethylene (XLPE), which has become the most widely used insulation material for power cables [5]. Compared with HV XLPE cables, 10 kV XLPE cables are more prevalent in the power distribution system, especially in urban power grids. The cable maintenance has become difficult due to the complicated underground environmental conditions in cities. As the part of the cable with a complex structure and relatively weak insulation, the cable joint is more likely to explode than the cable main body due to long-term usage and lack of maintenance [6–8]. The defects in the cable joint are the main cause of explosion accidents. Many studies have been conducted to investigate the electric field distribution and aging process of main cable body, whereas studies of the cable joint are inadequate due to its complicated structure and multiple installation procedures [9,10]. This paper focuses on the electric field analysis around the 10 kV cable joint to determine the characteristics of the field distribution corresponding to each typical defect. The electric field was then measured to locate and identify the defect in the cable joint.

Cable joint defects are more likely to occur on the interfaces between the XLPE and silicone rubber layers, caused by various factors during the manufacturing and installation processes [11–14]. Four typical defects were analyzed: air void defect, water film defect, metal debris defect, and metal needle defect. The air void defects are created under the following two scenarios. The XLPE main insulation could be dented during the manufacturing process. The cable joint installer may accidently cut the XLPE material with a knife or other tool during the removal of the external semi-conductive layer of the cable. Water film defects are created by water leaking into a small gap in the interface when the XLPE is not fully sealed by a silicon rubber tube during installation. Metal debris defects are created by the remaining metal tips on the interface that were not properly cleaned up during installation. The metal needle penetrates the silicon rubber tube due to external forces and causes defects. Many researchers have developed different methods to detect the defects in the cable joints. Zhang et al. developed a thermal-probability-density-based method to detect the internal defects of power cable joints [15]. Yang et al. proposed a new method for determining the connection resistance of the compression connector in a cable joint and evaluated the crimping process defects using coupling field analysis [16,17]. Zhu et al. evaluated the thermal effect of different laying modes on XLPE insulation and estimated the cable ampacity [18]. Most existing methods adopted the thermal or ultrasonic signal as the indicator to detect the internal defects of cable joints [19,20]. However, these signals are the derivative consequences of the insulation material breakdown and susceptible to interferences in the environment, while the electric field intensification is the original cause of the insulation material breakdown. Therefore, electric field analysis around the cable joint can identify the types of defects by categorizing the characteristics of electric field distribution with relatively less influence from the external noises.

The charge simulation method (CSM) is used to calculate the electric field around the cable joint surface because the CSM significantly reduces the computational complexity due to the axis-symmetrical geometry of the cable joint [21]. CSM was first applied in HV electric field calculation by Singer in 1974 [22]. Malik then modified the CSM application using the least square method to minimize calculation errors [23]. Takuma et al. applied CSM to calculate the electric fields with multi-dielectric material conditions [24]. This paper compared the calculation results with experimental results to validate the simulation model and the electric field measurement method. The electric field distribution was measured based on the Pockels effect to reduce the electromagnetic interference [25,26]. The distance from the probe to the cable joint surface was precisely controlled by a ball screw structure to obtain the same measurement position as the calculation model.

Electrical trees develop inside dielectric material when internal defects distort the electric field distribution, and the maximum electric field exceeds the critical value of the field strength [27–29]. Electrical trees create conductive paths in the XLPE main insulation and increase the risk of failure. Rompe and Weizel first introduced the formula to obtain the time-dependent voltage, current, and resistance during the discharge process [30]. In 1998, Champion et al. simulated the process of an electrical tree spreading along with orthogonal mesh based on the traditional dielectric breakdown model [31]. Zhou et al. then proposed that the thermal aging of silicon rubber led to the formation of an electrical tree by performing a dielectric withstand test on silicone rubber [32]. Since the electrical tree propagation is a stochastic process creating different trajectories even the electric field distributions remain the same, the random walk theory is more consistent with the physical phenomena in the cable joint, when compared with other approaches. Therefore, the random walk theory is combined with CSM to model the electrical tree growth in a cable joint with defects.

Random walk theory describes a sequence of random events and the probabilities of states at each time related to the previous time. The probability of all next possible states is determined by electric field vectors close to the electrical tree. Noskov et al. developed the electrical tree propagation model based on random walk theory in the rod–plane system to evaluate the self-consistent dynamic characteristics of the electrical tree [33,34].

The CSM was combined with random walk theory to describe the stochastic process of electrical tree propagation. The electrical tree propagation processes were repeated multiple times to analyze the pattern of the electrical tree trajectories with four types of defects. The characteristics of electric field distribution under the presence of defects were also investigated. The electric field was measured to locate and identify the defects, then predict the pattern of electrical tree growth in a 10 kV XLPE cable joint to prevent further failures and accidents.

2. Materials and Methods

2.1. Cable Joint Structure

A schematic of a 10 kV XLPE cable joint structure is shown in Figure 1.

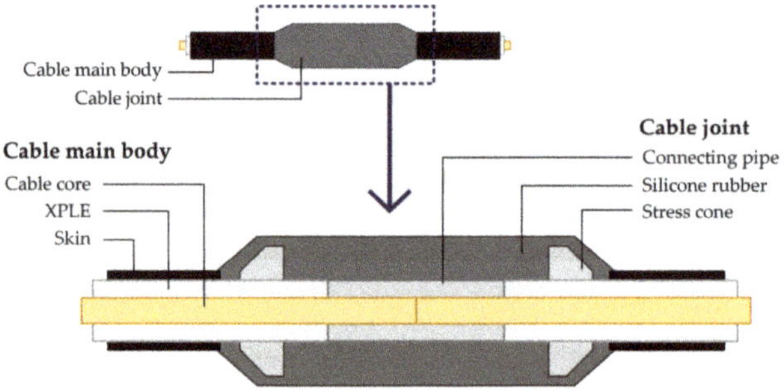

Figure 1. Schematic of a 10 kV cross-linked polyethylene (XLPE) cable and the joint.

2.2. Charge Simulation Based Electric Field Calculation

Electric field distributions around the cable joint were calculated based on the CSM. The CSM simulates the field by a number of discrete charges placed outside the calculation region to solve Poisson's equations with boundary conditions satisfied numerically [35].

$$\begin{array}{ll} \nabla^2 u = \frac{\partial^2 \varphi}{\partial x^2} + \frac{\partial^2 \varphi}{\partial y^2} = \frac{\rho}{\varepsilon} & \text{Poisson's equation} \\ \varphi(x,y)\big|_{\Gamma} = f_1(\Gamma) & \text{Dirichlet boundary condition} \\ \frac{\partial \varphi}{\partial n}\Big|_{\Gamma} = f_2(\Gamma) & \text{Neumann boundary condition} \end{array} . \tag{1}$$

The ring charges, line charges, and point charges were input in the XLPE main insulation, cable core, and defects, respectively, to fit the axis-symmetrical geometry of the joint. The discretized equations are formulated as follows to calculate the values of charges according to the Dirichlet and Neumann boundary conditions [22].

$$[P][Q] = [V]. \tag{2}$$

P is the $m \times n$ coefficient matrix, Q is the $n \times 1$ unknown simulating charge matrix, and V is the $m \times 1$ electric potential matrix at the contour points.

The charge distributions were shown in the following sections to satisfy the potential continuity (Dirichlet) and normal flux density continuity (Neumann) on the interfaces between different materials. The model was built on the MATLAB platform.

2.2.1. Charges Distribution Close to the Cable Joint and Its Accessories

The distribution of all simulation charges close to the cable joint and its accessories is shown in Figure 2.

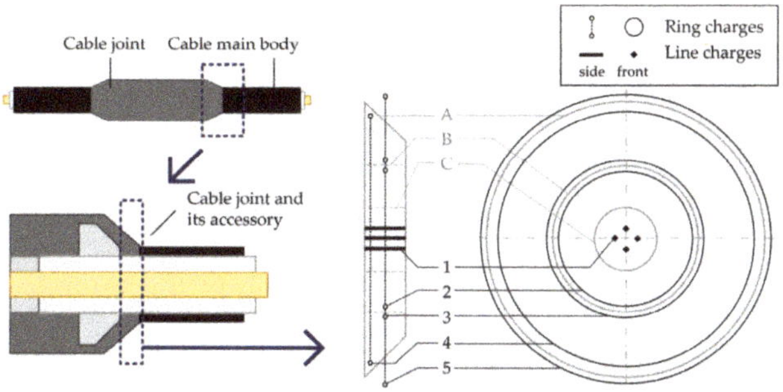

Figure 2. Distribution of simulation charges close to the junction of the cable and the joint. Simulation charges: 1, Q_{core}, line charges in the cable core; 2, $Q_{XLPE.O}$, ring charges at the outer side of the XLPE layer; 3, $Q_{SiR.I}$, ring charges at the inner side of the silicone rubber; 4, $Q_{SiR.O}$, ring charges at the outer side of the silicone rubber; and 5, Q_A, ring charges close to the surface of the silicone rubber. Boundaries: A, the surface of the silicone rubber; B, the interface of silicone rubber and XLPE layer; and C, the interface of the XLPE layer and cable core.

The potential of the interface of cable core and XLPE layer is calculated as

$$\begin{gathered}\sum_{j=1}^{n_{core}} P_{i,j}Q_{core(j)} + \sum_{j=1}^{n_{SiR.I}} P_{i.j}Q_{SiR.I(j)} = \varphi_{core},\\ i = 1,2,\ldots,n_{core},\end{gathered} \tag{3}$$

where n_{core} is the number of line charges Q_{core} in the cable core, $n_{SiR.I}$ is the number of ring charges $Q_{SiR.I}$ at the inner side of the silicone rubber, P is the potential coefficient, and φ_{core} is the potential of the cable core.

The potential continuity boundary condition on the interface of the XLPE layer and silicone rubber, and the surface of the silicone rubber are calculated by Equations (4) and (5), respectively:

$$\begin{gathered}\sum_{j=1}^{n_{XLPE.O}} P_{i,j}Q_{XLPE.O(j)} - \sum_{j=1}^{n_{SiR.I}} P_{i.j}Q_{SiR.I(j)} = 0,\\ i = 1,2,\ldots,n_{XLPE.O},\\ n_{XLPE.O} = n_{SiR.I};\end{gathered} \tag{4}$$

$$\begin{gathered}\sum_{j=1}^{n_{SiR.O}} P_{i,j}Q_{SiR.O(j)} - \sum_{j=1}^{n_{A}} P_{(i,j)}Q_{A(j)} = 0,\\ i = 1,2,\ldots,n_{SiR.O},\\ n_{SiR.O} = n_{A},\end{gathered} \tag{5}$$

where $n_{XLPE.O}$ is the number of line charges $Q_{XLPE.O}$ at the inner side of the XLPE layer, $n_{SiR.O}$ is the number of ring charges $Q_{SiR.O}$ at the outer side of the silicone rubber, and n_A is the number of ring charges Q_A close to the surface of the silicone rubber.

The field strength boundary conditions on the interface of the XLPE layer and the silicone rubber, and the surface of the silicone rubber are calculated by Equations (6) and (7), respectively:

$$\varepsilon_{XLPE}E_{XLPE} - \varepsilon_{SiR}E_{SiR} = \sum_{j=1}^{n_{XLPE.O}} \varepsilon_{XLPE}P_{i,j}Q_{XLPE.O(j)} - \sum_{j=1}^{n_{SiR.I}} \varepsilon_{SiR}P_{i.j}Q_{SiR.I(j)} = 0, \quad i = 1,2,\ldots,n_{XLPE.O}; \tag{6}$$

$$\varepsilon_{SiR}E_{SiR} - \varepsilon_{A}E_{A} = \sum_{j=1}^{n_{SiR.O}} \varepsilon_{SiR}P_{i,j}Q_{SiR.O(j)} - \sum_{j=1}^{n_{A}} \varepsilon_{A}P_{(i,j)}Q_{A(j)} = 0, \quad i = 1,2,\ldots,n_{SiR.O}, \tag{7}$$

where ε_{XLPE}, ε_{SiR}, and ε_{A} are the permittivity of XLPE, silicone rubber, and air, respectively.

2.2.2. Charges Distribution Close to the Stress Cone

The distribution of all simulation charges close to the stress cone shown in Figure 3.

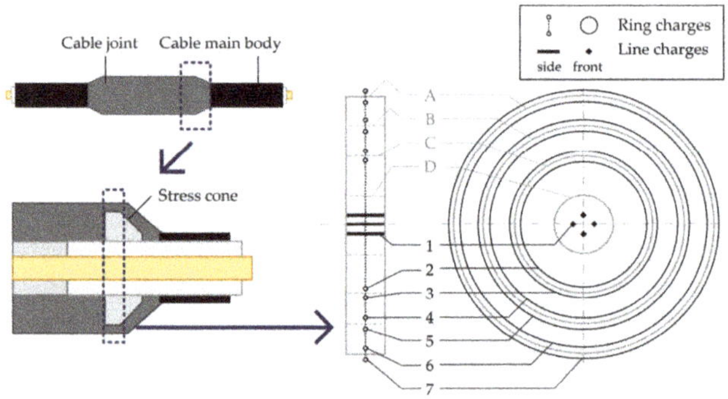

Figure 3. Distribution of simulation charges close to the stress cone. Simulation charges: 1, Q_{core}, line charges in the cable core; 2, $Q_{XLPE.O}$, ring charges at the outer side of the XLPE layer; 3, $Q_{SC.I}$, ring charges at the inner side of the stress cone; 4, $Q_{SC.O}$, ring charges at the outer side of the stress cone; 5, $Q_{SiR.I}$, ring charges at the inner side of the silicone rubber; 6, $Q_{SiR.O}$, ring charges at the outer side of the silicone rubber; and 7, Q_{A}, ring charges close to the surface of the silicone rubber. Boundaries: A, the surface of the outer layer of the silicone rubber; B, the interface of the silicone rubber and the stress cone; C, the interface of the stress cone and the XLPE layer; and D, the interface of the XLPE layer and the cable core.

The potential of the interface of cable core and XLPE layer is calculated as:

$$\sum_{j=1}^{n_{core}} P_{i,j}Q_{core(j)} + \sum_{j=1}^{n_{SC.I}} P_{i,j}Q_{SC.I(j)} = \varphi_{core}, \quad i = 1,2,\ldots,n_{core}, \tag{8}$$

where $n_{SC.I}$ is the number of ring charges $Q_{SC.I}$ at the inner side of the stress cone.

The potential continuity boundary conditions on the interfaces of the XLPE layer–stress cone, stress cone–silicone rubber, and air–silicone rubber are satisfied by Equations (9)–(11) respectively:

$$\sum_{j=1}^{n_{XLPE.O}} P_{i,j}Q_{XLPE.O(j)} - \sum_{j=1}^{n_{SC.I}} P_{i,j}Q_{SC.I(j)} = 0, \quad i = 1,2,\ldots,n_{XLPE.O}, \quad n_{XLPE.O} = n_{SC.I}; \tag{9}$$

$$\sum_{j=1}^{n_{SC.O}} P_{i,j}Q_{SC.O(j)} - \sum_{j=1}^{n_{SiR.I}} P_{i,j}Q_{SiR.I(j)} = 0, \quad i = 1,2,\ldots,n_{SC.O}, \quad n_{SC.O} = n_{SiR.I}; \tag{10}$$

$$\sum_{j=1}^{n_{SiR.O}} P_{i,j}Q_{SiR.O(j)} - \sum_{j=1}^{n_A} P_{i,j}Q_{A(j)} = 0, \quad i = 1,2,\ldots,n_{SiR.O}, \quad n_{SiR.O} = n_A, \tag{11}$$

where $n_{SC.O}$ is the number of ring charges $Q_{SC.O}$ at the outer sides of the stress cone.

The field strength boundary conditions on the interfaces between XLPE layer–stress cone, stress cone–silicone rubber interfaces, and air-silicone rubber are satisfied by Equations (12)–(14), respectively:

$$\varepsilon_{XLPE}E_{XLPE} - \varepsilon_{SC}E_{SC} = \sum_{j=1}^{n_{XLPE.O}} \varepsilon_{XLPE}P_{i,j}Q_{XLPE.O(j)} - \sum_{j=1}^{n_{SC.I}} \varepsilon_{SC}P_{i,j}Q_{SC.I(j)} = 0, \quad i = 1,2,\ldots,n_{XLPE.O}; \tag{12}$$

$$\varepsilon_{SC}E_{SC} - \varepsilon_{SiR}E_{SiR} = \sum_{j=1}^{n_{SC.O}} \varepsilon_{SC}P_{i,j}Q_{SC.O(j)} - \sum_{j=1}^{n_{SiR.I}} \varepsilon_{SiR}P_{i,j}Q_{SiR.I(j)} = 0, \quad i = 1,2,\ldots,n_{SC.O}; \tag{13}$$

$$\varepsilon_{SiR}E_{SiR} - \varepsilon_{A}E_{A} = \sum_{j=1}^{n_{SiR.O}} \varepsilon_{SiR}P_{i,j}Q_{SiR.O(j)} - \sum_{j=1}^{n_A} \varepsilon_{A}P_{i,j}Q_{A(j)} = 0, \quad i = 1,2,\ldots,n_{SiR.O}, \tag{14}$$

where ε_{SC} is the permittivity of the stress cone material.

2.2.3. Charges Distribution Close to the Connecting Pipe Inside the Cable Joint

The distribution of all simulation charges close to the connecting pipe shown in Figure 4.

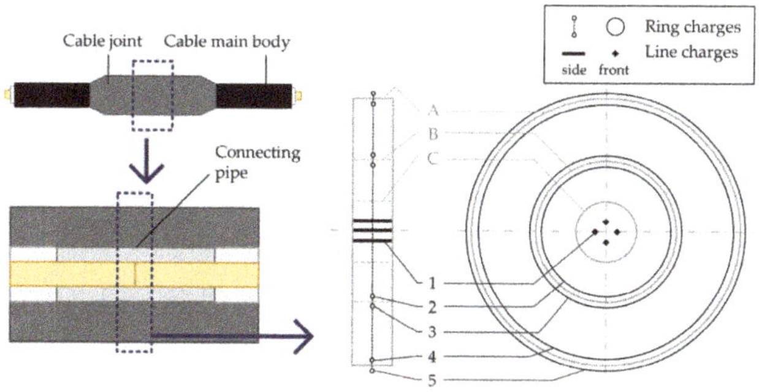

Figure 4. Distribution of simulation charges inside the cable joint. Simulation charges: 1, Q_{core}, line charges in the cable core; 2, $Q_{pipe.O}$, ring charges at the outer side of the connecting pipe; 3, $Q_{SiR.I}$, ring charges at the inner side of the silicone rubber; 4, $Q_{SiR.O}$, ring charges at the outer side of the silicone rubber; and 5, Q_A, ring charges close to the surface of the silicone rubber. Boundaries: A, the surface of the silicone rubber; B, the interface of silicone rubber and the connecting pipe; and C, the interface of connecting pipe and cable core.

The potential of the interface of the cable core and the connection pipe layer is calculated by:

$$\sum_{j=1}^{n_{core}} P_{i,j}Q_{core(j)} + \sum_{j=1}^{n_{SiR.I}} P_{i,j}Q_{SiR.I(j)} = \varphi_{core}. \quad i = 1, 2, \ldots, n_{core}. \tag{15}$$

The potential continuity boundary conditions on the interfaces between of the connecting pipe–silicone rubber and the silicone rubber-air are calculated by Equations (16) and (17), respectively:

$$\sum_{j-1}^{n_{pipe.O}} P_{i,j}Q_{pipe.O(j)} - \sum_{j=1}^{n_{SiR.I}} P_{i,j}Q_{SiR.O(j)} = 0, \quad i = 1, 2, \ldots, n_{pipe.O}, \quad n_{pipe.O} = n_{SiR.I}; \tag{16}$$

$$\sum_{j=1}^{n_{SiR.O}} P_{i,j}Q_{SiR.O(j)} - \sum_{j=1}^{n_A} P_{i,j}Q_{A(j)} = 0, \quad i = 1, 2, \ldots, n_{SiR.O}, \quad n_{SiR.O} = n_A, \tag{17}$$

where $n_{pipe.O}$ is the number of ring charges $Q_{pipe.O}$ at the outer side of the connection pipe.

The field strength boundary conditions on the interface of connection pipe and silicone rubber are calculated by:

$$\varepsilon_{pipe}E_{pipe} - \varepsilon_{SiR}E_{SiR} = \sum_{j=1}^{n_{pipe.O}} \varepsilon_{pipe}P_{i,j}Q_{pipe.O(j)} - \sum_{j=1}^{n_{SiR.I}} \varepsilon_{SiR}P_{i,j}Q_{SiR.O(j)} = 0, \quad i = 1, 2, \ldots, n_{pipe.O}; \tag{18}$$

$$\varepsilon_{SiR}E_{SiR} - \varepsilon_A E_A = \sum_{j=1}^{n_{SiR.O}} \varepsilon_{SiR}P_{i,j}Q_{SiR.O(j)} - \sum_{j=1}^{n_A} \varepsilon_A P_{i,j}Q_{A(j)} = 0, \quad i = 1, 2, \ldots, n_{SiR.O}, \tag{19}$$

where ε_{pipe} is the permittivity of the connection pipe material.

2.2.4. Charges Distribution Close to Four Types of Defects

Four main kinds of defects occur on the interface of the XLPE layer and the silicone rubber: ware air voids, water film, metal debris, and metal needles shown in Figure 5.

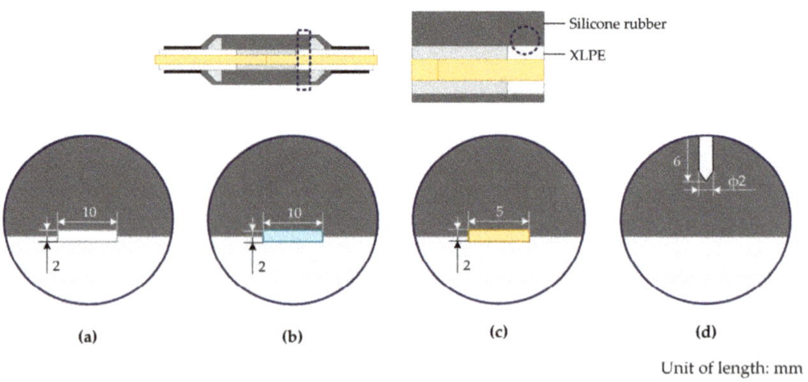

Figure 5. Defects between the XLPE layer and the silicone rubber: (**a**) air void, (**b**) water film, (**c**) metal debris, and (**d**) metal needle.

For non-conductive defects, such as air voids or water films, a series of point charges are placed close to the interface of the defects and other materials (Figure 6), and the potential continuity and field strength boundary conditions must be satisfied.

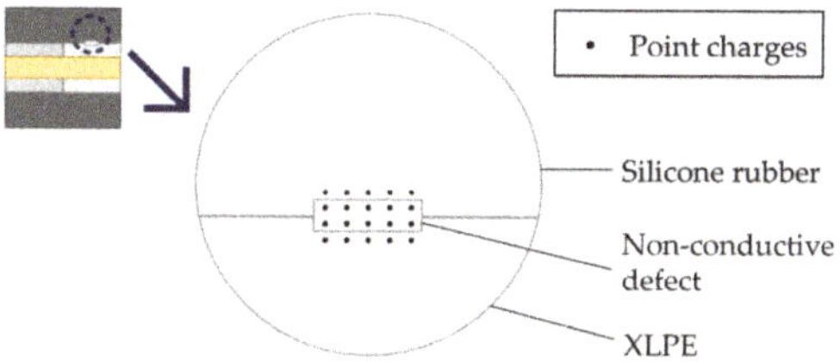

Figure 6. Charges distribution of non-conductive defects.

The potential continuity equations are:

$$\sum_{j=1}^{n_{D.SiR}} P_{i,j}Q_{D.SiR(j)} - \sum_{j=1}^{n_{SiR.D}} P_{i,j}Q_{SiR.D(j)} = 0, \quad i = 1,2,\ldots,n_{D.SiR}, \quad n_{D.SiR} = n_{SiR.D}; \tag{20}$$

$$\sum_{j=1}^{n_{D.XLPE}} P_{i,j}Q_{D.XLPE(j)} - \sum_{j=1}^{n_{XLPE.D}} P_{i,j}Q_{XLPE.D(j)} = 0, \quad i = 1,2,\ldots,n_{D.XLPE}, \quad n_{D.XLPE} = n_{XLPE.D}, \tag{21}$$

where $n_{D.SiR}$ and $n_{D.XLPE}$ are the numbers of the point charges $Q_{D.SiR}$ and $Q_{D.XLPE}$ at the rubber and XLPE sides inside the defect, respectively; and $n_{SiR.D}$ and $n_{XLPE.D}$ are the numbers of the point charges $Q_{SiR.D}$ and $Q_{XLPE.D}$ at the defect side inside the silicone rubber and XLPE layer, respectively.

The field boundary equations are:

$$\varepsilon_D E_D - \varepsilon_{SiR} E_{SiR} = \sum_{j=1}^{n_{D.SiR}} \varepsilon_D P_{i,j}Q_{D.SiR(j)} - \sum_{j=1}^{n_{SiR.D}} \varepsilon_{SiR} P_{i,j}Q_{SiR.D(j)} = 0, \quad i = 1,2,\ldots,n_{D.SiR}; \tag{22}$$

$$\varepsilon_D E_D - \varepsilon_{XLPE} E_{XLPE} = \sum_{j=1}^{n_{D.XLPE}} P_{i,j}Q_{D.XLPE(j)} - \sum_{j=1}^{n_{XLPE.D}} P_{i,j}Q_{XLPE.D(j)} = 0, \quad i = 1,2,\ldots,n_{D.XLPE}, \tag{23}$$

where ε_D is the permittivity of the defect, such as the permittivity of air (ε_A) or water (ε_W).

Metal debris (Figure 7) has floating potential [36]. Hence, the calculation of metal debris is different from non-conductive defects.

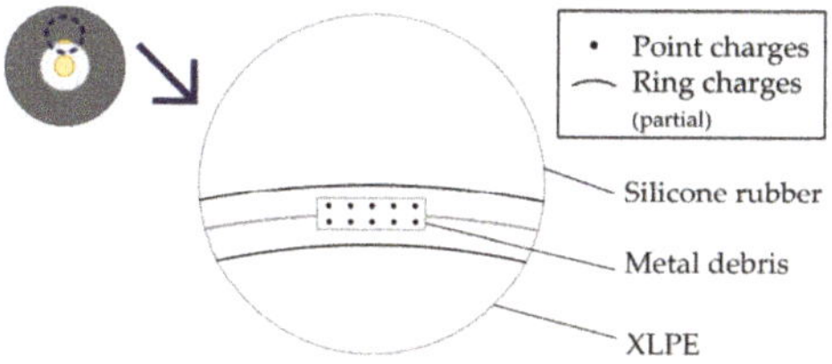

Figure 7. Charges distribution of metal debris.

To calculate the potential of the metal debris, the potential boundary equations are:

$$\sum_{j=1}^{n_M} P_{i,j}Q_{M(j)} + \sum_{j=1}^{n_{SiR.I}} P_{i,j}Q_{SiR.I(j)} = \varphi_M, \quad i = 1,2,\ldots,n_M; \tag{24}$$

$$\sum_{j=1}^{n_M} P_{i,j}Q_{M(j)} + \sum_{j=1}^{n_{XLPE.O}} P_{i,j}Q_{XLPE.O(j)} = \varphi_M, \quad i = 1,2,\ldots,n_M; \tag{25}$$

where φ_M is the unknown potential of the metal debris and n_M is the number of point charges Q_M inside the metal debris.

2.3. Electrical Tree Propagation Model

In this model, the trajectory of the electrical tree was composed of a sequence of point charges (Figure 8a).

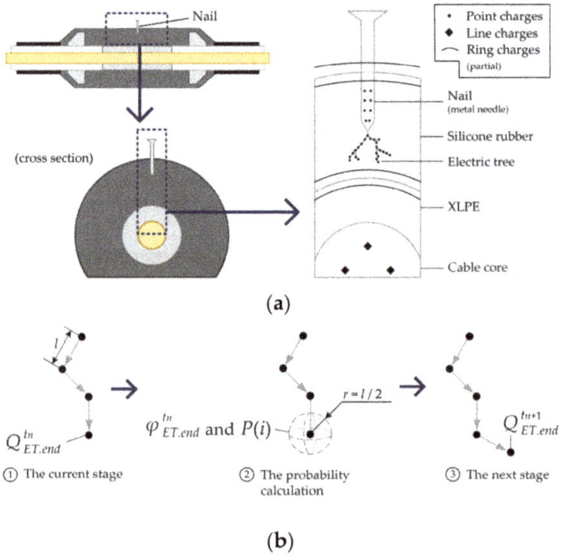

Figure 8. Simulation model of the electrical tree: (**a**) charge distribution inside the metal needle and the electrical tree and (**b**) simulation of electrical tree development process.

Since the formation of the electrical tree is a time-varying process, the potential φ of each position inside the silicone rubber is calculated by:

$$\varphi^{t_n} = \sum_{i=1}^{n_{MN}} P_i Q_{MN(i)} + \sum_{i=1}^{n_{ET}} P_i Q_{ET(i)} + \sum_{i=1}^{n_A} P_i Q_{A(i)} + \sum_{i=1}^{n_{XLPE.O}} P_i Q_{XLPE.O(i)}, \tag{26}$$

where n_{MN} and n_{ET} are the numbers of point charges Q_{MN} and Q_{ET} inside the metal needle and on the electrical tree, respectively; and the superscript t_n represents the nth state. The instant t_n is related to the initial time t_0 and the time step Δt, i.e.,

$$t_n = t_0 + \Delta t. \tag{27}$$

In this model, the length of the electrical tree propagation step is l. Thus, the probability of electrical tree development is related to the potential $\varphi_{ET.end}^{t_n}$ on a sphere with point charge $Q_{ET.end}^{t_n}$ based on random walk theory at the end of the electrical tree (Figure 8b). The calculation for potential $\varphi_{ET.end}^{t_n}$ is shown in Equation (26).

The number of possible development directions is n_{dir}. For the *i*-th direction, its probability *P*(*i*) is calculated by [33,34]:

$$P(i) = \frac{\tau\left[\varphi_{ET.end(i)}^{t_n}\right]}{\sum_{j=1}^{n_{dir}} \varphi_{ET.end(j)}^{t_n}}, \tag{28}$$

where $\varphi_{ET.end(j)}^{t_n}$ is the potential on the sphere in the *j*-th direction and $\tau(\varphi)$ is a piecewise function:

$$\tau(\varphi) = \begin{cases} \varphi, \varphi > E_{dielectr} l \\ 0, \varphi \leq E_{dielectr} l \end{cases}, \tag{29}$$

where $E_{dielectr}$ is the dielectric strength of the silicone rubber. In other words, for electrical tree development the electric strength on the sphere must exceed the dielectric strength of the silicone rubber.

The program flow diagram is shown in Figure 9.

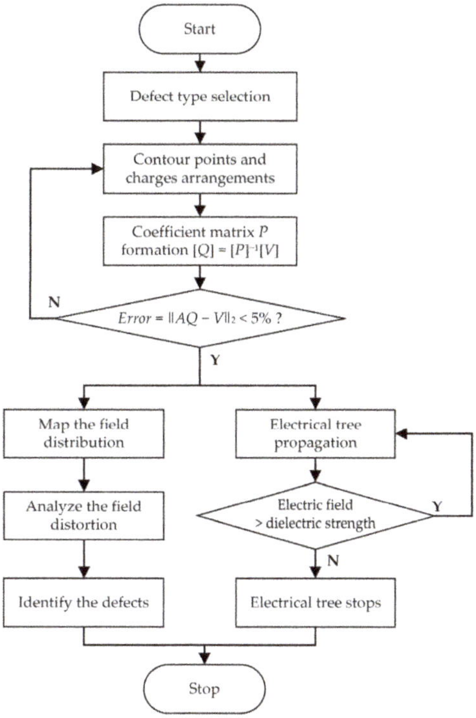

Figure 9. The program flow chart for the simulation process.

3. Results

3.1. Simulation Results with Four Types of Defects

3.1.1. Electric Field Distribution with the Air Void Defect

Electric field distributions with the air void defect are shown in Figure 10. Figure 10a,b shows the location of the defect in the cylindrical cable joint structure. Figure 10c displays the electric field distribution in the front view of the cable joint. Figure 10d shows the electric field distribution with an air void defect along the measurement line in Figure 8a. Figure 10e presents the electric field strength on the cable joint surface along the measurement lines in Figure 10a,b.

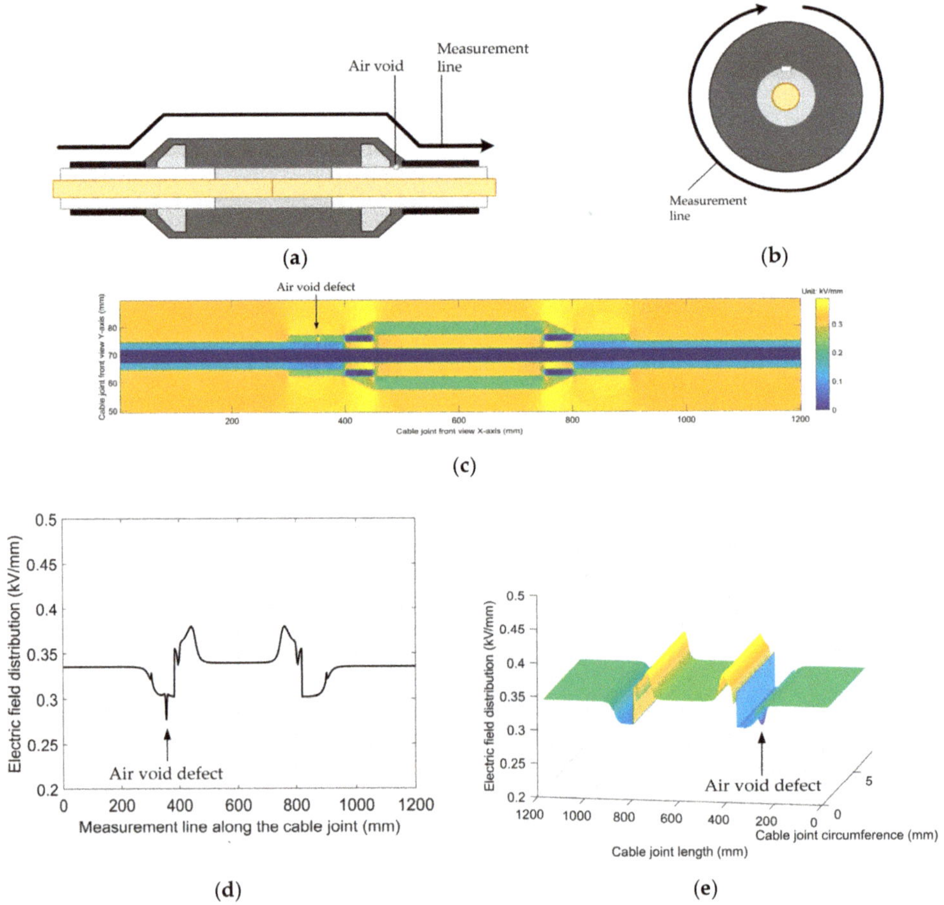

Figure 10. Electric field distributions with the air void defect: (**a**) front view of defect and field measurement line in the cable joint, (**b**) cross-section view of defect and field measurement line of the cable joint, (**c**) front view of electric field distribution of the cable joint, (**d**) front view of electric field distribution along measurement line, and (**e**) electric field strength on the cable joint surface.

Figure 10c shows that the electric field strength increased in the air void defect area due to the small relative permittivity of air, and the electric field around the defect decreased (Figure 10d). Figure 10e illustrates that the electric field strength on the cable joint surface was distorted and evidently decreased in the area close to the air void defect.

3.1.2. Electric Field Distribution with the Water Film Defect

Electric field distributions with the water film defect are shown in Figure 11. Figure 11a,b shows the location of the defect in the cylindrical cable joint structure. Figure 11c displays the electric field distribution from the front view of the cable joint. Figure 11d shows the electric field distribution with the water film defect along the measurement line shown in Figure 11a. Figure 11e presents the electric field strength on the cable joint surface along the measurement lines from Figure 11a,b.

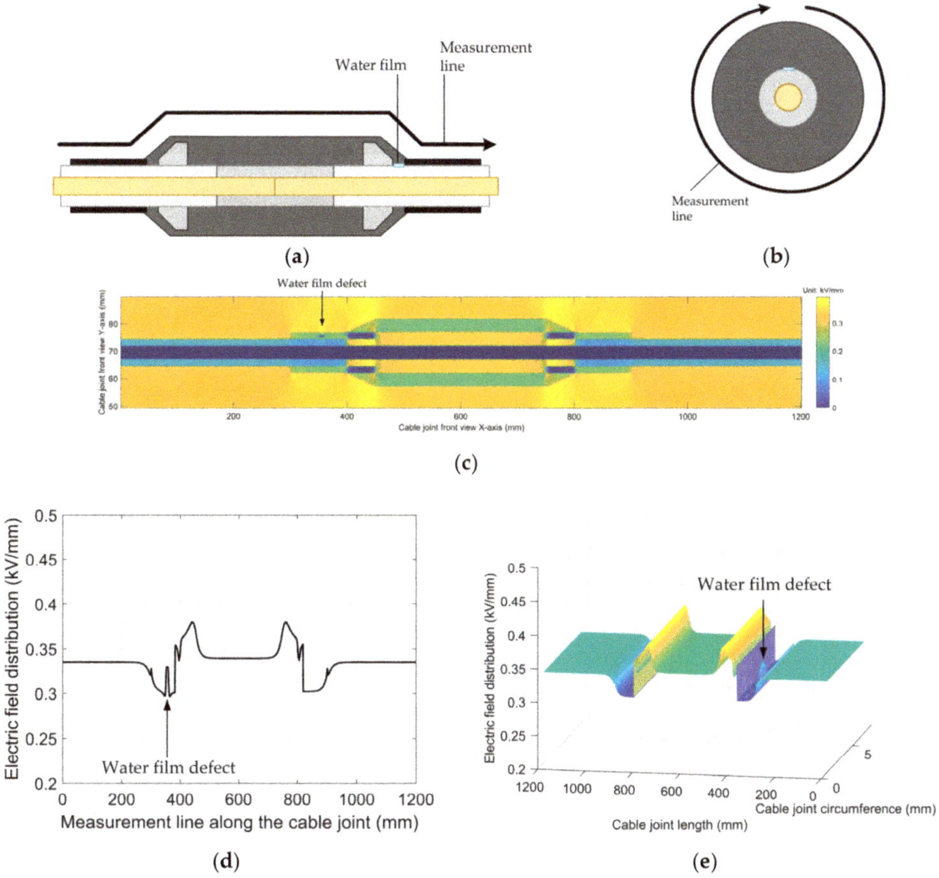

Figure 11. Electric field distributions with the water film defect: (**a**) defect and field measurement line from the front view of the cable joint, (**b**) cross-section view of defect and field measurement line of the cable joint, (**c**) electric field distribution from the front view of the cable joint, (**d**) electric field distribution along the front view measurement line, and (**e**) electric field strength on the cable joint surface.

Figure 11c shows that the electric field strength decreased in the water film defect area due to the large relative permittivity of water, and the electric field around the defect increased (Figure 11d). Figure 11e illustrates that the electric field strength on the cable joint surface was distorted and evidently increased in the area close to the water film defect.

3.1.3. Electric Field Distribution with the Metal Debris Defect

Electric field distributions with the metal debris defect are shown in Figure 12. Figure 12a,b shows the location of the defect in the cylindrical cable joint structure. Figure 12c displays the electric field distribution from the front view of the cable joint. Figure 12d shows the electric field distribution with a metal debris defect along the measurement line from Figure 12a. Figure 12e presents the electric field strength on the cable joint surface along the measurement lines from Figure 12a,b.

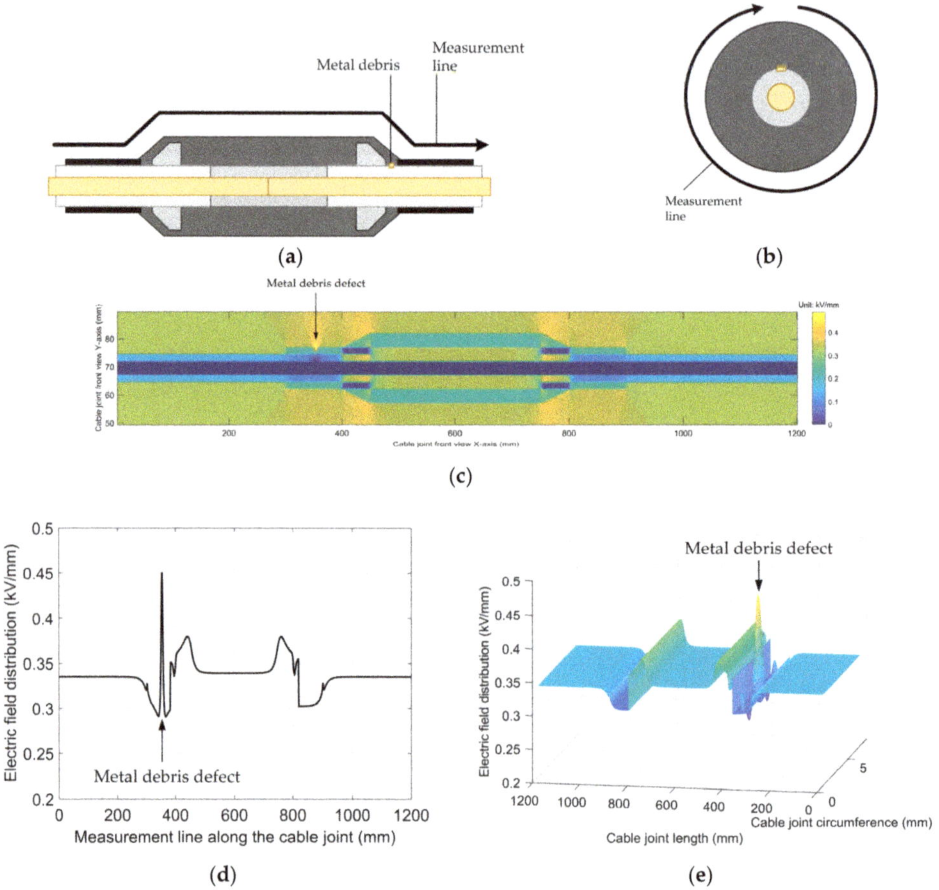

Figure 12. Electric field distributions with the metal debris defect: (**a**) defect and field measurement line from the front view of the cable joint, (**b**) defect and field measurement line of the cross-section of the cable joint, (**c**) electric field distribution from the front view of the cable joint, (**d**) electric field distribution along the front view measurement line, and (**e**) electric field strength on the cable joint surface.

Figure 12c shows that the electric field strength decreased in the metal debris defect area due to the conductive property of the metal debris, and the electric field around the defect increased (Figure 12d). Figure 12e illustrates that the electric field strength on the cable joint surface was distorted and evidently increased in the area close to the metal debris defect.

3.1.4. Electric Field Distribution with the Metal Needle Defect

Electric field distributions with the metal needle defect are shown in Figure 13. Figure 13a,b shows the location of the defect in the cylindrical cable joint structure. Figure 13c displays the front view of the electric field distribution of the cable joint. Figure 13d shows the electric field distribution with the metal needle defect along the measurement line from Figure 13a. Figure 13e presents the electric field strength on the cable joint surface along the measurement lines from Figure 13a,b.

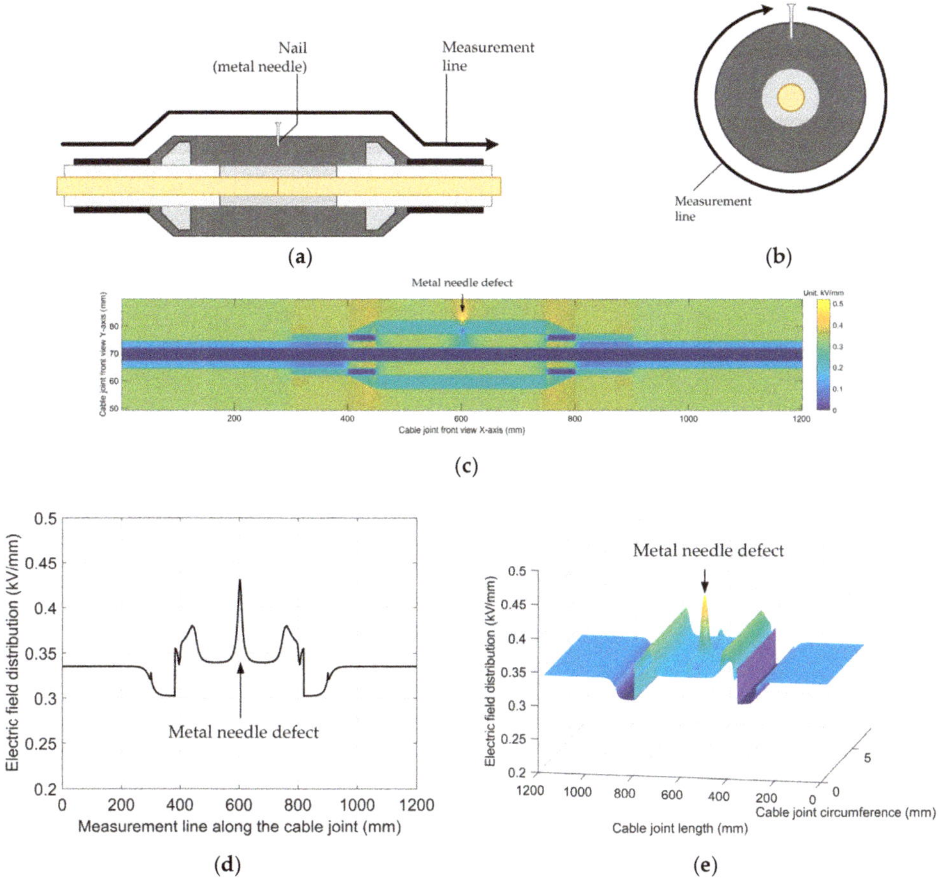

Figure 13. Electric field distributions with the metal needle defect: (**a**) front view of the defect and field measurement line of the cable joint, (**b**) cross-section view of the defect and field measurement line of the cable joint, (**c**) electric field distribution from the front view of the cable joint, (**d**) electric field distribution along the front view measurement line, and (**e**) electric field strength on the cable joint surface.

Figure 13c shows that the electric field strength decreased in the metal needle defect area due to the conductive property of the metal needle, and the electric field around the defect increased (Figure 13d). Figure 13e shows that the electric field strength on the cable joint surface was distorted and evidently increased in the area close to the metal needle defect.

3.1.5. Electric Field Comparison with Four Types of Defects

The pattern of electric field distributions with four types of defects were analyzed based on the field distortion magnitude and tendency along the measurement lines on the cable joint surface. The characteristics of the field distributions for all types of defects are shown in Table 1.

Table 1. Comparison of electric field distributions of four types of defects.

Type of Defect	Field Distortion Tendency	Field Distortion Magnitude (%)
Air void	Decrease	9.3
Water film	Increase	11.8
Metal debris	Increase	26.2
Metal needle	Increase	29.7

Table 1 shows that the field distortions of metal debris and needle were more severe than the field distortions caused by an air void and water film. The field distortion tendencies of air voids and water films were opposite. The types and locations of defects in the cable joint could be identified and located based on the characteristics of the measured electric field.

3.2. *Experiment Results of Four Types of Defects*

Electric field distribution along the measurement lines on the cable joint surface was measured using a probe based on the Pockels effect to reduce electromagnetic interference. The probe was installed on the ball screw structure to maintain a 10 mm distance from the top of the probe to the cable joint surface. The experiment schematic and setup are shown in Figures 14 and 15, respectively. The supply voltage of the cable core was set to 7 kV AC (Line to ground).

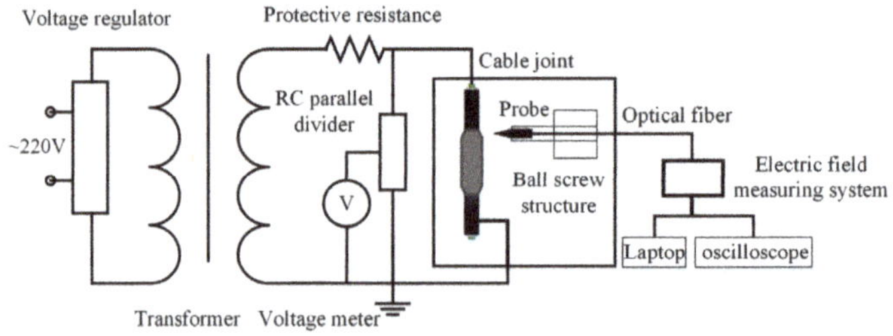

Figure 14. Electric field measurement system schematic.

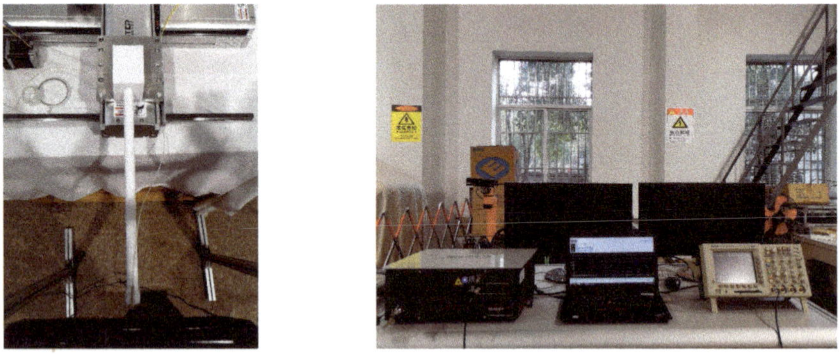

(**a**) (**b**)

Figure 15. Electric field measurement system setup: (**a**) the probe on the ball screw structure and (**b**) the electric field measurement equipment.

3.2.1. Artificial Defects in the Cable Joint

For the experiment, four typical defects were created to imitate the defects caused by external forces during cable joint manufacturing and installation processes (Figure 16).

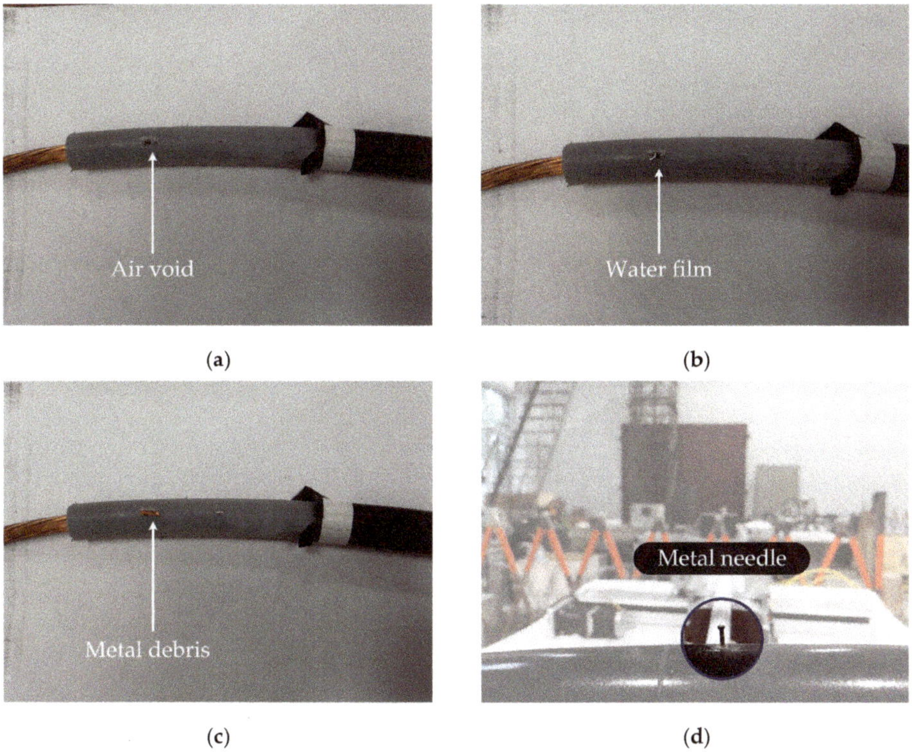

Figure 16. Artificial defects in the cable joint: (**a**) air void, (**b**) water film, (**c**) metal debris, and (**d**) metal needle.

3.2.2. Electric Field Measurement Results

The electric field distribution was measured at discrete points around the cable joint surface. The location of the probe was accurately controlled by the ball screw structure to ensure that the distance between all the measurement points was 10 mm. The consecutive points were interpolated in the discrete measurement points to obtain the continuous plots in Figure 17 for four types of defects. The sizes and locations of the artificial defects were the same as the sizes and locations of the defects in the simulation model to enable comparison of the results. The electric field measurements were repeated five times at the same location close to the defect for each type of defect. The number of discrete points with more than 30% variance was less than 4% during the measurement. The maximum variances of the electric field strength from five repetitions are shown in Table 2.

It was observed that the variances of the repetitions were relatively small, and the number of repetitions did not significantly affect the accuracy of the measurement.

Figure 17 illustrates that electromagnetic interference and noise in the environment affects the accuracy of the electric field measurement. The accumulated errors between the simulation model and experiment results are shown in Table 3. From the experiment, it was concluded that the air void and water film defects could not be detected when their sizes were less than 5 mm^3 because the field distortions caused by the air void and water film were relatively small and could be confused with the interference. The conductive metal defects could be identified and located when their sizes were larger than 2 mm^3 due to the severe electric field distortion they produced.

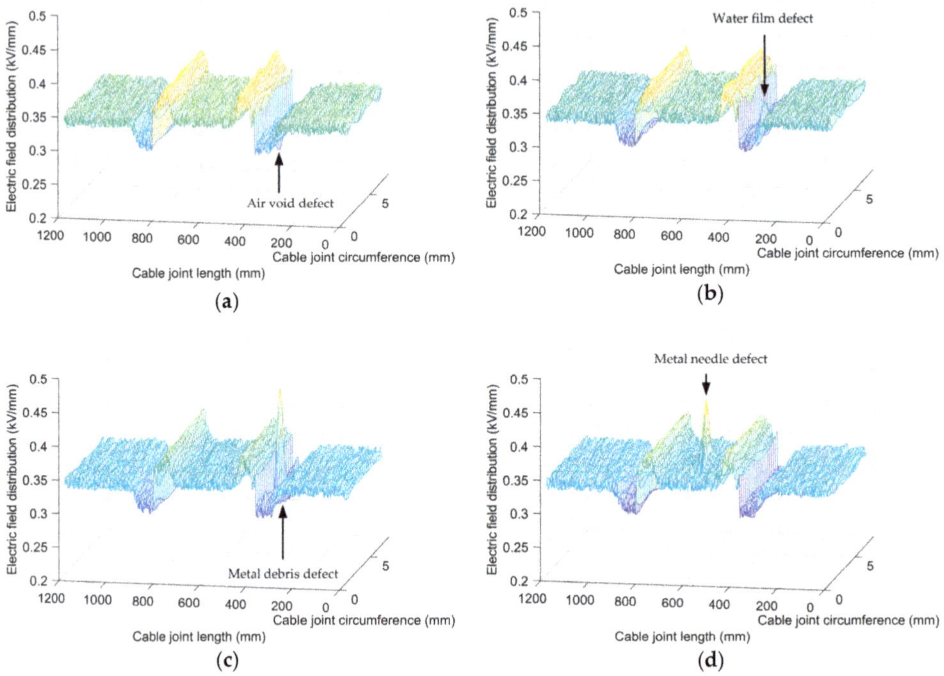

Figure 17. Electric field measurement results around the cable joint surface with (**a**) the air void defect, (**b**) the water film defect, (**c**) the metal debris defect, and (**d**) the metal needle defect.

Table 2. The maximum variances of the electric field strength from five repetitions.

Type of Defect	Maximum Variances of Field Strength (kV^2/mm^2)
Air void	0.00015
Water film	0.00021
Metal debris	0.00018
Metal needle	0.00022

Table 3. Accumulated errors between the simulation and experiment results.

Type of Defect	Accumulated Error (%)
Air void	12.1
Water film	11.7
Metal debris	18.4
Metal needle	19.1

3.3. Simulation Results of Electrical Tree Propagation

The electrical tree propagation processes were simulated close to the defects, using the same types, locations, and sizes as in the electric field calculation model. The electrical tree initiated at the location with the maximum electric field strength and propagated for 9.2 min. The propagation processes were repeated 50 times to analyze the patterns of the tree trajectories.

Figure 18a shows that the average length of the electrical tree trajectories around the air void defects was 2.2 mm. The average length of the trajectories was small because the electric field around the air void defect decreased (Table 1) and the field distortion was mild. Figure 18b shows that the average length of the electrical tree trajectories around the water film defects was 4.7 mm. The average length of the trajectory increased because the electric field increased around the water film defect (Table 1). Figure 18c,d indicate that the average lengths of the electrical tree trajectories around the metal debris and needle defects were 7.5 mm and 11.2 mm, respectively, due to the severe electric field distortion around the defects. The electrical tree trajectories around the conductive metal defects tended to approach the cable core because the vertical field vector to the cable core was significantly larger than the field vectors in other directions during the tree propagation process.

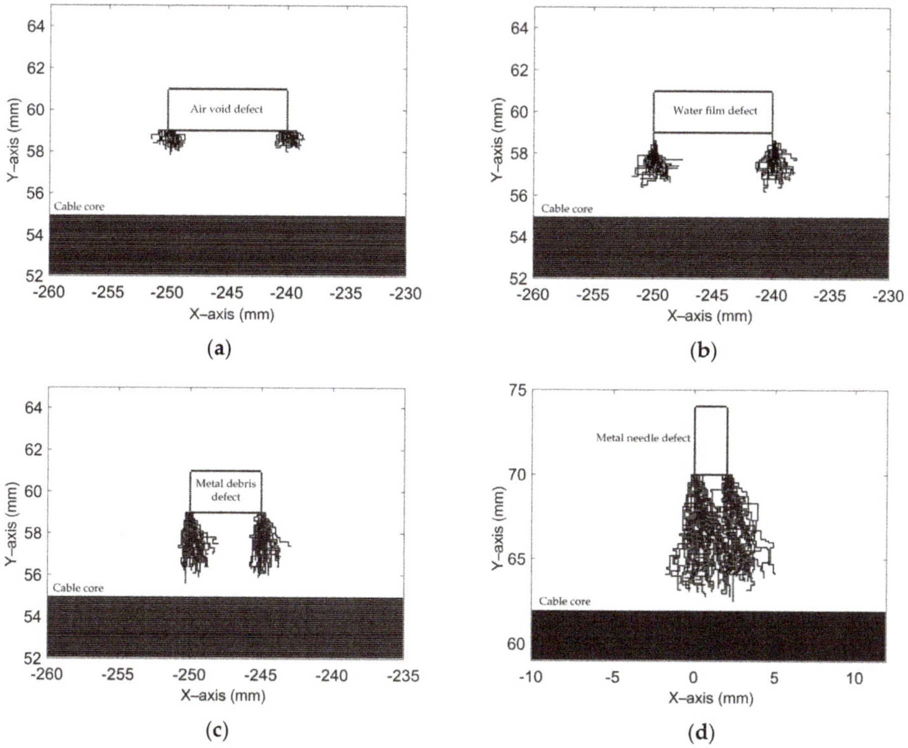

Figure 18. Electrical tree distribution after propagation with (**a**) the air void defect, (**b**) the water film defect, (**c**) the metal debris defect, and (**d**) the metal needle defect.

4. Conclusions

The number of cable failure accidents continuously increased due to the defects produced during the manufacturing and aging processes, this paper provided a defect detection method for a 10 kV XLPE cable and predicted the aging rate based on the characteristics of the electric field distribution. The electric field distribution was calculated and measured around the defective cable surface. The simulation and calculation results were compared to validate the charge-simulation-based electric field model and the electric field measurements. The electrical tree propagation processes were simulated based on CSM and random walk theory to analyze the pattern of electrical tree trajectories.

1. Four typical defects could be identified in the cable joint based on their unique electric field distribution characteristics. Electric field intensification was the original cause of insulation material breakdown, and the electric field measurement with Pockels effect reduced environmental interferences. A feature library of defects could be built for the operating personnel to locate and identify the internal and external defects.
2. The electric field distortion magnitudes were 9.3%, 11.8%, 26.2%, and 29.7% for the defects of air void, water film, metal debris, and metal needle, respectively. The air and water defects of large size caused less field distortion when compared with metal defects of small size. Therefore, air void and water film defects were difficult to detect when they were smaller than 5 mm^3 because their field distortions were relatively small and might be concealed by interference.
3. The combination of CSM and random walk theory could accurately describe the electrical tree propagation in cable joints. CSM was used to calculate the instantaneous electric field and charge distributions at each step of tree propagation. Random walk theory describes the stochastic property of electrical tree growth by determining the tree propagation direction based on the electric field analysis and the probabilistic model. The electrical tree model could predict the aging rate of the insulation material and prevent the failure accident of the cable joint.
4. The average length of the electrical tree trajectories around a water film defect was larger than those around an air void defect due to the opposite field distortion tendencies around the defect. The electrical trees around metal debris and needles were more inclined to approach the cable core and cause main insulation breakdown compared with other types of defects.

Author Contributions: Conceptualization: J.H.; Methodology: J.H.; Software: J.H. and K.H.; Validation: K.H. and L.C.; Formal Analysis: J.H. and K.H.; Investigation: J.H.; Experiment: L.C.; Writing—Original Draft Preparation: J.H.; Writing—Review and Editing: J.H., K.H., and L.C.

Funding: This research was supported by the National Natural Science Foundation of China (Grant No. 51807028), the Basic Research Program of Jiangsu Province (Grant No. BK20170672).

Conflicts of Interest: The authors declare no conflict of interest.

Nomenclature

$E_{dielectr}$	Dielectric strength of silicone rubber (unit: kV/mm)
l	Length step of electrical tree development (unit: mm)
n_A	Number of ring charges in the air close to the silicone rubber
n_{core}	Number of line charges in the cable core
$n_{D.SiR}$	Number of point charges in the defects close to the silicone rubber interface
$n_{D.XLPE}$	Number of point charges in the defects close to the XLPE layer
n_{dir}	Number of possible directions of electrical tree development
n_{ET}	Number of point charges in the electrical tree
n_M	Number of point charges in the metal debris
n_{MN}	Number of point charges in the metal needle
$n_{pipe.O}$	Number of ring charges at the outer side of the connecting pipe
$n_{SC.I}$	Number of ring charges at the inner side of the stress cone
$n_{SC.O}$	Number of ring charges at the outer side of the stress cone
$n_{SiR.D}$	Number of point charges in the silicone rubber close to the defect interface
$n_{SiR.I}$	Number of ring charges at the inner side of the silicone rubber
$n_{SiR.O}$	Number of ring charges at the outer side of the silicone rubber
$n_{XLPE.D}$	Number of point charges in the XLPE layer close to the defect interface
$n_{XLPE.O}$	Number of ring charges at the outer side of the XLPE layer
$P(i)$	Probability of the i-th possible direction of electrical tree development
$P_{i,j}$	Potential coefficient (unit: kV/pC)
Q_A	Ring charges in the air close to the silicone rubber (unit: pC)

Q_{core}	Line charges in the cable core (unit: pC)
$Q_{D.SiR}$	Point charges in the defects close to the silicone rubber interface (unit: pC)
$Q_{D.XLPE}$	Point charges in the defects close to the XLPE layer (unit: pC)
Q_{ET}	Point charges in the electrical tree (unit: pC)
$Q^{t_n}_{ET.end}$	Point charge at the end of the electrical tree at n-th stage (current stage, unit: pC)
$Q^{t_{n+1}}_{ET.end}$	Point charge at the end of the electrical tree at $(n + 1)$-th stage (next stage, unit: pC)
Q_M	Point charges in the metal debris (unit: pC)
Q_{MN}	Point charges in the metal needle (unit: pC)
$Q_{pipe.O}$	Ring charges at the outer side of the connecting pipe (unit: pC)
$Q_{SC.I}$	Ring charges at the inner side of the stress cone (unit: pC)
$Q_{SC.O}$	Ring charges at the outer side of the stress cone (unit: pC)
$Q_{SiR.D}$	Point charges in the silicone rubber close to the defect interface (unit: pC)
$Q_{SiR.I}$	Ring charges at the inner side of the silicone rubber (unit: pC)
$Q_{SiR.O}$	Ring charges at the outer side of the silicone rubber (unit: pC)
$Q_{XLPE.D}$	Point charges in the XLPE layer close to the defect interface (unit: pC)
$Q_{XLPE.O}$	Ring charges at the outer side of the XLPE layer (unit: pC)
r	Radius of the sphere centered on $Q^{t_n}_{ET.end}$, $r = l/2$ (unit: mm)
Δt	Time step of electrical tree development (unit: s)
t_0	Initial state of electrical tree development (unit: s)
t_n	The n-th state (current state) of electrical tree development (unit: s)
t_{n+1}	The $(n + 1)$-th state (next state) of electrical tree development (unit: s)
ε_A	Permittivity of air (unit: pF/m)
ε_D	Permittivity of dielectric of non-conductive defect (unit: pF/m)
ε_{SC}	Permittivity of material of stress cone (unit: pF/m)
ε_{pipe}	Permittivity of material of connecting pipe (unit: pF/m)
ε_{SiR}	Permittivity of silicone rubber (unit: pF/m)
ε_W	Permittivity of water (unit: pF/m)
ε_{XLPE}	Permittivity of XLPE (unit: pF/m)
φ_{core}	Potential of cable core (unit: kV)
$\varphi^{t_n}_{ET.end}$	Potential of the sphere surface with $Q^{t_n}_{ET.end}$ as the center and $l/2$ as the radius(unit: kV)
φ_M	Potential of metal debris (unit: kV)

References

1. Blodgett, R.B.; Fisher, R.G. Insulations and Jackets for Cross-Linked Polyethylene Cables. *IEEE Trans. Power App. Syst.* **1963**, *82*, 971–980. [CrossRef]
2. McKean, A.L.; Oliver, F.S.; Trill, S.W. Cross-Linked Polyethylene for Higher Voltages. *IEEE Trans. Power App. Syst.* **1967**, *PAS-86*, 1–10. [CrossRef]
3. Pack, A.V. Service Aged Medium Voltage Cables – A Critical Review of Polyethylene Insulated Cables. In Proceedings of the Conference Record of the 2004 IEEE International Symposium on Electrical Insulation, Indianapolis, IN, USA, 19–22 September 2004.
4. Zhang, C.; Li, C.; Zhao, H.; Han, B.A. Review on the Aging Performance of Direct Current Cross-linked Polyethylene Insulation Materials. In Proceedings of the 2015 IEEE 11th International Conference on the Properties and Applications of Dielectric Materials (ICPADM), Sydney, NSW, Australia, 19–22 July 2015.
5. Thue, W.A. *Electrical Power Cable Engineering*, 3rd ed.; CRC Press: Boca Raton, FL, USA, 2011.
6. Luo, H.; Cheng, P.; Liu, H.; Kang, K.; Yang, F.; Yang, Q. Investigation of Contact Resistance Influence on Power Cable Joint Temperature based on 3-D Coupling Model. In Proceedings of the 2016 IEEE 11th Conference on Industrial Electronics and Applications (ICIEA), Hefei, China, 5–7 June 2016.
7. Ruan, J.; Liu, C.; Tang, K.; Huang, D.; Zheng, Z.; Liao, C. A Novel Approach to Estimate Temperature of Conductor in Cable Joint. In Proceedings of the 2015 IEEE International Magnetics Conference (INTERMAG), Beijing, China, 11–15 May 2015.
8. Colavitto, A.; Contin, A.; Vicenzutti, A.; Sulligoi, G.; McCandless, M. Impact of Harmonic Pollution in Junctions between DC Cables with Different Insulating Technologies: Electrical and Thermal Analyses. In Proceedings of the 2019 IEEE Milan PowerTech, Milan, Italy, 23–27 June 2019.

9. Kubota, T.; Takahashi, Y.; Hasegawa, T.; Noda, H.; Yamaguchi, M.; Tan, M. Development of 500-kV XLPE Cables and Accessories for Long Distance Underground Transmission Lines-Part II: Jointing Techniques. *IEEE Trans. Power Deliv.* **1994**, *9*, 1750–1759. [CrossRef]
10. Takeda, N.; Lzumi, S.; Asari, K.; Nakatani, A.; Noda, H.; Yamaguchi, M.; Tan, M. Development of 500-kV XLPE Cables and Accessories for Long-distance Underground Transmission Lines. IV. Electrical Properties of 500-kV Extrusion Molded Joints. *IEEE Trans. Power Deliv.* **1996**, *11*, 635–643. [CrossRef]
11. Yang, H.; Liu, L.; Sun, K.; Li, J. Impacts of Different Defects on Electrical Field Distribution in Cable Joint. *J. Eng.* **2019**, *2019*, 3184–3187. [CrossRef]
12. Song, M.; Jia, Z. Calculation and Simulation of Mechanical Pressure of XLPE-SR Surface in Cable Joints. In Proceedings of the 2018 12th International Conference on the Properties and Applications of Dielectric Materials (ICPADM), Xi'an, China, 20–24 May 2018.
13. Chen, C.; Liu, G.; Lu, G.; Wang, J. Influence of Cable Terminal Stress Cone Install Incorrectly. In Proceedings of the 2009 IEEE 9th International Conference on the Properties and Applications of Dielectric Materials, Harbin, China, 19–23 July 2009.
14. Chen, C.; Liu, G.; Lu, G.; Wang, J. Mechanism on Breakdown Phenomenon of Cable Joint with Impurities. In Proceedings of the 2009 IEEE 9th International Conference on the Properties and Applications of Dielectric Materials, Harbin, China, 19–23 July 2009.
15. Zhang, L.; LuoYang, X.; Le, Y.; Yang, F.; Gan, C.; Zhang, Y. A Thermal Probability Density-Based Method to Detect the Internal Defects of Power Cable Joints. *Energies* **2018**, *11*, 1674. [CrossRef]
16. Yang, F.; Liu, K.; Cheng, P.; Wang, S.; Wang, X.; Gao, B.; Fang, Y.; Xia, R.; Ullah, I. The Coupling Fields Characteristics of Cable Joints and Application in the Evaluation of Crimping Process Defects. *Energies* **2016**, *9*, 932. [CrossRef]
17. Yang, F.; Zhu, N.; Liu, G.; Ma, H.; Wei, X.; Hu, C.; Wang, Z.; Huang, J. A New Method for Determining the Connection Resistance of the Compression Connector in Cable Joint. *Energies* **2018**, *11*, 1667. [CrossRef]
18. Zhu, W.; Zhao, Y.; Han, Z.; Wang, X.; Wang, Y.; Liu, G.; Xie, Y.; Zhu, N. Thermal Effect of Different Laying Modes on Cross-Linked Polyethylene (XLPE) Insulation and a New Estimation on Cable Ampacity. *Energies* **2019**, *12*, 2994. [CrossRef]
19. Wang, Z.; Li, H.; Li, Y. PD Detection in XLPE Cable Joint based on Electromagnetic Coupling and Ultrasonic Method. In Proceedings of the 2011 International Conference on Electrical and Control Engineering, Yichang, China, 16–18 September 2011.
20. Huang, Z.; Zeng, Y.; Wang, R.; Ma, J.; Nie, Z.; Jin, H. Study on Ultrasonic Detection System for Defects inside Silicone Rubber Insulation Material. In Proceedings of the 2018 International Conference on Power System Technology (POWERCON), Guangzhou, China, 6–8 November 2018.
21. Jin, W.; Wang, H.; Kuffel, E. Application of the Modified Surface Charge Simulation Method for Solving Axial Symmetric Electrostatic Problems with Floating Electrodes. In Proceedings of the 1994 4th International Conference on Properties and Applications of Dielectric Materials (ICPADM), Brisbane, Queensland, Australia, 3–8 July 1994.
22. Singer, H.; Steinbigler, H.; Weiss, P. A Charge Simulation Method for the Calculation of High Voltage Fields. *IEEE Trans. Power Appl. Syst.* **1974**, *PAS-93*, 1660–1668. [CrossRef]
23. Malik, N.H. A Review of the Charge Simulation Method and Its Applications. *IEEE Trans. Electr. Insul.* **1989**, *24*, 3–20. [CrossRef]
24. Takuma, T.; Kawamoto, T.; Fujinami, H. Charge Simulation Method with Complex Fictitious Charges for Calculating Capacitive-Resistive Fields. *IEEE Trans. Power App. Syst.* **1981**, *PAS-100*, 4665–4672. [CrossRef]
25. Murooka, Y.; Nakano, T.; Takahashi, Y.; Kawakami, T. Modified Pockels Sensor for Electric-field Measurements. *IEE Proc. Sci. Meas. Technol.* **1994**, *141*, 481–485. [CrossRef]
26. Long, F.; Zhang, J.; Xie, C.; Yuan, Z. Application of the Pockels Effect to High Voltage Measurement. In Proceedings of the 2007 8th International Conference on Electronic Measurement and Instruments, Xi'an, China, 16–18 August 2007.
27. Talaat, M. Influence of Transverse Electric Fields on Electrical Tree Initiation in Solid Insulation. In Proceedings of the 2010 Annual Report Conference on Electrical Insulation and Dielectic Phenomena, West Lafayette, IN, USA, 17–20 October 2010.

28. Song, W.; Zhang, D.; Wang, X.; Lei, Q. Characteristics of Electrical Tree and Effects of Barriers on Electrical Tree Propagation under AC Voltage in LDPE. In Proceedings of the 2009 IEEE 9th International Conference on the Properties and Applications of Dielectric Materials, Harbin, China, 19–23 July 2009.
29. Wang, W.; Chen, S.; Yang, K.; He, D.; Yu, Y. The Relationship between Electric Tree Aging Degree and the Equivalent Time-frequency Characteristic of PD Pulses in High Voltage Cable. In Proceedings of the 2012 IEEE International Symposium on Electrical Insulation, San Juan, PR, USA, 10–13 June 2012.
30. Fujiwara, O. An Analytical Approach to Model Indirect Effect Caused by Electrostatic Discharge. *IEICE Trans. Commun.* **1996**, *E79-B*, 483–489.
31. Champion, J.V.; Dodd, S.J. Modelling Partial Discharges in Electrical Trees. In ICSD'98. In Proceedings of the 1998 IEEE 6th International Conference on Conduction and Breakdown in Solid Dielectrics, Vasteras, Sweden, 22–25 June 1998.
32. Zhou, Y.; Zhang, Y.; Zhang, L.; Guo, D.; Zhang, X.; Wang, M. Electrical tree Initiation of Silicone Rubber after Thermal Aging. *IEEE Trans. Dielectr. Electr. Insul.* **2016**, *23*, 748–756. [CrossRef]
33. Noskov, M.D.; Malinovski, A.S.; Sack, M.; Schwab, A.J. Self-consistent Modeling of Electrical Tree Propagation and PD Activity. *IEEE Trans. Dielectr. Electr. Insul.* **2000**, *7*, 725–733.
34. Noskov, M.D.; Malinovski, A.S.; Sack, M.; Schwab, A.J. Modelling of Partial Discharge Development in Electrical Tree Channels. *IEEE Trans. Dielectr. Electr. Insul.* **2003**, *10*, 425–434. [CrossRef]
35. Zhou, P.B. *Numerical Analysis of Electromagnetic Fields*, 1st ed.; Springer Verlag: Berlin, Germany, 1993.
36. El-Kishky, H.; Gorur, R.S. Electric Potential and Field Computation along AC HV Insulators. *IEEE Trans. Dielectr. Electr. Insul.* **1994**, *1*, 982–990. [CrossRef]

energies

MDPI

Article

Thermal Effect of Different Laying Modes on Cross-Linked Polyethylene (XLPE) Insulation and a New Estimation on Cable Ampacity

WenWei Zhu [1,2,†], YiFeng Zhao [1,†], ZhuoZhan Han [1], XiangBing Wang [2], YanFeng Wang [2], Gang Liu [1,*], Yue Xie [1,*] and NingXi Zhu [1]

1 School of Electric Power, South China University of Technology, Guangzhou 510640, China
2 Grid Planning Research Center, Guangdong Power Grid Co., Ltd., Guangzhou 510000, China
* Correspondence: liugang@scut.edu.cn (G.L.); epxieyue@mail.scut.edu.cn (Y.X.)
† These authors contributed equally to this work.

Received: 4 July 2019; Accepted: 1 August 2019; Published: 3 August 2019

Abstract: This paper verifies the fluctuation on thermal parameters and ampacity of the high-voltage cross-linked polyethylene (XLPE) cables with different insulation conditions and describes the results of a thermal aging experiment on the XLPE insulation with different operating years in different laying modes guided by Comsol Multiphysics modeling software. The thermal parameters of the cables applied on the models are detected by thermal parameter detection control platform and differential scanning calorimetry (DSC) measurement to assure the effectivity of the simulation. Several diagnostic measurements including Fourier infrared spectroscopy (FTIR), DSC, X-ray diffraction (XRD), and breakdown field strength were conducted on the treated and untreated specimens in order to reveal the changes of properties and the relationship between the thermal effect and the cable ampacity. Moreover, a new estimation on cable ampacity from the perspective on XLPE insulation itself has been proposed in this paper, which is also a possible way to judge the insulation condition of the cable with specific aging degree in specific laying mode for a period of time.

Keywords: thermal effect; cable; XLPE; laying modes; Comsol Multiphysics; thermal parameters; cable ampacity

1. Introduction

Thermoplastics are widely used in the insulation of power cables, such as polyvinyl chloride (PVC), low density polyethylene (LDPE), and cross-linked polyethylene (XLPE), which possess higher physical, electrical, and heat-resistant properties compared with oil-paper insulation [1]. The PVC insulation is commonly applied in low-voltage and medium-voltage cables, of which the permissible long-term load temperature is 70 °C [2]; the LDPE power cables can be operated at higher voltage levels by adding special additives [3]. In addition, the permissible long-term load temperature is 75 °C and the permissible maximum short-circuit temperature is 130 °C. Due to the crosslinking treatment of PE, the permissible long-term load temperature and the permissible maximum short-circuit temperature have been elevated to 90 and 250 °C, respectively [4]. The XLPE cables have been widely applied in high-voltage (HV) and extra-high voltage (EHV) transmission systems due to the its high performance on electricity and thermal properties. Therefore, it is meaningful to expand the research frontiers about the XLPE cables in the actual operation to elevate the permissible long-term load capacity and extend their service life.

XLPE is a high-molecular polymer with crystal and amorphous phase, which is subjected mostly to the thermal effect because the crystal structure of XLPE is prone to be affected by the melting and cooling process [5]. It is well evidenced that long-term thermal effect can change the morphology of the insulation distinctly and lead to degradation on the insulation [6,7]. In fact, the influence of thermal effect on polymers is a complicated process and the insulation properties can be improved in some certain situations [8]. Much research has been carried out on this subject. The inverse temperature effect is presented to indicate that the degradation of semi-crystalline polymers is obvious at the low temperatures and a significant recovery at elevated temperatures [9]. In the study about the non-isothermal melt-crystallization kinetics of polymers [10,11], it is well known that the ability of XLPE to crystallize is highly dominated by the cooling rate and the current melting condition of XLPE, because different melting and cooling rates can change the form of spherulites, the size and distribution level of which determine some specific conductive properties of XLPE [12,13]. Therefore, it is meaningful to investigate the changes of properties on XLPE with different melting and cooling rates.

During its practical operation, cable ampacity is influenced by many factors, such as the kind of laying modes and cable specifications [14,15]. Although the factor of the laying mode has been taken into account in International Electrotechnical Commission (IEC) and Institute of Electrical and Electronics Engineers (IEEE) standards in the calculation of cable ampacity, the influence of cable aging on the cable ampacity was rarely concerned. It is universal to calculate the XLPE cable ampacity by the thermal-circuit method and simulation software, which follows the basic rule of calculating the maximum current through the conductor corresponding to the steady-state temperature of 90 °C in different environments [16]. The precondition of the cable ampacity calculation is setting the volumetric thermal capacity and thermal resistance of XLPE as fixed values [17–19]. To a large extent, these methods neglect the changes on XLPE insulation itself caused by the different aging factors and the influence of the heating and cooling process to the material during the cable operating condition, which would totally change the morphology of XLPE [20–22]. It can be inferred that the thermal resistance and thermal capacity of XLPE in the thermal-circuit model are variable for the reason that the microstructure of XLPE is changing with many factors, such as heat and electricity. Moreover, papers [23,24] have found that the measured values of thermal resistance and thermal capacity of XLPE are different distinctly from the IEC standard. Further, it can be considered that the cable ampacity is a fluctuating value rather than a constant as to the same specification XLPE cable in the same environment with the passage of time. Therefore, it is meaningful to pursue a new way to assess the cable ampacity.

This paper has focused on two aspects of the XLPE insulation in order to reveal the changes on XLPE properties and the relationship between thermal effect and cable ampacity. The first aspect concerns the verification of fluctuation on thermal parameters and ampacity of the cable with different insulation conditions. The second aspect provides a new estimation on cable ampacity from the perspective of the changes of XLPE insulation condition, considering the factors of the different operating years and the different laying modes of the cables. Such correlations between the cable insulation and the cable ampacity should be researched further for a better understanding of the cable ampacity and the aging mechanism of XLPE.

2. Experimental

2.1. Preparation of XLPE Specimens

Two retired high-voltage AC XLPE cables with service years of 15 and 30, and a spare high-voltage AC XLPE cable were selected in this paper. For convenience, they are named by their service year: XLPE-0, XLPE-15, and XLPE-30. Some critical parameters of these cables are listed in Table 1.

Overheated operation has not been reported for these two retired cables, which means that the temperature in the insulation layer remained below 90 °C during the cable operation. Each cable insulation was peeled parallel to the conductor surface, and the tape-like XLPE peels were obtained. Peels near the inner semi-conductive layer were taken as the specimens, because these positions of the insulation endured the most severe electrical and thermal stresses. These obtained specimens were all cleaned by alcohol to remove the surface impurities.

Table 1. Critical parameters of the cables.

Cable	V_L	M_I	M_C	d_I	O_P
XLPE-0	110/63.5	XLPE	Cu	18.5	1998-
XLPE-15	110/63.5	XLPE	Cu	18.5	1999–2015
XLPE-30	110/63.5	XLPE	Cu	18.5	1985–2015

V_L—voltage level in kV, M_I—insulation material, M_C—conductor material, d_I—insulation thickness in mm, O_P—operation period. Other parameters such as cross-linking method are unknown.

2.2. Cable Thermal Parameters and Ampacity Measurement

By means of the high-voltage cable thermal parameter detection control platform, as illustrated in Figure 1, the three cables with the length of 1.5 m were taken to test its ampacity under the air-laying mode. The critical thermal parameters and ampacity of these three cables are listed in Table 2, where volumetric thermal capacity δ of XLPE was measured by differential scanning calorimetry (DSC) at the temperature of 30 °C; thermal resistance R of XLPE and thermal conductivity λ of XLPE were deducted by thermal circuit model and its Matlab program, and results are shown in the Appendix A; ambient temperature θ_O and conductor steady-state temperature θ_C were measured by thermocouples, and the ampacity under the air-laying mode I_A was tracked by the Matlab program.

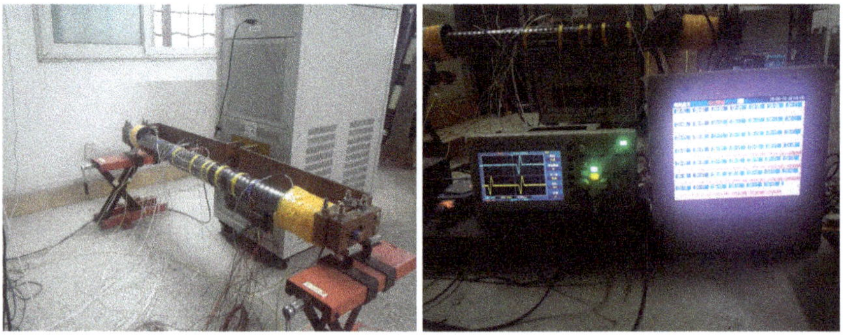

(**a**) (**b**)

Figure 1. High-voltage cable thermal parameter detection control platform. (**a**) Ampacity experimental cable and AC constant current source, (**b**) thermal compensation system and paperless recorder.

Table 2. Critical thermal parameters and ampacity of the cables.

Cable	δ (J/K·m³)	R (K·m/W)	λ (W/K·m)	θ_O (°C)	θ_C (°C)	I_A (A)
XLPE-0	1.94×10^6	3.23	0.31	17.7	89.6	1244.07
XLPE-15	1.96×10^6	3.45	0.29	17.7	89.7	1222.99
XLPE-30	2.07×10^6	5.26	0.19	17.7	90.2	1096.47

δ—volumetric thermal capacity of cross-linked polyethylene (XLPE), R—thermal resistance of XLPE, λ—thermal conductivity of XLPE, θ_O—ambient temperature, θ_C—conductor steady-state temperature and I_A—ampacity under the air-laying mode.

From the results in Table 2, it can be verified that the thermal parameters and ampacity fluctuated with different conditions (thermal history, operating years or material characteristic) of the XLPE insulation. Therefore, it is meaningful to provide new estimations on cable ampacity to improve the current research on cable ampacity.

The structural model and thermal parameters in Table 2 of these cables with the same specification is the foundation of the subsequent analysis on simulation of different laying modes based on Comsol Multiphysics. The configuration of the experimental cables is shown in Figure 2.

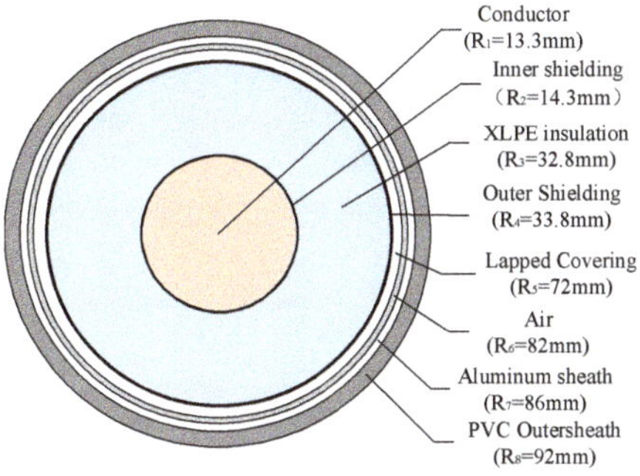

Figure 2. Configuration of the ampacity experimental cable and structural model used for Comsol Multiphysics. R_x is the corresponding radius of each layer.

2.3. Simulation of Different Laying Modes of Cables

Comsol Multiphysics was adopted to simulate different laying modes of the XLPE cables. The thermal parameters of the three cables listed in Table 2 were extracted to construct the simulation model, respectively, and we found that the simulation results had the same changing tendency. The permanents of XLPE-0 were used to construct the following simulation model. The establishment of thermal field model and determination of the boundary conditions are not the highlight in this paper, the detail can be referred to literature [25–27].

Firstly, we constructed the structural model (Figure 2) under the air-laying mode on the Comsol Multiphysics and we found that the simulative ampacity approached to the practical experiment, which was 1152 A. Therefore, it is assured that the following simulative results are effective.

Secondly, we defined cable ampacity under air-laying mode as reference cable ampacity I_R (I_R = 1200 A) and constructed the structural models into three different laying modes, including pipeline, ground, and tunnel-laying modes based on Comsol Multiphysics, as shown in Figure 3.

Thirdly, we applied the 1.2 I_R and the 0 I_R into the simulated cables, which were laid in pipeline, underground, and in the tunnel, respectively, and obtained the simulative melting and cooling curves of three different laying modes that are shown in Figure 4a,b.

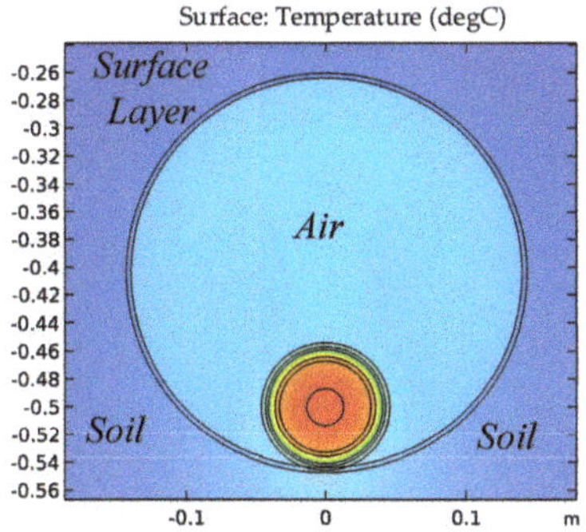

(**a**) Pipeline-laying mode

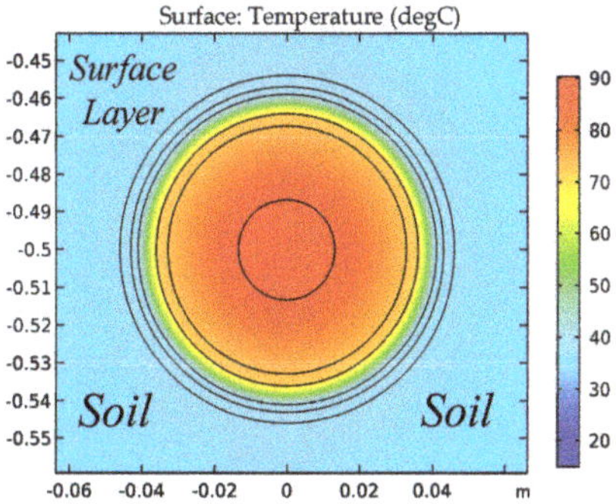

(**b**) Ground-laying mode

Figure 3. *Cont.*

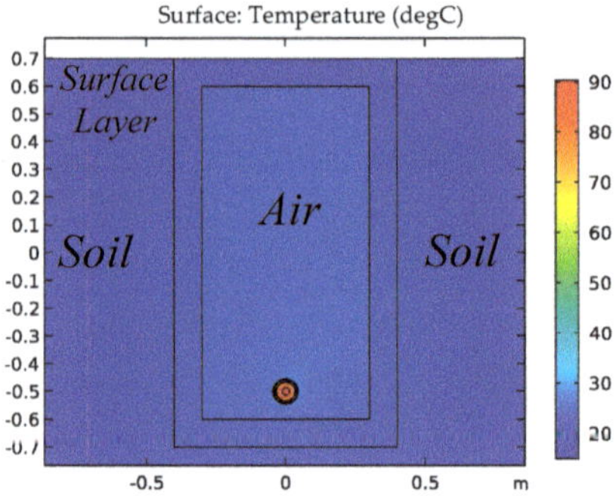

(**c**) Tunnel-laying mode

Figure 3. Schematic diagrams of the structural models of three different laying modes based on Comsol Multiphysics.

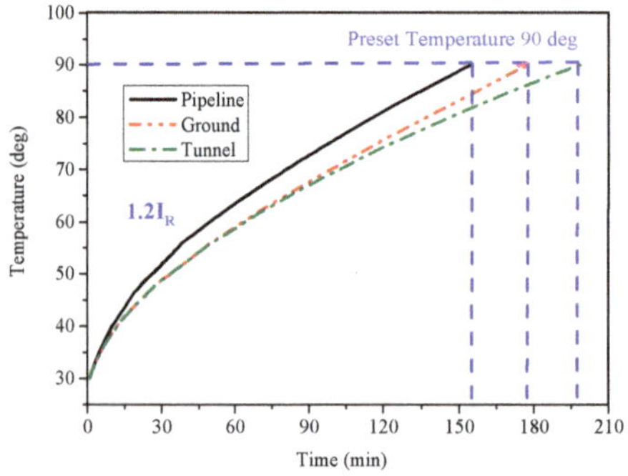

(**a**) Simulative melting curves

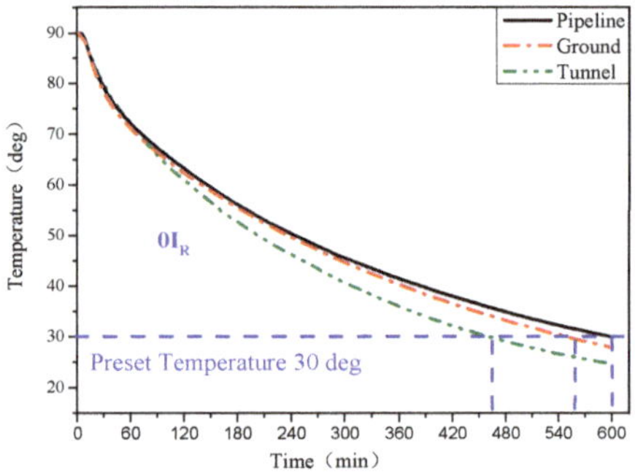

(**b**) Simulative cooling curves

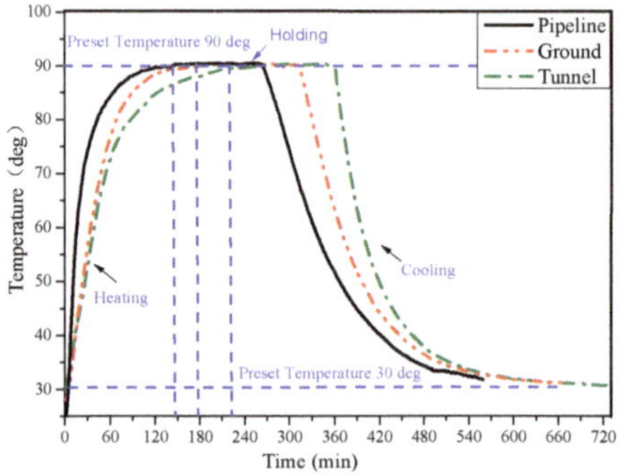

(**c**) Practical measured curves

Figure 4. Simulative melting and cooling curves and practical measured curves of three different laying modes.

2.4. Thermal Aging

The cable specimens with different operation years were formed into a group, and there were four groups in total. The first group was the untreated group which played a role of reference. The second group simulated the cable laying in the pipeline with the lowest heat dissipation efficiency. The third group simulated the cable laying under the ground with the moderate heat dissipation efficiency. The last group simulated the cable laying in the tunnel with the highest heat dissipation efficiency.

Thermal aging experiments were performed on the last three groups of the specimens. Three modified aging ovens were used to control the heating and cooling process, which fitted the corresponding simulative curves, and a 2 h constant temperature phase of 90 °C between the heating and cooling phases were added to the thermal aging. The three phases, including heating, holding, and cooling, were repeated 80 times in order to enlarge the differences among the specimens. The practical temperature curves of the three different laying modes are shown in Figure 4c, measured by the thermo-couple sensors.

2.5. Diagnostic Measurements

The Fourier transform infrared spectroscopy (FTIR), the differential scanning calorimetry (DSC) measurement, X-ray diffraction (XRD), and the breakdown field strength measurement were adopted to analyze the properties changes on the specimens after the thermal effects.

Micro-structure changes on the specimens were analyzed by the VERTEX 70 infrared spectrometer manufactured by German. Each specimen was tested at 32 scans in the range of 600~3600 cm^{-1} with a resolution of 0.16 cm^{-1} and the signal-to-noise ratio of 55,000:1. The obtained spectra were analyzed by OPUS software.

Thermal properties changes on the specimens were analyzed by the DSC NETZSCH-DSC 214 instrument manufactured by German. Five milligram specimens were prepared to test, with the program of two heating phases and a cooling phase under nitrogen atmosphere to avoid thermal degradation. The temperature was increased from 25 to 140 °C at a rate of 10 °C/min and maintained at 140 °C for 5 min, and then cooled to 25 °C. This scanning was repeated twice per measurement, and the first cooling and second heating phases were analyzed in this paper.

Crystal structure changes on the specimens were analyzed by Bruker D8 ADVANCE X-ray diffractometer manufactured by Germany. The experimental interval of Bragg angle was $2\theta = 5°–90°$ by step size of 0.02° with 0.1 s/step scan rate. The obtained data were analyzed by software of DIFFIAC plus XRD Commander.

Electric properties changes of the specimens were analyzed by the ZJC-100 kV voltage breakdown tester manufactured by China. Each specimen was cut into 50 × 50 × 0.5 mm to test, with the voltage rising rate of 1 kV/s under the transformer oil. The valid AC breakdown field strength of each specimen was measured 5 times to obtain the average breakdown field strength.

3. Results and Discussions

In the process of the thermal effects, the features of each laying mode from the heating and cooling curves were observed in the results of Figure 4. Among the three laying modes, the pipeline mode possesses the quick heating and slow cooling features, the ground mode possesses the moderate heating and slow cooling features, and the tunnel mode possesses the slow heating and quick cooling features. These differences led to the different statuses among the specimens with different operating years. In the following parts, we will focus on the connection among the changes of micro-structure, crystal structure, and external electrical property of each specimen in order to reveal the relationship between the thermal effects and the cable ampacity.

3.1. Result of FTIR Spectroscopy Measurement

The oxidation process of XLPE under the thermal effects can generate major by-products of thermal-oxidation, such as carbonyl groups and unsaturated groups, which can signify the aging status of the specimens. Figure 5 is the FTIR spectra of each specimen whose wavenumber and absorbance represent the kind and content of the corresponding group. The wavelengths of 720, 1471, 2856, and 2937 cm^{-1} are all caused by the vibration of the methylene band ($-CH_2-$). Absorption peaks ranging from 1700 cm^{-1} to 1800 cm^{-1} can be considered as the thermo-oxidative products [3]. Among of them, carboxylic acid absorption appears at 1701 cm^{-1}, ketone absorption locates at 1718 cm^{-1}, and aldehyde absorption situates at 1741 cm^{-1}. The peak at 1635 cm^{-1} is assigned to the unsaturated groups absorption, which can indicate the decomposition process.

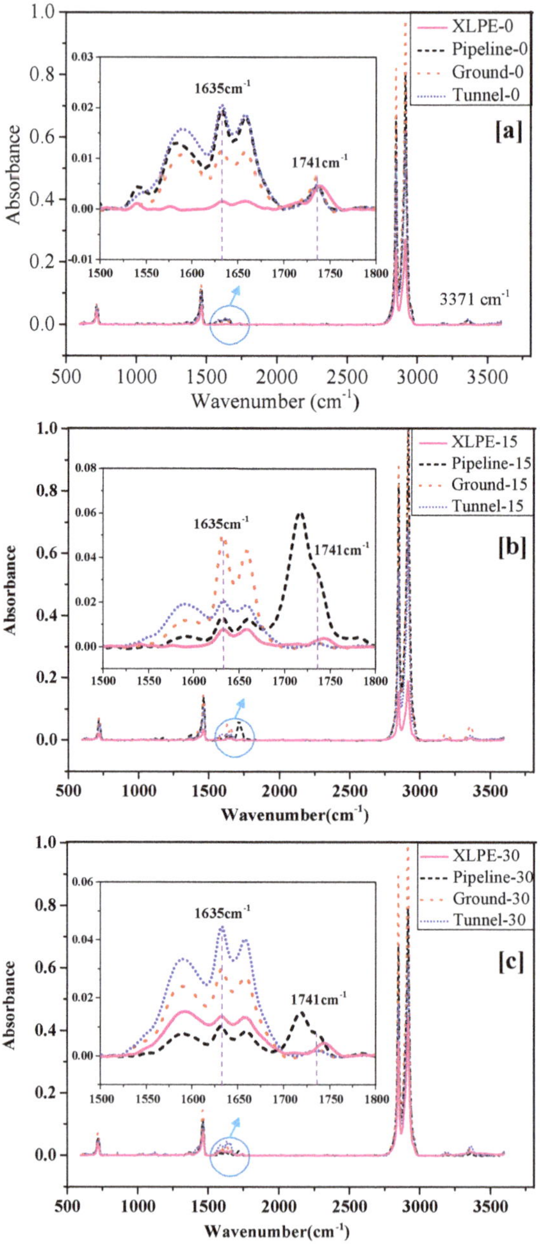

Figure 5. Fourier transform infrared spectroscopy (FTIR) spectra of the untreated specimens and the treated specimens under thermal aging of three different laying modes. (**a**) the spare cable; (**b**) the cable with service years of 15; (**c**) the cable with service years of 30.

From Figure 5a, it is clear that some peaks in the range of 1500 to 1700 cm^{-1} are present after the thermal effects. These peaks reflect a slight decomposition process on the backbone of the XLPE macromolecules. This phenomenon can be associated with the thermal activation to the XLPE molecular chain, which, as a consequence of the adequate motion of the molecular chain, leads to the generation of a certain quantity of small chain segments, free polar groups, etc. On the other aspect, there are no excessive displacement in the position of the peaks in the range of 1700 to 1800 cm^{-1} among the specimens after thermal aging, which signifies the process of thermo-oxidation is moderate. Therefore, it can be considered that the different thermal aging modes mainly have activated the motion of macromolecular chains, but hardly cause the oxidative degradation in regard to the spare cable.

From Figure 5b, we can see the similar phenomenon happening on the specimens except for Pipeline-15. Significant peaks in the range of 1700 to 1800 cm^{-1} are presented in Pipeline-15, which is responsible for the aggravation of oxidative degradation. Moreover, there was a pronounced increase in peaks in the range of 1600 to 1650 cm^{-1} in Ground-15, which indicates a dominant process of chains scission occurs under the thermal effect of ground mode.

From Figure 5c, we can notice that quantities of small molecular chains have emerged on the untreated one (XLPE-30) due to the long-term operation in actual condition. After the thermal aging with different laying modes, it can be deduced that the thermal effects have sped up the decomposition process on the molecular backbone of Ground-0 and Tunnel-0, and a certain quantity of broken molecular chains have already transformed into oxidative groups in Pipeline-30. This indicates severe oxidative degradation occurs in Pipeline-30. It may be admitted that the potential to resist oxidation of XLPE-30 was weakened compared with XLPE-0.

In order to quantify the situation of oxidation and decomposition under the thermal effects, carbonyl index and unsaturated band index were chosen for research. The definition of these two indexes are as follows [3]:

$$CI = I_{1741}/I_{1471}, \quad (1)$$

$$UBI = I_{1635}/I_{1471}, \quad (2)$$

where carbonyl index (*CI*) is the relative intensities of the carbonyl band at 1741 cm^{-1} (aldehyde absorption) to the methylene band at 1471 cm^{-1}; unsaturated band index (*UBI*) is the relative intensities of the unsaturated group at 1635 cm^{-1} to the methylene band at 1471 cm^{-1}.

The two indexes are shown in Figure 6. It can be stated that the carbonyl index is decreased in all the specimens compared with the untreated ones, except for Pipeline-15 and Pipeline-30, and the unsaturated band index is increased except for Tunnel-15, Pipeline-15, and Pipeline-30. That means the thermal effects aggravate basically the decomposition process of the specimens to generate a certain quantity of broken molecular chains. The oxidation process was followed by chains scission and formation of smaller chain segments, which would easily react with oxygen (O_2) to transform into oxidative groups. For Pipeline-15 and Pipeline-30, the broken molecular chain segments are prone to transform into oxidative groups, probably due to the heating and cooling features of pipeline mode. With regard to Tunnel-15, the two indexes are decreased, which symbolizes the optimization of the micro-structure. As we demonstrated above, the status of the degradation and the stability of the specimens after thermal effects are listed in Table 3.

Table 3. Status of the degradation and stability of the specimens after thermal effects.

Specimen	Pipeline	Ground	Tunnel
XLPE-0	Improved	Improved	Improved
XLPE-15	Degraded	Maintained	Improved
XLPE-30	Degraded	Improved	Maintained

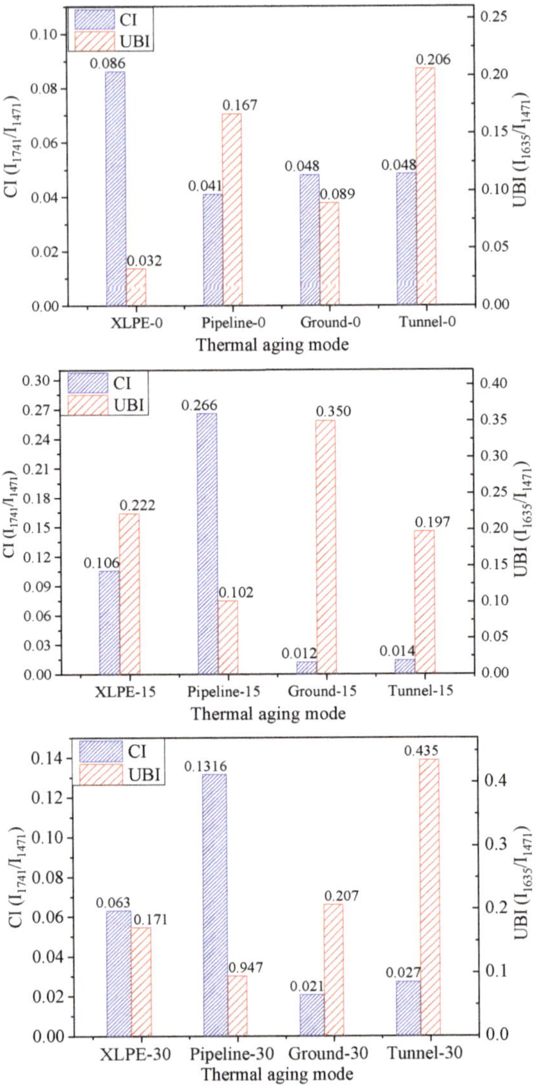

Figure 6. Carbonyl index, *CI*, and unsaturated band index, *UBI*, of the untreated specimens and the treated specimens under thermal aging of three different laying modes.

3.2. Result of DSC Measurement

With the program running, two consecutive scans were conducted on each specimen. The first scan began by heating from 30 to 140 °C and then followed by cooling from 140 to 30 °C. The second scan was excerpted in the same way as the first scan. The thermos-gram of the first heating phase is commonly used for observing the thermal history and the current crystalline condition of the specimens, which can also be analyzed by XRD measurement in Section 3.3. In order to avoid long arguments, two phases were analyzed, including the first cooling and the second heating phase, which can be analyzed on the ability to crystallize and the quality of the crystal structure of the specimens after the thermal effects.

Figures 7 and 8 show the thermos-grams of the two phases corresponding to the operating years of each cable. The obtained parameters are listed in Table 4, where T_c is the crystallizing peak temperature, ΔH_c is the enthalpy of crystallization, T_m is the melting peak temperature, and $\Delta T = T_m - T_c$ is the degree of supercooling, which is proportional inversely to the crystallization rate [22].

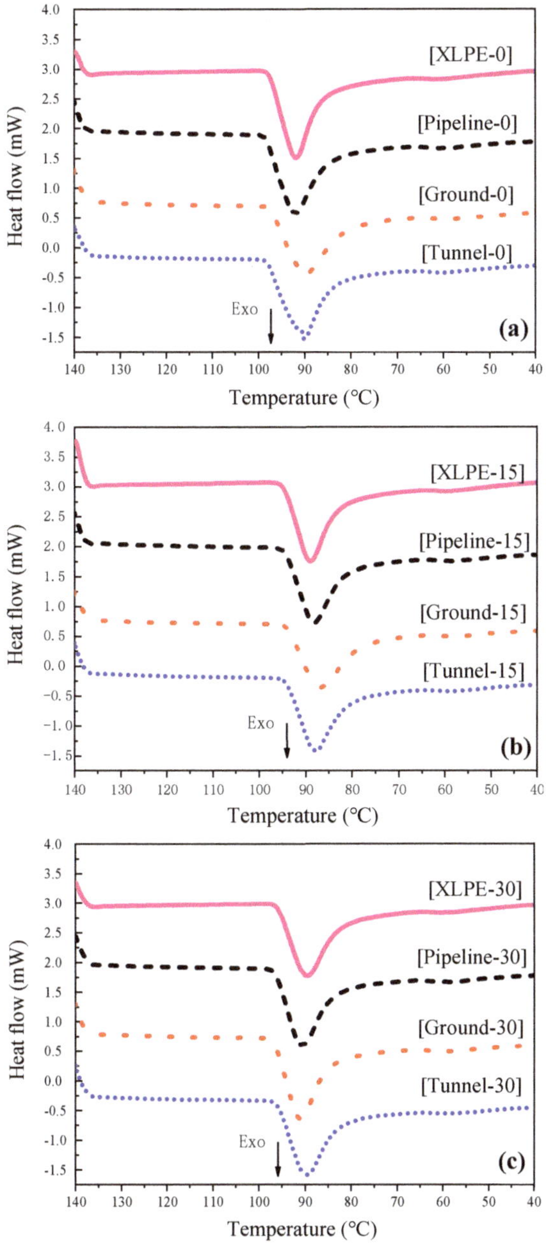

Figure 7. Heat flow as a function of measurement temperature in the first cooling of differential scanning calorimetry (DSC). (**a**) the spare cable; (**b**) the cable with service years of 15; (**c**) the cable with service years of 30.

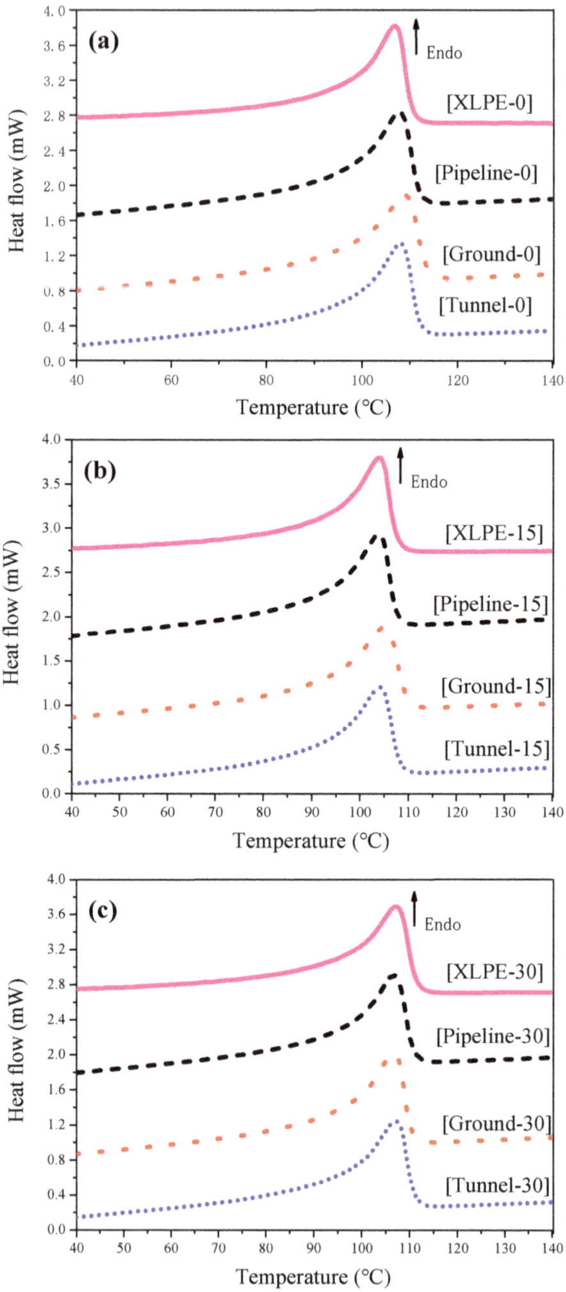

Figure 8. Heat flow as a function of measurement temperature in the second heating of differential scanning calorimetry (DSC). (**a**) the spare cable; (**b**) the cable with service years of 15; (**c**) the cable with service years of 30.

Table 4. Parameters obtained from first cooling and second heating phases.

Specimen	T_c (°C)	ΔH_c (J/g)	T_m (°C)	ΔH_f (J/g)	ΔT (°C)
XLPE-0	91.8	−99.0	106.9	100.8	15.1
Pipeline-0	92.0	−108.8	107.6	116.6	15.6
Ground-0	89.7	−102.8	109.0	111.1	19.3
Tunnel-0	90.0	−110.8	108.0	117.2	18.0
XLPE-15	89.0	−93.4	104.0	98.7	15.0
Pipeline-15	88.2	−99.6	103.8	108.1	15.6
Ground-15	86.9	−93.4	105.0	101.9	18.1
Tunnel-15	87.8	−99.1	104.1	101.1	16.3
XLPE-30	89.5	−103.9	107.2	103.4	17.7
Pipeline-30	90.6	−103.0	106.6	110.5	16.0
Ground-30	90.9	−105.1	106.6	111.3	15.7
Tunnel-30	89.8	−104.5	107.1	106.4	17.0

T_c—crystallizing peak temperature, ΔH_c—enthalpy of crystallization, T_m—melting peak temperature, ΔH_f—enthalpy of fusion and ΔT—super-cooling degree.

In Figure 7a, it is clearly observed that the exothermic peaks of Ground-0 and Tunnel-0 appear at a slight lower temperature and the shapes become broader, compared to XLPE-0 and Pipeline-0, in the first cooling phase. In Figure 8a, it can be found that melting peaks of all the treated specimens move towards higher temperatures, but the endothermic peaks are expanded. That means that although the thermal effects disperse the crystalline region, especially for Ground-0 and Tunnel-0, the main crystal structure of each treated specimen is improved.

In Figure 7b, exothermic peaks of all the treated specimens displace slightly toward lower temperatures. In Figure 8b, the melting peaks locate at higher temperatures for Ground-15 and Tunnel-15 but at lower temperature for Pipeline-15. The endothermic peaks become broader especially for Pipeline-15. It can be stated that the crystal structure becomes more deteriorated for a long time in the mode of the pipeline.

In Figure 7c or Figure 8c, the exothermic peaks of the treated specimens move towards a higher temperature and the endothermic peaks move towards a lower temperature, which means the thermal effects disrupt the distribution of crystal structure [28]. Therefore, the ability to crystallize is increased, but the quality is declined relatively. Among the specimens, Ground-30 presents the highest crystallization performance.

In addition, the DSC endotherms indicate a range of melting processes that can be related to the crystallinity and lamellar thickness variations. The crystallinity and average lamellar thickness [3,8] can be calculated by Formulas (1) and (2). The calculating formulas are as follows:

$$\chi(\%) = {}^{\Delta H_f}/_{\Delta H_f^0} \times 100, \tag{3}$$

$$T_m = T_{m0}(1 - {}^{2\sigma_e}/_{\Delta H_m L}), \tag{4}$$

where $\chi(\%)$ is crystallinity; ΔH_f^0 is the enthalpy of fusion of an ideal polyethylene crystal per unit volume; T_m is the observed melting temperature (K) of lamellar of thickness L; T_{m0} is the equilibrium melting temperature of an infinitely thick crystal; σ_e is the surface-free energy per unit area of basal face; ΔH_m is the enthalpy of fusion of an ideal polyethylene crystal per unit volume; and L is the lamellar thickness. The used values for calculation were as follows: T_{m0} = 414.6 K, ΔH_m = 2.88 × 10^8 J/m^3, and σ_e = 93 × 10^{-3} J/m^2. Changes on crystallinity and lamellar thickness with different thermal aging modes are depicted in Figure 9.

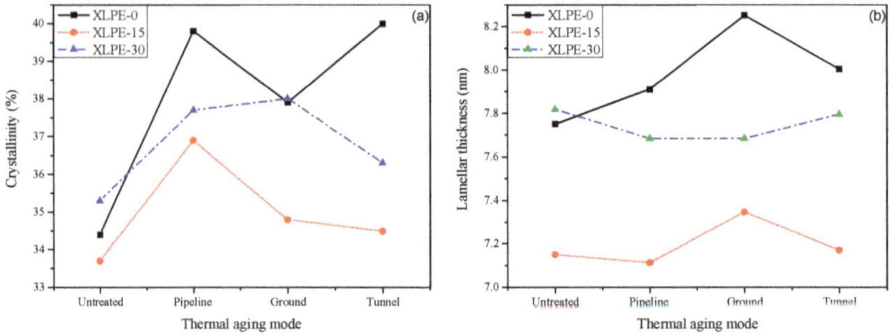

Figure 9. Changes on crystallinity and lamellar thickness of the main endothermic peak with different thermal aging modes: (**a**) Crystallinity; (**b**) lamellar of thickness.

We can observe that the crystallinity of all the specimens are increased in various degrees after the thermal aging in Figure 9a. This phenomenon is ascribed to adequate movement of the molecular chains and the generation of short chain segments, which are inclined to form the secondary crystal [3]. It is hard to judge the status of crystal structure just from the crystallinity. Therefore, the analysis should be combined with the change of lamellar of thickness obtained at the main endothermic peak shown in Figure 9b. We can find that both Pipeline-15 and Pipeline-30 have a high crystallinity but the lamellar thickness of main endothermic peaks are low, which indicates the crystal structures are mainly made up of secondary crystal. On the contrary, the Ground-0 and the Ground-15 have a relatively low crystallinity with solid lamellar of thickness, which means the crystal structures are firm and compact. Analogously, we can deduce the status of crystal structure of each specimen as shown in Table 5.

Table 5. Status of the crystal structure of the specimens after thermal effects.

Specimen	Pipeline	Ground	Tunnel
XLPE-0	Improved	Improved	Improved
XLPE-15	Degraded	Improved	Improved
XLPE-30	Degraded	Improved	Maintained

3.3. Result of XRD Measurement

The changes on the crystal structure of each specimen before and after the thermal effect were analyzed by X-ray diffractometer, to observe the influence on the crystalline phase of each layer under the different aging conditions.

Figure 10 displays the X-ray spectrum of each position in cable insulation before and after the accelerated aging test. It can be observed that two main crystalline peaks of each specimen appear at $2\theta = 21.22°$ and $2\theta = 23.63°$, which correspond to the (110) and (200) lattice planes. There is a one small peak observed at $2\theta = 36.5°$ which corresponds to the (020) lattice plane as Miller demonstrates [29]. There is no excessive displacement in the position of the peaks or in their splitting among the specimens, but the intensity and the shape of the peaks are different. It is indicated that accelerated aging hardly produces any new crystalline phase in the crystal structure, but results in the changes on the crystallinity and the grain size of the specimens.

From Figure 10, it can be noticed remarkably that a turnover phenomenon occurs in the thermal aging of tunnel mode, which is reflected by the fact that the crystal of the (110) lattice plane modifies into the (200) lattice plane. It is commonly assumed that the (200) lattice plane corresponds to the deformed spherulites, whose optical axes are oriented parallel to the radial direction [29]. This transformation indicates the shape of spherulite is associated with the heating and cooling process under different thermal effects. Therefore, we may admit the crystal structure changes on XLPE should be in the light of the features of different laying modes. For more precision on the objective to analyze the crystal structure changes among the specimens, the crystallinity percentage and the grain size are introduced by Formulas (5) and (6).

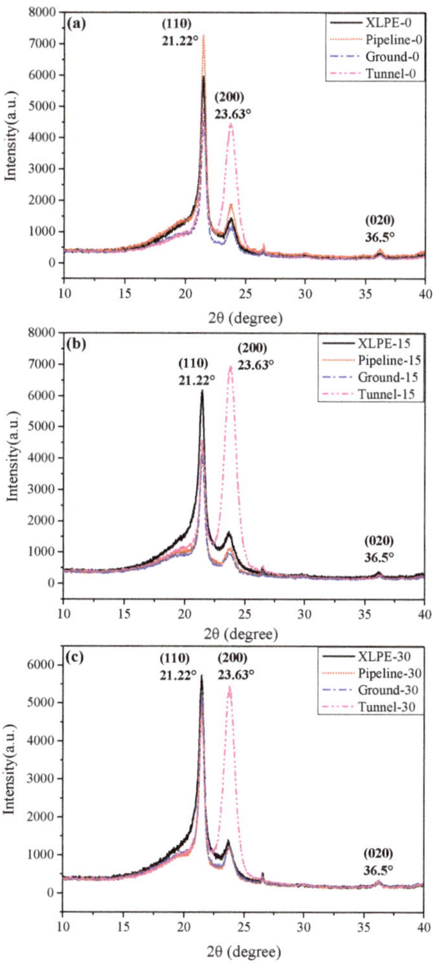

Figure 10. X-ray diffraction (XRD) spectrum of the untreated specimens and the treated specimens under thermal aging of three different laying modes. (**a**) the spare cable; (**b**) the cable with service years of 15; (**c**) the cable with service years of 30.

The crystallinity percentage can be calculated by the Hinrichsen method [30]. The X-ray spectrum of each specimen is fitted by Gaussian functions. Figure 11 displays the corresponding three Gauss fit peaks obtained by using the original 9.1. The calculating process is given as follows:

$$\chi(\%) = (S_2 + S_3)/(S_1 + S_2 + S_3) \times 100, \quad (5)$$

where χ (%) is crystallinity percentage, S_1 is the area of the amorphous halo, S_2 is the area of the main crystallization peak at $2\theta = 21.22°$, and S_3 is the area of the secondary crystallization peak at $2\theta = 23.63°$.

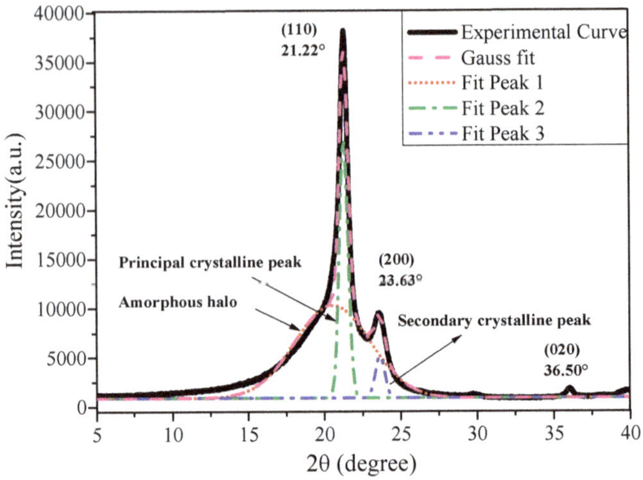

Figure 11. Gaussian fitting of the crystalline peaks and the amorphous halo by the Hinrichsen method.

In addition, the grain size of the different diffraction peaks corresponding to different crystal lamellar can be calculated by Scherrer equation [31]. The calculating formula is as follows:

$$D_{hkl} = (K{\cdot}\lambda_X)/(\beta\cos\theta), \tag{6}$$

where D_{hkl} is the grain size (A) perpendicular to the (hkl) crystal face; λ_X is experimental X-ray wavelength (nm), which is 0.15418 nm; β is the broadening of the diffraction peak (khl) (Rad) producing by grain refinement; and K is 0.89 when β is the full width at half maximum of the diffraction peak.

Basing on Formulas (5) and (6), we present the crystallinity percentage and the grain size perpendicular to the (110) and (200) crystal faces in Figure 12. With regard to XLPE-0 in Figure 12a, the crystallinity was lifted up in varying degrees by the disruption of the crystalline order after the thermal effect in each laying mode. Especially for Tunnel-0, a significant increase in crystallinity and distinct shrink in grain size perpendicular to the (200) crystal face indicate that the spherulites are tightly distributed in the insulation and the crystal structure is relatively perfect, even though most of the spherulites were deformed by thermal effect.

With regard to XLPE-15 in Figure 12b, it shows that the changes of the crystallinity and the grain sizes of Ground-15 and Tunnel-15 are similar to the corresponding ones of XLPE-0. The decrease in crystallinity and increase in crystallite size of the principal crystalline area for Pipeline-15 reflect that the spaces among the crystal structures are expanded and the integral crystal structure is not closely arranged.

With regard to XLPE-30 in Figure 12c, minor differences in the crystallinity and the grain sizes between Pipeline-30 and Ground-30 demonstrate that the influence of the features in pipeline and ground modes on the current crystalline condition is not pronounced. For Tunnel-30, we can also find high crystallinity with tight, deformed spherulites, as shown in Figure 12a,b. As we demonstrated above, the status of the current crystal structure of the specimens after thermal effects are listed in Table 6.

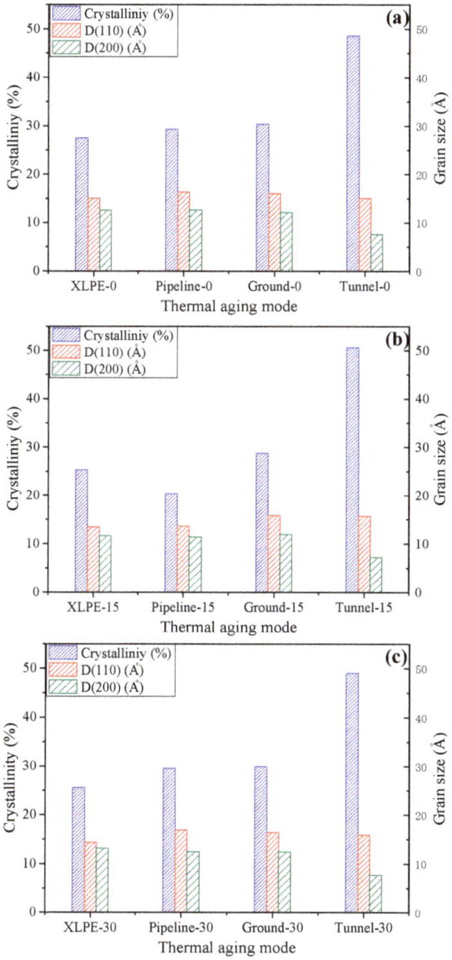

Figure 12. Changes on crystallinity and grain size perpendicular to the (110) and (200) crystal faces. (**a**) the spare cable; (**b**) the cable with service years of 15; (**c**) the cable with service years of 30.

Table 6. Status of the current crystal structure of the specimens after thermal effects.

Specimen	Pipeline	Ground	Tunnel
XLPE-0	Improved	Improved	Improved
XLPE-15	Degraded	Improved	Improved
XLPE-30	Maintained	Maintained	Improved

3.4. Results of the Breakdown Field Strength Measurement

The breakdown field strength test was conducted for each specimen and the valid average values are listed in Table 7. It is obvious that the electric properties were improved in varying degrees among XLPE-0 specimens after the different thermal effects. This phenomenon is probably attributed to the flexible movement of the molecular chain which activates the development of crystalline order and the annealing effect [6].

Table 7. Breakdown field strength (kV/mm).

Specimen	Untreated	Pipeline	Ground	Tunnel
XLPE-0	66.106	69.273	68.860	71.325
XLPE-15	73.489	63.941	75.640	81.650
XLPE-30	68.382	65.463	67.609	69.600

With regard to XLPE-15 and XLPE-30, the electric properties were improved under thermal effects in the ground- and tunnel-laying modes, but were degraded in the pipeline-laying mode. It may be considered that the features of the pipeline-laying mode are not relatively suitable for the aged cables with high loading for a long time.

3.5. Assessment of the Structural and Electrical Properties of XLPE under Thermal Effects

From the results of four diagnostic measurements, the correlations between structural and electrical properties were connected to assess the insulation condition.

For the spare cable under the three different thermal effects, the change of temperature activates the adequate movement of the molecular chain, leading to higher crystallinity and electrical performance of the treated specimens. The increase in crystallinity among the specimens is probably ascribed to the scission of the molecules traversing the amorphous regions, followed by rearrangement of the chains imperfectly crystallized at the manufacturing step to increase crystallinity [32]. Moreover, the improvement of crystal structure has a positive influence on the electrical breakdown and the resistivity of the insulation [33].

For the cable with service years of 15, thermal effect in tunnel mode presented a higher performance in structural and electrical properties, which can be identified by the diminution of oxygen-containing groups, improvement of crystalline morphology, and elevation of breakdown field strength. Properties indicators from the diagnostic measurements show a certain degradation happening on thermal effect in pipeline mode with high carbonyl content, inferior crystal structure, and drop in electrical breakdown.

For the cable with service years of 30, the flexibility of the molecular chain and the potential to recrystallize were weakened during long-term practical operation. Thermal effect in pipeline mode reflects degradation from the aspects of carbonyl content, crystal structure, and breakdown field strength compared with XLPE-30. A slight improvement of properties occurs on thermal effect in ground mode and maintaining of properties occurs on thermal effect in tunnel mode.

In this case, compared with the untreated ones, the status of the specimens under different thermal effects can be assessed comprehensively in Table 8.

Table 8. Comprehensive assessment on the specimens under the thermal effects.

Specimen	Pipeline	Ground	Tunnel
XLPE-0	Improved (greatly)	Improved (greatly)	Improved (greatly)
XLPE-15	Degraded (severely)	Maintained	Improved (greatly)
XLPE-30	Degraded (slightly)	Improved (slightly)	Maintained

3.6. A Proposal of New Estimation on Cable Ampacity

With the combination of the reference cable ampacity I_R and the results of diagnostic measurements, the relationship between the cable ampacity and the insulation properties was revealed according to the connection among the change of micro-structure, crystal structure, and external electrical property of each specimen. A new estimation on cable ampacity can be proposed from the perspective of changes of XLPE insulation properties to prolong the cables' service life. The specific regulation strategy of cable ampacity is as follows:

From the results of Table 8, the improvement and maintenance of overall insulation properties indicate that the margin of reference cable ampacity I_R can be lifted properly to 1.2 I_R. Analogously, the degradation of overall insulation properties indicates that the reference cable ampacity I_R should not be lifted to 1.2 I_R or should be decreased properly based on I_R.

4. Conclusions

The verification of fluctuation on thermal parameters and ampacity of the cable, with different insulation conditions and thermal effects of different laying modes, on XLPE insulation, with different operating years, was analyzed in this paper. Moreover, a new estimation on cable ampacity in regard to the specific aging status of XLPE cable with a specific laying mode was proposed. The following regulation strategy of cable ampacity based on I_R can be made to prolong the cables' service life from the perspective of XLPE insulation itself:

1. For the spare cable, it has a high potential to adapt to thermal effect of the three laying modes, so the margin of ampacity can be elevated to 1.2 I_R.
2. For the cable which had operated for 15 years, the margin of ampacity can be elevated to 1.2 I_R in the tunnel, should not be lifted to 1.2 I_R for a long time in ground mode, and should be decreased properly in the pipeline based on I_R.
3. For the cable which had operated for 30 years, the margin of ampacity can be elevated to 1.2 I_R in ground mode, should be reduced from 1.2 I_R in tunnel mode, but the margin of ampacity should not be elevated in the pipeline based on I_R.

In the future, more work will be devoted to assessing the relationship between the cable ampacity and XLPE insulation properties in the cable actual working condition. Establishing more effective methods on estimating cable ampacity and prolonging the cables service life are of great significance to the electric power industry.

Author Contributions: This paper is a result of the collaboration of all co-authors. G.L. and Y.X. conceived and designed the study. W.Z. and Y.Z. established the model and drafted the manuscript. Z.H. refined the language and provided statistical information. N.Z. helped with the corrections. X.W. and Y.W. designed and performed the experiments. All authors read and approved the final manuscript.

Funding: This research was funded by Guangdong Power Grid Co., Ltd. Grid Planning Research Center, grant number No. GDKJXM20172797.

Acknowledgments: The authors gratefully acknowledge Guangdong Power Grid Co., Ltd. Grid Planning Research Center and this research is truly supported by the foundation: Technical Projects of China Southern Power Grid (No. GDKJXM20172797).

Conflicts of Interest: The authors declare no conflicts of interest.

Appendix A

The Matlab program of calculating the ampacity under the air-laying mode of the cables is shown as follows (XLPE-0):

```
function y = XLPE-0()
global k dki dko Rk Ck Rk1 Ck1 l p i A B P T1u E To I h2 R12 number number1 T1 number2 C0 Tou r0p R11
load('XLPE-0.mat');           %introduce the matrix (column 1: conductor temperature;
                              column 2: current)
I = XLPE(:,2);                %reading the current in the matrix
T1 = XLPE(:,1);               %reading the temperature in the matrix
t1 = 0:60:54900;              %calculating time range (total data volume in the matric)
hold on                       %keeping the drawing interface
```

```
number = 30;                                    %insulation hierarchical Number
number1 = number + 1;                           %lapped covering
number2 = number + 2;                           %outer-sheath
d1 = 26.6*10^-3;                                %conductor diameter
d2 = 65.6*10^-3;                                %insulation diameter
d6 = 92.0*10^-3;                                %outer-sheath diameter
h1 = (d2-d1)/(2*number);                        %insulation hierarchical thickness
h2 = 3.0*10^-3;                                 %outer-sheath thickness
a = 0.00393;                                    %resistance temperature coefficient of copper conductor
p1 = 3.23;                                      %thermal resistance coefficient of insulation layer
p6 = 3.5;                                       %thermal resistance coefficient of outer sheath
Dc = 344.312*10^4;                              %volumetric thermal capacity of Cu
DPE = 194.12*10^4;                              %volumetric thermal capacity of XLPE
D6 = 242.54*10^4;                               %volumetric thermal capacity of outer-sheath
r0p = 3.482e-5;                                 %DC Resistance of Conductor at 20 °C
R11 = 0.1095;                                   %thermal resistance of lapped covering
dki = zeros(1,number);                          %inner diameter array of each insulation layer
dko = zeros(1,number);                          %outer diameter array of each insulation layer
Rk = zeros(1,number);                           %thermal resistance array of each insulation layer
Ck = zeros(1,number);                           %thermal capacity array of each insulation layer
Rk1 = zeros(1,number2);                         %thermal resistance array
Ck1 = zeros(1,number2);                         %thermal capacity array
A = zeros(number2,number2);                     %A matrix
B = zeros(number2,number2);                     %A matrix
P = zeros(number2,1);                           %P matrix
E = zeros(1,number2);                           %temperature initial matrix
T1c = zeros(1,915);                             %calculating point number
R12 = 0.0299;                                   %outer-sheath thermal resistance
C0 = Dc*pi*0.25*d1*d1;                                  %conductor thermal capacity
C12 = D6*pi*0.25*(d6^2-(d6-2*h2)^2);                    % outer-sheath thermal capacity
for k = 1:number
    dki(k) = d1 + 2*h1*(k-1);
    dko(k) = d1 + 2*h1*k;
    Rk(k) = p1/(2*pi)*log(1 +
2*h1/dki(k));                                           %thermal resistance of each insulation layer
Ck(k) = DPE*pi*0.25*(dko(k)^2-dki(k)^2);                %thermal capacity of each insulation layer
end
C11 = 5549.4;                                           %lapped covering thermal capacity
for n = 1:number2                                       %array of initial cable internal temperature
E(n) = 17.7;
end
T1u = 17.7;                                             %initial conductor temperature
for u = 0:60:54900;                                     %time horizon
for l = 1:number2
    if(l==1)
        Rk1(l) = Rk(l);
        Ck1(l) = Ck(l) + C0;
          else if(l==number1)
        Rk1(l) = R11;
        Ck1(l) = C11;
          else if(l==number2)
```

```
            Rk1(l) = R12;
            Ck1(l) = C12;
        else
            Rk1(l) = Rk(l);
            Ck1(l) = Ck(l);
        end
    end
    for i = 1:number2
    for j = 1:number2
    if(j==i)
       if(j==1)
         A(i,j) = -(Ck1(i)*Rk1(i))^-1;B(i,j)=Ck1(i)^-1;
       else
              A(i,j) = -Ck1(i)^-1*(Rk1(i-1)^-1+Rk1(i)^-1);B(i,j)=Ck1(i)^-1;
       end
    else if(j==I + 1)
         A(i,j) = (Ck1(i)*Rk1(i))^-1;B(i,j)=0;
    else if(j==i-1)
         A(i,j) = (Ck1(i)*Rk1(j))^-1;B(i,j)=0;
    else
         A(i,j) = 0;B(i,j) = 0;
    end
    end
    end
    tt = [u,u+60];
    Tou = To(u/60+1);                              %initial surface temperature
    i = I(u/60 + 1);                               %initial current
    T1c(u/60 + 1) = E(1);                          %calculating conductor temperature
    rp = r0p*(1+a*(T1u-20));                       %DC resistance of conductor
    x = pi*400e-7/rp;
    Y = x^2/(192 + 0.8*x^2);                       %skin effect factor of conductor
    r = rp.*(1 + Y);                               %AC resistance of conductor
    p = i^2*r;                                     %heating power of conductor
    [t,x] = ode23t(@odefun3,tt,E);                 %solution of differential equation
    E = x(end,:);                      %updating the initial temperature array for the next period
    T1u = E(end,1);                    % updating the conductor temperature array for the next period
    end
    y = T1c';                                      %calculating conductor temperature
    plot(t1,y,'g')                                 %plotting curves
    hold on
    end
    function dx = odefun3(t,x)                     %P array solution
    global A B P p Tou R12 number2
    dx = zeros(number2,1);
    for m = 1:number2
         if(m==1)
       P(m) = p;
              else if(m==number2)
              P(m) = Tou/R12;
         else
              P(m) = 0;
         end
    end
    dx = A*x + B*P;
    end
```

The calculating result of the XLPE-0 cable ampacity through the Matlab program above is shown in Figure A1.

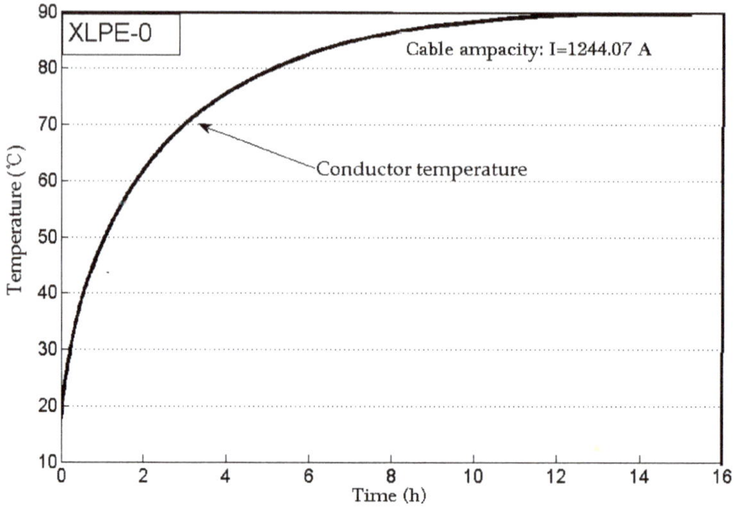

Figure A1. The calculating result of the XLPE-0 cable ampacity.

References

1. Orton, H. Power cable technology review. *High Volt. Eng.* **2015**, *41*, 1057–1067. [CrossRef]
2. Shwehdi, M.H.; Morsy, M.A.; Abugurain, A. Thermal aging tests on XLPE and PVC cable insulation materials of Saudi Arabia. In Proceedings of the IEEE Conference on Electrical Insulation and Dielectric Phenomena, Albuquerque, NM, USA, 19 November 2003; pp. 176–180. [CrossRef]
3. Liu, X.; Yu, Q.; Liu, M.; Li, Y.; Zhong, L.; Fu, M. DC electrical breakdown dependence on the radial position of specimens within HVDC XLPE cable insulation. *IEEE Trans. Dielectr. Electr. Insul.* **2017**, *24*, 1476–1486. [CrossRef]
4. Ouyang, B.; Li, H.; Li, J. The role of micro-structure changes on space charge distribution of XLPE during thermo-oxidative ageing. *IEEE Trans. Dielectr. Electr. Insul.* **2017**, *24*, 3849–3859. [CrossRef]
5. Diego, J.A.; Belana, J.; Orrit, J.; Cañadas, J.C.; Mudarra, M.; Frutos, F.; Acedo, M. Annealing effect on the conductivity of xlpe insulation in power cable. *IEEE Trans. Dielectr. Electr. Insul.* **2011**, *18*, 1554–1561. [CrossRef]
6. Xie, Y.; Zhao, Y.; Liu, G.; Huang, J.; Li, L. Annealing Effects on XLPE Insulation of Retired High-Voltage Cable. *IEEE Access.* **2019**. [CrossRef]
7. Xie, Y.; Liu, G.; Zhao, Y. Rejuvenation of Retired Power Cables by Heat Treatment. *IEEE Trans. Dielectr. Electr. Insul.* **2019**, *26*, 668–670. [CrossRef]
8. Celina, M.; Gillen, K.T.; Clough, R.L. Inverse temperature and annealing phenomena during degradation of crosslinked polyolefins. *Polym. Degrad. Stab.* **1998**, *61*, 231–244. [CrossRef]
9. Kalkar, A.K.; Deshpande, A.A. Kinetics of isothermal and non-isothermal crystallization of poly (butylene terephthalate) liquid crystalline polymer blends. *Polym. Eng. Sci.* **2010**, *41*, 1597–1615. [CrossRef]
10. Wang, Y.; Shen, C.; Li, H.; Qian, L.; Chen, J. Nonisothermal melt crystallization kinetics of poly (ethylene terephthalate)/clay nanocomposites. *J. Appl. Polym. Sci.* **2010**, *91*, 308–314. [CrossRef]
11. Xie, A.S.; Zheng, X.Q.; Li, S.T.; Chen, G. The conduction characteristics of electrical trees in XLPE cable insulation. *J. Appl. Polym. Sci.* **2010**, *114*, 3325–3330. [CrossRef]
12. Xie, A.; Li, S.; Zheng, X.; Chen, G. The characteristics of electrical trees in the inner and outer layers of different voltage rating XLPE cable insulation. *J. Phys. D Appl. Physic* **2009**, *42*, 125106–125115. [CrossRef]
13. Insulated Conductors Committee of the IEEE Power Engineering Society. *IEEE GUIDE for Soil Thermal Resistivity Measurements*; IEEE Std 442, Reaffirmed 2003; IEEE: Piscataway, NJ, USA, 1981.

14. Frank, D.W.; Jos, V.R.; George, A.; Bruno, B.; Rusty, B.; James, P.; Marcio, C.; Georg, H. *A Guide for Rating Calculations of Insulated Cables*; Cigré TB # 640; Cigré: Paris, France, 2015.
15. International Electrotechnical Commission. *Calculation of the Current Rating of Electric Cables*; IEC Press: Geneva, Switzerland, 2006.
16. Wang, P.; Liu, G.; Ma, H. Investigation of the Ampacity of a Prefabricated Straight-Through Joint of High Voltage Cable. *Energies* **2017**, *10*, 2050. [CrossRef]
17. Meng, X.K.; Wang, Z.Q.; Li, G.F. Dynamic analysis of core temperature of low-voltage power cable based on thermal conductivity. *Can. J. Electr. Comput. Eng.* **2006**, *39*, 59–65. [CrossRef]
18. Del Pino Lopez, J.C.; Romero, P.C. Thermal effects on the design of passive loops to mitigate the magnetic field generated by underground power cables. *IEEE Trans. Power Deliv.* **2011**, *26*, 1718–1726. [CrossRef]
19. Sharad, P.A.; Kumar, K.S. Application of surface-modified XLPE nanocomposites for electrical insulation-partial discharge and morphological study. *Nanocomposites* **2017**, *3*, 30–41. [CrossRef]
20. Andjelkovic, D.; Rajakovic, N. Influence of accelerated aging on mechanical and structural properties of cross-linked polyethylene (XLPE) insulation. *Electr. Eng.* **2001**, *83*, 83–87. [CrossRef]
21. Fu, Q.; Liu, J.P.; He, T.B. Non-isothermal Crystallization Behavior and Kinetics of Metallocene Short Chain Branched Polyethylen. *Chem. Res.* **2002**, *6*, 1183–1188. [CrossRef]
22. Olsen, R.; Anders, G.J.; Holboell, J.; Gudmundsdottir, U.S. Modelling of dynamic transmission cable temperature considering soil-specific heat, thermal resistivity, and precipitation. *IEEE Trans. Power Deliv.* **2013**, *28*, 1909–1917. [CrossRef]
23. Han, Y.J.; Lee, H.M.; Shin, Y.J. Thermal aging estimation with load cycle and thermal transients for XLPE-insulated underground cable. In Proceedings of the IEEE Conference on Electrical Insulation and Dielectric Phenomenon (CEIDP), Fort Worth, TX, USA, 1 October 2017; pp. 205–208. [CrossRef]
24. Xiaobin, C.; Zhixing, Y.I.; Kui, C.; Xianyi, Z.; Jia, Y. Laying mode and laying spacing for single-core feeder cable of high speed railway. *China Railw. Sci.* **2015**, 85–90. [CrossRef]
25. Chen, Y.; Duan, P.; Cheng, P.; Yang, F. Numerical calculation of ampacity of cable laying in ventilation tunnel based on coupled fields as well as the analysis on relevant factors. In Proceedings of the IEEE Conference on Intelligent Control and Automation, Shenyang, China, 29 June–4 July 2014; pp. 3534–3538. [CrossRef]
26. Wang, Y.; Chen, R.; Li, J.; Grzybowski, S.; Jiang, T. Analysis of influential factors on the underground cable ampacity. In Proceedings of the IEEE Conference on Electrical Insulation Conference, Annapolis, MD, USA, 5–8 June 2011; pp. 430–433. [CrossRef]
27. Wang, P.; Ma, H.; Liu, G. Dynamic Thermal Analysis of High-Voltage Power Cable Insulation for Cable Dynamic Thermal Rating. *IEEE Access.* **2019**, *7*, 56095–56106. [CrossRef]
28. Xu, Y.; Luo, P.; Xu, M.; Sun, T. Investigation on insulation material morphological structure of 110 and 220 kv xlpe retired cables for reusing. *IEEE Trans. Dielectr. Electr. Insul.* **2014**, *21*, 1687–1696. [CrossRef]
29. Bin, Y.; Adachi, R.; Tong, X. Small-angle HV light scattering from deformed spherulites with orientational fluctuation of optical axes. *Colloid Polym. Sci.* **2004**, *282*, 544–554. [CrossRef]
30. Li, J.; Li, H.; Wang, Q. Accelerated inhomogeneous degradation of XLPE insulation caused by copper-rich impurities at elevated temperature. *IEEE Trans. Dielectr. Electr. Insul.* **2016**, *23*, 1789–1797. [CrossRef]
31. He, K.; Chen, N.; Wang, C.; Wei, L.; Chen, J. Method for determining crystal grain size by X-ray diffraction. *Cryst. Res. Technol.* **2018**, *53*, 1700157. [CrossRef]
32. Rabello, M.S.; White, J.R. The role of physical structure and morphology in the photodegradation behavior of polypropylene. *Polym. Degrad. Stab.* **1997**, *56*, 55–73. [CrossRef]
33. Martini, H.; Zhao, S.; Friberg, A.; Jabri, Z. Influence of electron beam irradiation on electrical properties of engineering thermoplastics. In Proceedings of the IEEE Conference on Electrical Insulation Conference (EIC), Montreal, QC, Canada, 19–22 June 2016; pp. 305–308. [CrossRef]

energies

MDPI

Article

Investigations on the Performance of a New Grounding Device with Spike Rods under High Magnitude Current Conditions

Abdul Wali Abdul Ali, Nurul Nadia Ahmad, Normiza Mohamad Nor *, Muhd Shahirad Reffin and Syarifah Amanina Syed Abdullah

Faculty of Engineering, Multimedia University, Cyberjaya 63100, Malaysia;
1122702929@student.mmu.edu.my (A.W.A.A.); nurulnadia.ahmad@mmu.edu.my (N.N.A.);
muhd_shahirad@yahoo.com (M.S.R.); synina@gmail.com (S.A.S.A.)
* Correspondence: normiza.nor@mmu.edu.my; Tel.: +603-8312-5387

Received: 26 February 2019; Accepted: 21 March 2019; Published: 23 March 2019

Abstract: In many publications, the characteristics of practical earthing systems were investigated under conditions involving fast-impulse currents of different magnitudes by field measurements. However, as generally known, in practice the transient current can normally reach several tens of kiloamperes. This paper therefore aimed to investigate the characteristics of a new electrode for grounding systems under high current magnitude conditions, and compare it with steady-state test results. The earth electrodes were installed in low resistivity test media, so that high impulse current magnitudes can be achieved. The effects of impulse polarity and earth electrode's geometry of a new earth electrode were also quantified under high impulse conditions, at high currents (up to 16 kA).

Keywords: earthing systems; electrode's geometry; fast-impulses; high-magnitude currents and impulse polarity

---

1. Introduction

Much work [1–18] has been carried out to characterize practical earthing systems under fast-impulse, high-magnitude current conditions. As generally known, experimental investigations on practical earthing systems under high impulse currents can provide more realistic results on the characteristics of earthing systems under high currents, in comparison to laboratory and computational methods. The first work on impulse characteristics by field measurement was carried out by Towne [1] in 1928, on galvanized-iron pipes with peak currents up to 900 A. In 1941, Bellaschi [2] had used deep-driven earth rods with current magnitudes between 2 kA to 8 kA. The following year, Bellaschi [3], completed further tests on 12 earth rods, with impulse currents of 400 A to 15.5 kA. There have been a lot more impulse tests [4–18], were carried out on practical earthing systems, where tit was found that the 'impulse resistance' values were found to be less than those measured for low voltage, low frequency currents [1–18]. A lot of improvements and suggestions in the impulse test methodology on practical earthing systems can also be seen in the last three decades [1–18]. Impulse tests have also been carried out on practical earthing systems considering various factors, namely; earth electrode geometry, soil resistivity, impulse polarity and voltage/current magnitudes [1–18]. For the study of grounding performance with the current magnitudes up to 5 kA, limited studies have been carried out on the effect of the earth electrode's configuration. A remarkable work was done in [4–6], where impulse tests were conducted on various practical earthing systems, with current magnitudes of more than 20 kA. It was found that the impulse resistance becomes less current dependent in high currents. For the effect of soil resistivity, it was reported in [19,20] that in a low resistivity test medium, it is possible that no ionization process occurs in the test since the gaps between the sand grains are filled with water, and little field enhancement is expected to occur since there is only a small dielectric difference between the soil and the air gaps. In addition, this could also be due to the fact the earthing systems may have become a conducting mass in low resistivity soil.

As for the effect of impulse polarity, as early as 1948, Petropoulos [21] found that for similar electrode dimensions and soil resistivity, the critical electric field, Ec which is the onset of ionization, and the breakdown voltage were found higher for negative compared with positive impulses. A few more studies followed with investigations on soil characterisation under high impulse current for both impulse polarities, where most of the studies were done by laboratory testing [19,20]. In the last few years, a few studies can be found on the effect of impulse polarity on practical earthing systems under high impulse conditions [17,19]. Since very limited studies can be found on the effect of impulse polarity on various grounding system configurations for the same soil resistivity, this paper presents the experimental results on new earth electrodes, combined with various electrode's configurations, under both impulse polarities.

This paper reports the investigation on the performance of a new grounding electrode, called a grounding device with spike rods (GDSR) under high-magnitude fast impulses, of both impulse polarities. The reason GDSR was designed and studied in the current paper is to follow up on the study performed by Petropoulos [21], where he found that electrodes fitted with spike rods have lower resistance than that electrode without the spike rods. For these reasons, further studies were performed, and presented in this paper. He [21] described the high field intensity of the spike rods which causes more current to flow. In this current work, it was realized and evident by field testing and measurements, which have not been implemented before. GDSR was also combined with other electrodes, which were installed at one site. A smaller impulse generator was also employed to observe the effectiveness of GDSR combined with other electrodes in the same soil resistivity at lower current magnitudes, below 2 kA. A new grounding electrode with spike rods was postulated to enhance the ionization process in soil and compared with conventional rod-electrodes. It was shown the resistance becomes less current dependent at high currents, which was found to be agree with previously published works [2–10,18–21].

2. Experimental Arrangement

In this study, eight earth electrode configurations were installed at the same soil site. Using the Wenner method, the soil resistivity of the outdoor test site was measured with earth tester. The RESAP module of Current Distribution, Electromagnetic Fields, Grounding and Soil Structure Analysis (CDEGS) was used to interpret the measured data into 2-layer soil models. The test site was purposely selected at a farming land, to obtain for low soil resistivity result, hence high current magnitudes can be achieved. It was computed that an upper layer and lower layer resistivity of 2.95 Ωm and 0.23 Ωm, respectively with 4.9 m depth for upper layer, and an infinite depth for lower layer.

2.1. Earth Electrodes

The test area contains six configurations; a vertical single rod electrode (see Figure 1), two parallel vertical rod electrodes (see Figure 2), three parallel vertical rod electrodes (see Figure 3), a GDSR (see Figure 4), a GDSR in parallel with vertical one rod electrode (see Figure 5), a GDSR in parallel with two vertical rod electrodes (see Figure 6). Each installed vertical rod electrode is 1.5 m long and has a 16 mm diameter. The interconnections between the vertical electrodes were done with copper strips, with a width of 2.5 cm, and length of 3 m, buried to a depth of 30 cm under the ground surface. For lower current impulse tests, two new configurations were laid, where a single rod electrode was installed at 50 cm depth, and a GDSR buried to a depth of 50 cm. Very little has been mentioned in literature on the recommended spacing between the vertical rods. In [22], it is stated that 'spacing of less than 3 m may not provide the most economical use of materials'. This shows that it is not effective to have the adjacent vertical rods too close to each other. On the other hand, [23] stated that the recommended spacing between rods should be at least 2 times the length of the rod. No specific study has focused on the effect of spacing changes on the mutual effects and performance of grounding systems. In the current study, the vertical rods were all arranged with a spacing between the rods of no less than twice the length of the electrode.

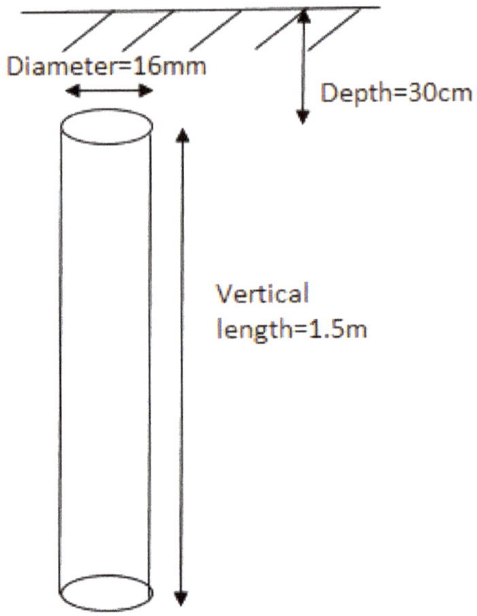

Figure 1. Configuration 1 with a single vertical rod electrode.

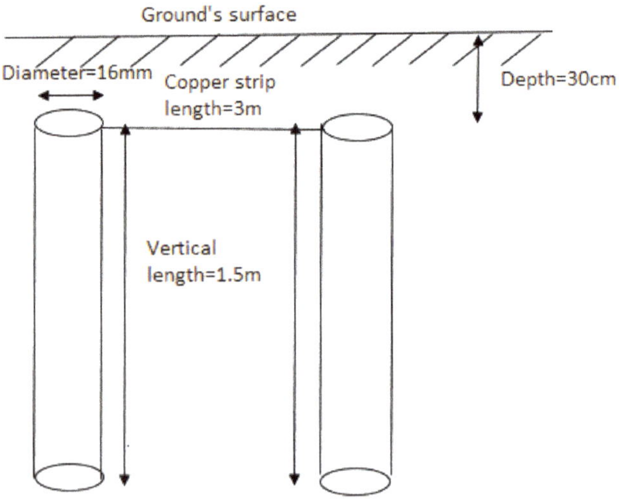

Figure 2. Configuration 2 with two parallel vertical rod electrodes.

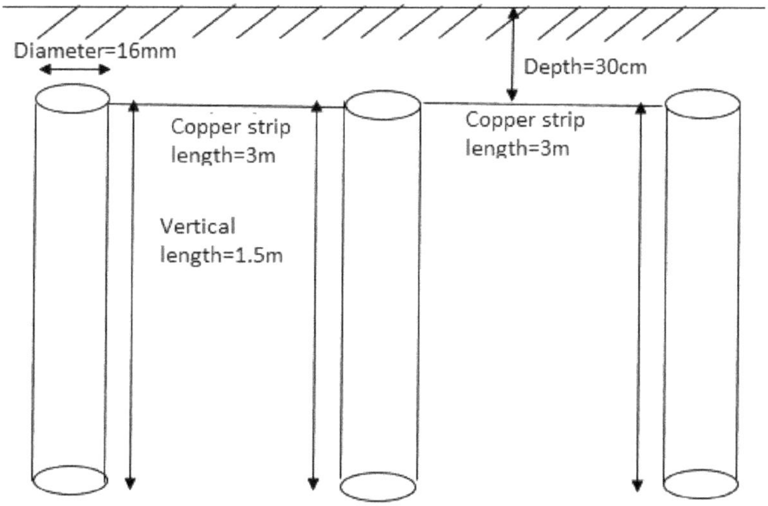

Figure 3. Configuration 3 with three parallel vertical rod electrodes.

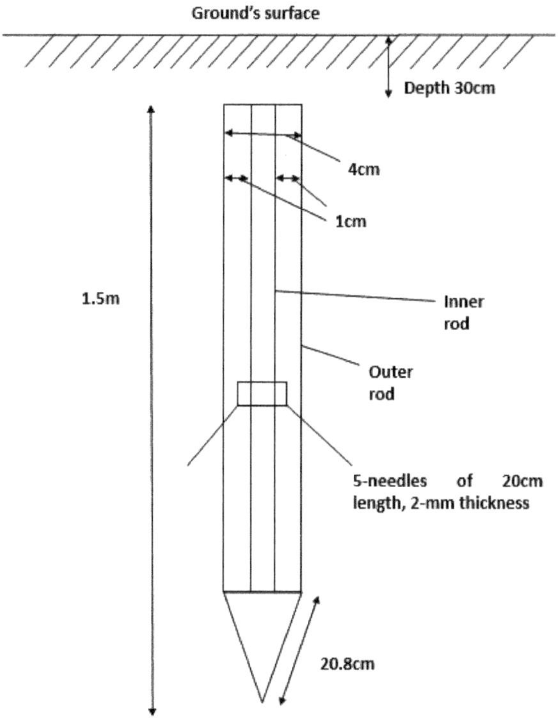

Figure 4. Configuration 4, a grounding device with spike rods.

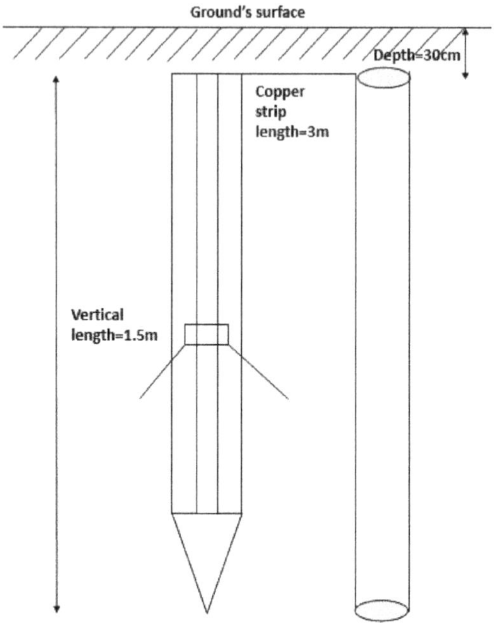

Figure 5. Configuration 5, a grounding device with spike rods and a vertical rod electrode.

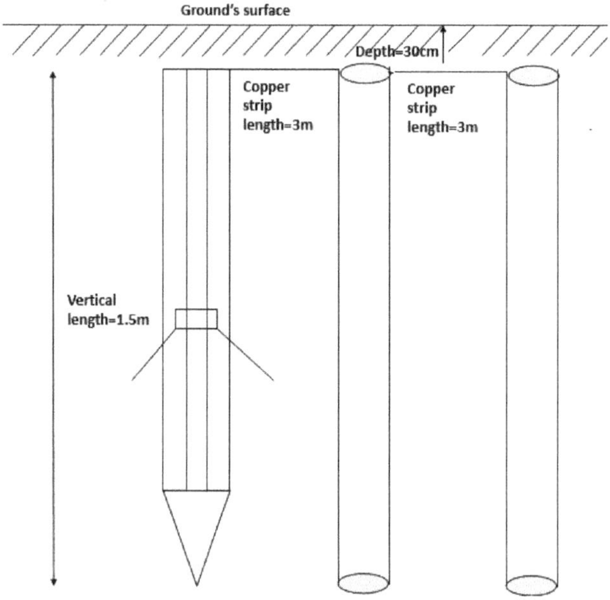

Figure 6. Configuration 6, a grounding device with spike rods and 2 vertical rod electrodes.

Figure 7 shows a detailed construction of a new electrode, GDSR. It comprises an outer shaft (110) and an inner shaft (120). The spike rods (123) are arranged on the body of the inner shaft (120), where the total number of spike rods is five. A pre-borehole was first made by using an auger, with a diameter of 3.8 cm. After completing the pre-borehole to a depth of at least 1.5 m, the grounding device with spike rods (100) is positioned in the hole, where the outer shaft (110) is subjected to stress and impact force while it is driven through hammering into the ground. The driving tip (122) has a pointed conical end to allow for easier penetration against obstructions encountered during driving of the rod. After that, using the provided winch (121), the inner shaft was turned in such a way that the grounding spike rods protrude out and pierce into the soil mass. GDSR was used in this study, since it was found by Petropoulos [21] that the earth electrodes attached with sharp points or needles have lower impulse resistance values than that without the sharp points. He [21] pointed out that by suitably shaping the electrodes with the sharp points, higher field intensities would be present, hence reducing the resistance values. For these reasons, a GDSR was used as part of the practical earthing systems. For checking the effectiveness of the shaping of various configurations, field measurements and testing at practical sites were carried out in this current work. A GDSR is postulated to cause and enhance ionization process in the soil, and discharge higher currents to the ground.

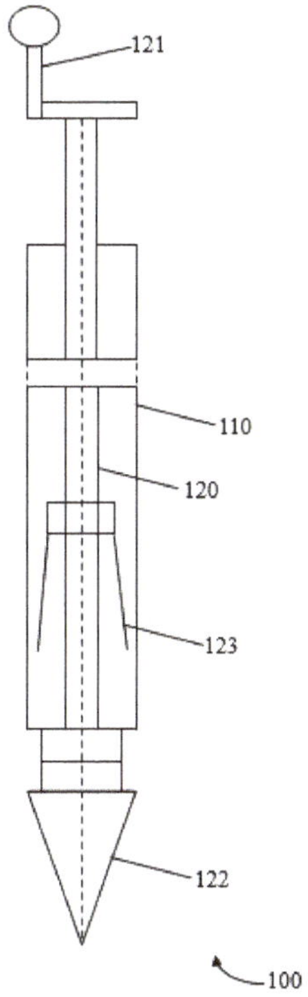

Figure 7. Construction of a grounding device with spike rods.

These eight earth electrode arrangements were used in this study for the reason that a vertical rod electrode is typically used locally, thus replicating the real conditions more closely. In order to ensure that the same test area was used, firstly, a single rod electrode was installed. Secondly, another vertical rod electrode was added, to make two parallel rod electrodes. It was then followed by the third rod electrode installed in a straight line arrangement, which gave us three parallel vertical rod electrodes. The first single rod that was installed earlier, as shown in Figure 1, was then removed, and replaced with a GDSR, giving the configuration shown in Figure 6. The third single rod electrode was then removed, whereby the configuration become as seen in Figure 5. Finally, the rod electrode was removed, thus leaving a GDSR only. For each configuration, FOP measurement and impulse tests were performed. Table 1 shows the steady-state earth resistance values, DC earth resistance value, RDC, determined with the Fall-of-Potential (FOP) method for all eight configurations. It can be seen from Table 1 that the RDC values of the earthing systems consisting of a GDSR are lower than that of conventional vertical rod electrodes. RDC for configurations 2 and 3 were found to be improved with the addition of vertical rod electrodes, with a decrease by 63% and 76.4% with the addition of one and two rod electrodes, respectively, to the single rod electrode, configuration 1. When a GDSR was used, RDC decreased by 75.5% for configuration 4 from configuration 1, 47.1% for configuration 5 from configuration 2, and 36% for configuration 6 from configuration 3. This indicates that a new electrode, the GDSR, is effective in reducing the RDC of earthing systems, which could be due the large cross sectional area of the GDSR', as compared to conventional electrode.

Table 1. Measured RDC for all configurations.

Configurations	Earthing Systems	RDC, Ω
1	A vertical single rod electrode	75.5
2	2 parallel vertical rod electrodes	27.6
3	3 parallel vertical rod electrodes	17.8
4	GDSR	18.5
5	GDSR in parallel with vertical one rod electrode	14.6
6	GDSR with spike rods in parallel with two vertical rods	11.4
7	A vertical single rod electrode, buried at a depth of 30 cm	313.2
8	A vertical grounding device with spike rods	85.2

2.2. *Experimental Test Set Up*

Figure 8 shows the field test arrangement consisting of an impulse generator, which needs to be mounted on a lorry, insulating rods, which are made of epoxy conduits to suspend the leads/copper mesh/coaxial cables, and isolate them from earth, DC converters and batteries to provide a power source to digital storage oscilloscopes (DSOs) and a laptop, and a diesel generator to power up the impulse generator. Separate DSOs were used to capture current and voltage measurements. A resistive divider with a ratio of 3890:1 was used to capture high voltage and a current transformer with a ratio of 0.01V/A was used for current measurements.

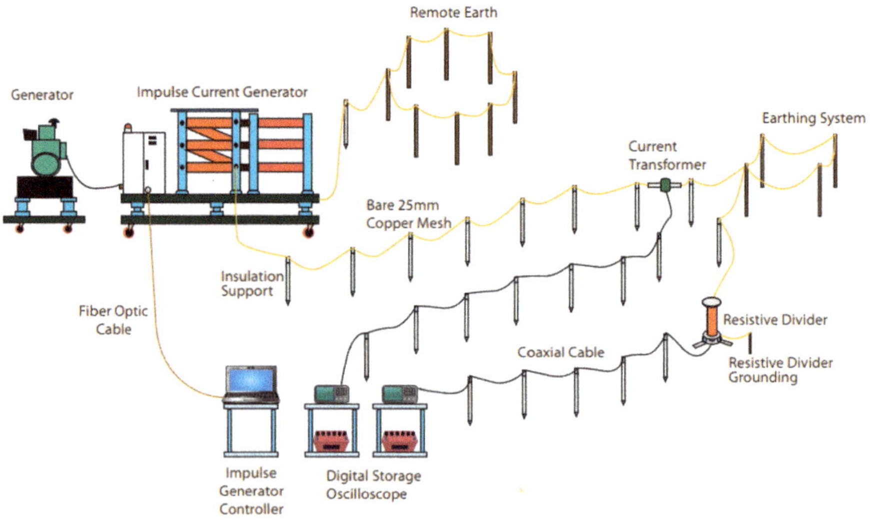

Figure 8. Test arrangement for field tests on practical grounding systems.

A remote or auxillary earth is needed to provide a return path for the discharge of high impulse currents to the ground during the measurements. In this study, the remote earth consists of 10 rods in a circular ring configuration (see Figure 9). These rods, are interconnected using copper mesh, arranged on top of the rod electrodes. Using a FOP method, RDC of the remote earth was measured, and found to be 4.8 Ω. This RDC value is acceptable since it is lower than that RDC of grounding systems under tests (see Table 1).

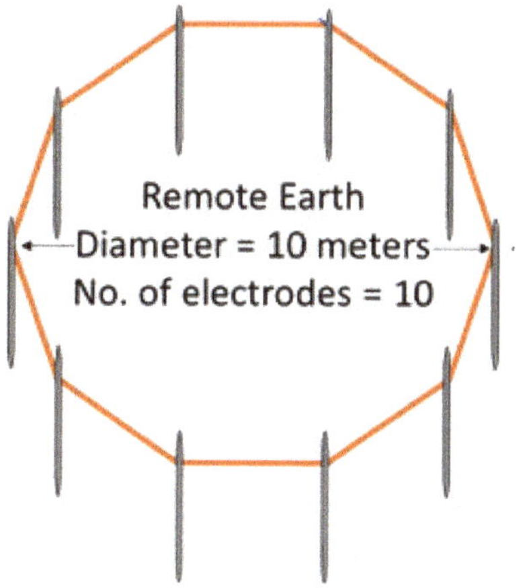

Figure 9. Remote earth for tests on various grounding systems.

3. Results and Analysis

This present work aims to quantify the effectiveness of a GDSR with other practical grounding systems under high magnitude impulse currents, above 5 kA, under both impulse polarities. The RDC values of grounding systems have been presented in Section 2. This allows comparison with the performance of grounding systems under high-magnitude impulse currents.

3.1. *Effect of Earth Electrode's Configurations*

Figures 10 and 11 show selected typical voltage and current traces for configuration 2 at charging voltages of 150 kV and 350 kV, respectively. The voltage and current traces of other configurations and voltage magnitudes were similar to those presented in Figures 10 and 11. In this work, the time to discharge to zero for current trace was measured. This parameter may provide information related to the effectiveness of the grounding systems in discharging a high current into the ground, where the faster the time taken to discharge for current trace, the better the grounding system is.

Figure 12 shows the measured time to discharge to zero for current trace at different applied voltage. For all configurations, (except for configuration 4), the time for current trace to discharge to zero were found to be independent of applied voltage. Configurations 3 and 6, with large sized grounding systems were found to have a faster time for the current trace to discharge to zero, which shows the good conductivity of the grounding systems. Generally, it is understood that the larger the size of a grounding system, the less time taken for current to discharge to zero, due to the fact more paths are available for the current to disperse. Some discrepancies in the results were also seen, where the time taken for current trace to discharge to zero for configuration 5 is higher than that configuration 1 and 2, despite the fact configuration 5 has a larger sized grounding system. This could be due to uncontrollable thermal and ionization processes in the soil, which have been highlighted in previously published work [1–3,19,20]. Another possibility is that this is also due to the inductive component, which can be significant under transient conditions. However, it is out of the scope of the current paper to come up with the equivalent circuit, and show evidence of an inductive effect for each configuration.

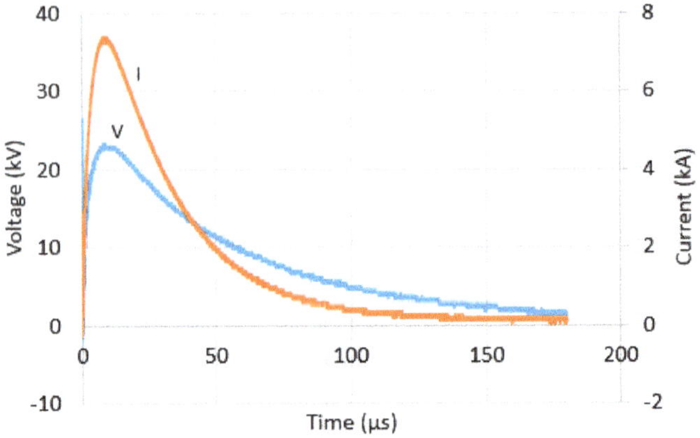

Figure 10. Voltage and current traces for configuration 2 at charging voltage of 150 kV under positive polarity.

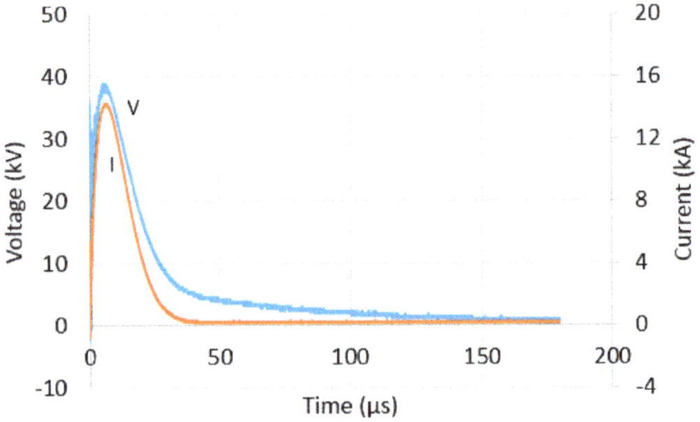

Figure 11. Voltage and current traces for configuration 2 at a charging voltage of 350 kV under positive polarity.

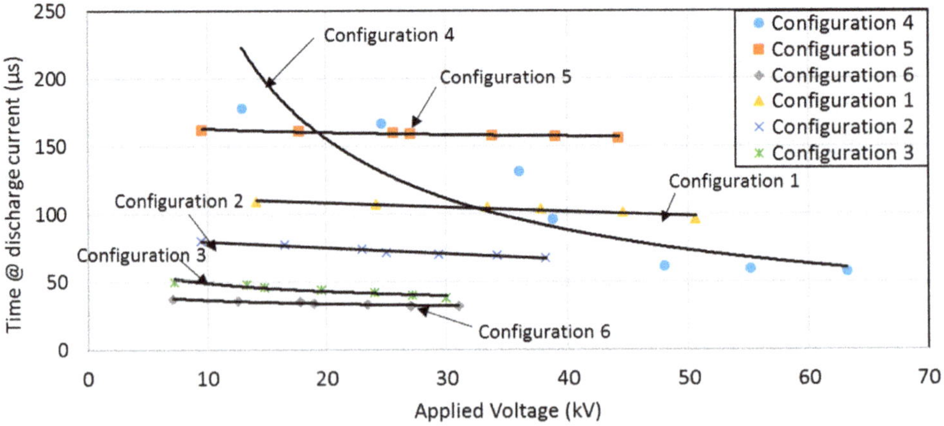

Figure 12. Time taken for current trace to discharge to zero vs. applied voltage for various grounding systems under positive impulse polarity.

In this study, the impulse resistance, $R_{impulse}$, was determined as the ratio of voltage at the peak current to the corresponding peak current. Figure 13 shows the measured $R_{impulse}$ for various grounding systems under high impulse currents, up to 16 kA. As can be seen, there is little dependence of $R_{impulse}$ on the current magnitudes (except for configuration 1). Configuration 1 had the highest RDC, and thus expectedly showed the highest reduction, as claimed in many prior publications [2–10,18–21] where the higher the DC earth resistance, the higher the current-dependent characteristics of earth resistance during the passage of high currents.

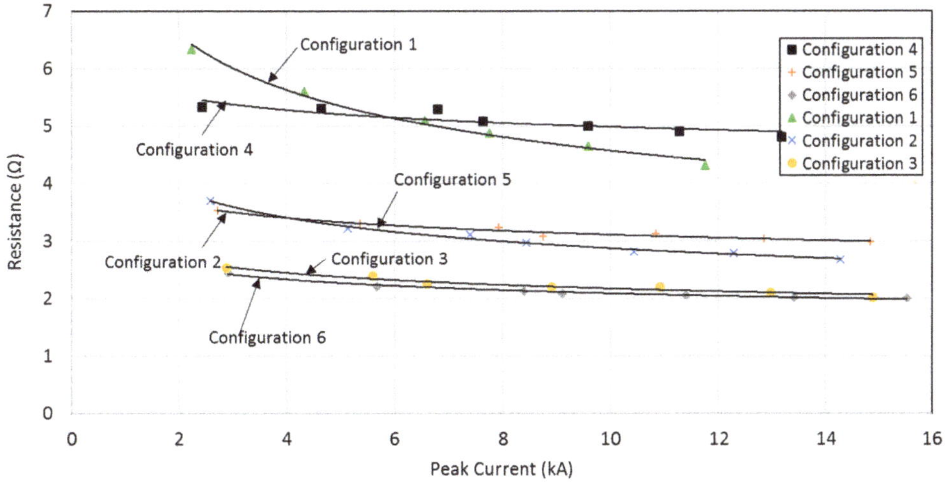

Figure 13. $R_{impulse}$ vs. peak current for various grounding systems under positive impulse polarity.

It was also noted that some differences in the $R_{impulse}$ with the current peak can be seen with the addition of GDSR at low current magnitudes, below 3 kA. However, at higher current magnitudes, little observable effect can be seen on $R_{impulse}$. This could be due to a full development of an ionization zone at high current, thus the performance of grounding systems has become independent of the grounding electrodes and current magnitudes.

3.2. *Effect of Low Current Magnitudes*

Due to current independence of earth resistance at high current magnitudes, as presented in Section A, experiments using a smaller impulse generator, which can generate up to 50 kV, 2 kA currents were performed to further investigate the grounding characteristics under lower current magnitude conditions. Impulse tests were conducted on four configurations; configurations 3 and 4, and another two new configurations, labelled as configurations 7 and 8. Configuration 7 is similar to configuration 1, but buried at 50 cm, and configuration 8 is similar to configuration 4, buried at 50 cm in the soil. Lesser depth, of 50 cm below the ground's surface, was used to obtain high resistances, hence low current magnitudes, for the vertical electrodes. RDC values were measured for both configurations 7 and 8, and found to be 313.2 Ω and 85.2 Ω, respectively. Using a similar test set up, remote earth and transducers as presented in Section 2.2, impulse tests were conducted on the four configurations using smaller impulse generator. Figure 14 shows voltage and current impulse shapes of configuration 7 at a charging voltage of 25 kV. Similar voltage and current traces were seen at different voltage magnitudes for configurations 7 and 8. However, faster voltage and current discharge times were seen for configuration 3 and 4, at various voltage magnitudes (see Figure 15), due to their low RDC, which thus provides better conduction of the grounding systems. When time to discharge to zero for current trace versus applied voltage was plotted for all four configurations under lower voltage magnitudes (Figure 16), it was noted that a reduction in time to discharge to zero for current traces with increasing applied voltage. This trend was not clearly observable at higher current magnitudes, presented earlier in Figure 12. A graph of $R_{impulse}$ was plotted for increasing current magnitudes (see Figure 17) for all four configurations. It can be seen that $R_{impulse}$ is decreasing significantly with current magnitude for configuration 7, with the highest RDC, 313.2 Ω. However, $R_{impulse}$ was found to be less current dependent for grounding systems with lower RDC, below 85 Ω.

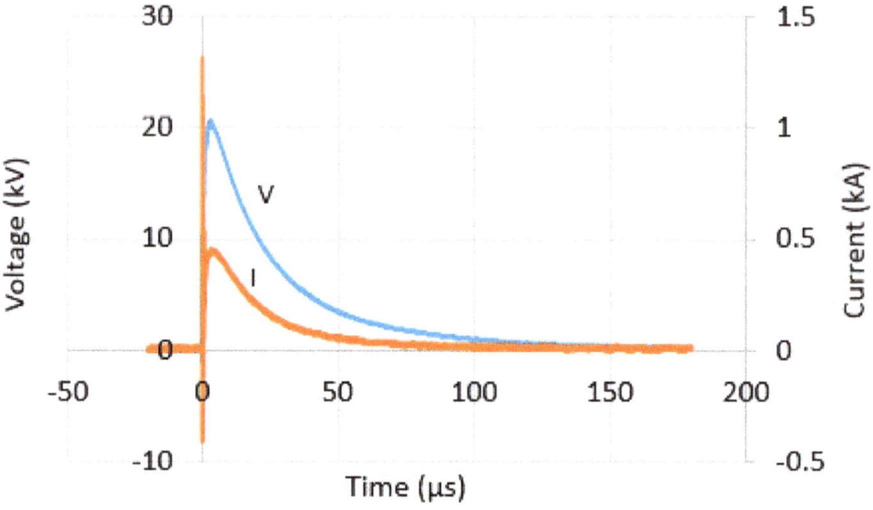

Figure 14. Voltage and current traces for configuration 7 at charging voltage of 25 kV under positive polarity.

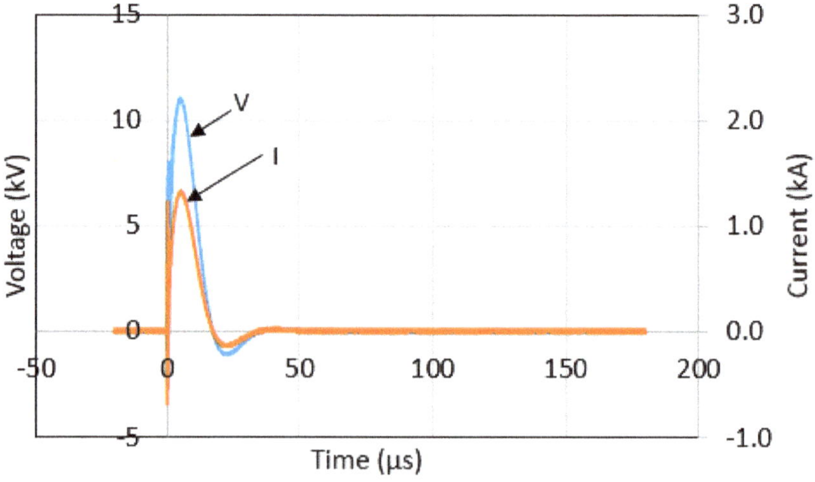

Figure 15. Voltage and current traces for configuration 8 at charging voltage of 25 kV under positive polarity.

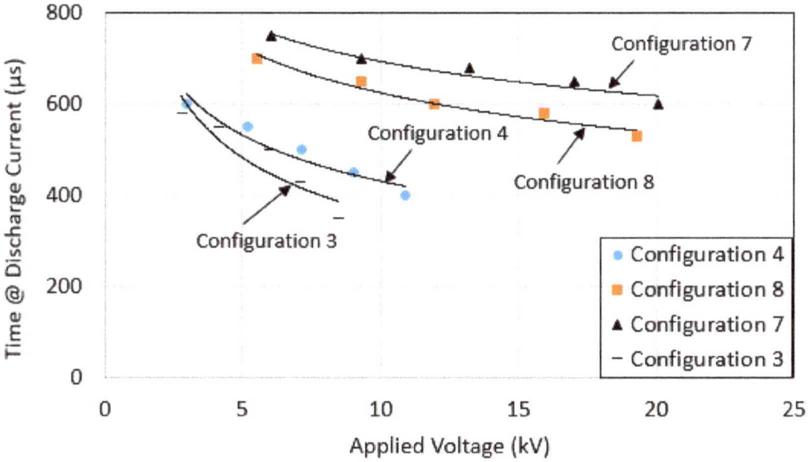

Figure 16. Time taken for current trace to discharge to zero vs. applied voltage for various grounding systems under low voltage magnitudes.

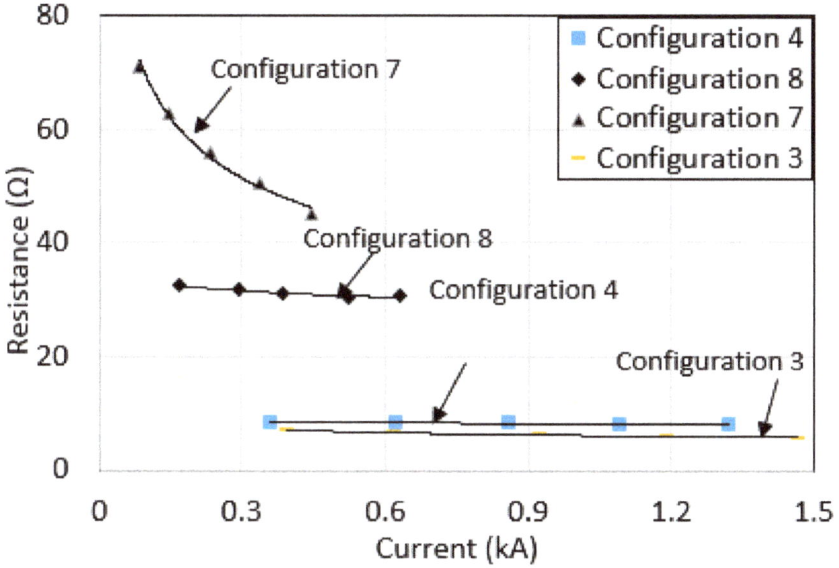

Figure 17. $R_{impulse}$ vs. peak current for various grounding systems under low voltage magnitudes.

3.3. *Effect of Impulse Polarity*

In a previously published work [18], it was noted that an effect of impulse polarity was seen in high resistivity soil. In this work, a low soil resistivity profile was used. Figures 18 and 19 show typical voltage and current traces for configuration 2 at charging voltages of 150 kV and 350 kV, respectively. Similar voltage and current traces were observed for various configurations and voltage magnitudes. As the voltage magnitudes were increased, a significant reduction in the time for current trace to discharge to zero was observed for all configurations of grounding systems (see Figure 20). This trend is found to be different than that observed for positive polarity (shown in Figure 12), where only configuration 4 has the reduction of time for current trace to discharge to zero when under positive impulse polarity. A faster time for current to discharge to zero for large sized grounding systems (configurations 3, 5 and 6) under negative impulse polarity was also noted. This could be influenced by the lower RDC in large grounding systems, thus discharging current at a faster time than that in smaller size of grounding systems. It was also noted that the trend of time taken for current trace to discharge to zero under negative polarity is more consistent, and similar to that found in other publications [19,20], than that found under positive impulse polarity. However, the inconsistent results for the time taken for current to discharge to zero under positive polarity are still not well understood.

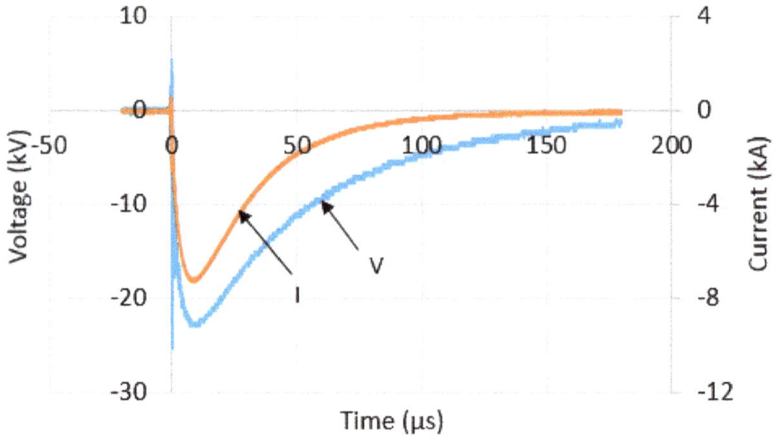

Figure 18. Voltage and current traces for configuration 2 at charging voltage of 150 kV under negative polarity.

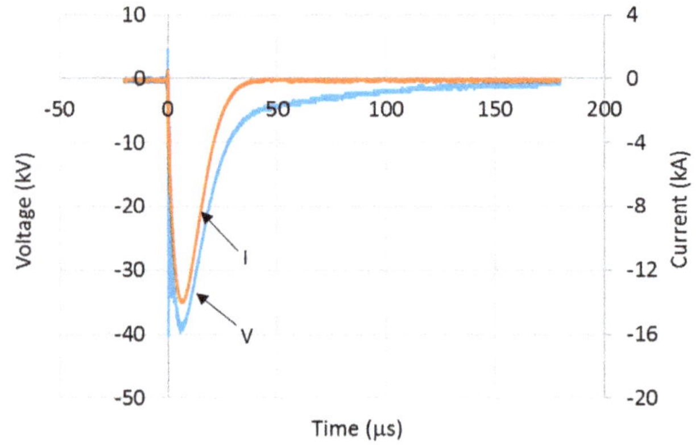

Figure 19. Voltage and current traces for configuration 2 at charging voltage of 350 kV under negative polarity.

When the voltage magnitudes under negative impulse polarity were increased, the $R_{impulse}$ was found to have small reduction with increasing current (see Figure 21). Configurations 3 and 6, with larger sized grounding systems were found to have the lowest $R_{impulse}$, which is a similar trend to that obtained under positive impulse polarity. These measured $R_{impulse}$ values with increasing currents were also plotted together under both impulse polarities for each configuration (see Figures 22–27). As can be seen, for configurations 1, 2, 3 and 4, a higher $R_{impulse}$ was recorded with negative impulses compared with positive impulses, as shown in Figures 22–25, respectively. A similar trend was seen in [20,21] whom conducted laboratory tests where $R_{impulse}$ values were found to be higher for negative impulses than for positive impulses. As generally known, a decrease in $R_{impulse}$ with increasing voltage indicates the ionization process in soil, which was thought to occur in the air voids within the soil [1–3,19,20]. Since it is expected that the discharges in air would require higher level voltages, and less currents for negative impulses compared to positive impulses, higher $R_{impulse}$ in negative impulses than positive impulses would occur. When Reffin et al. [18] performed experiments using field measurements on practical grounding systems installed in various soil conditions, they found that higher $R_{impulse}$ with negative impulses than that positive impulses for grounding systems with high RDC (62.6 Ω). On the other hand, $R_{impulse}$ was found to be independent of impulse polarity for low RDC (4.7 Ω). In this work, the electrodes were installed at the same site, with the same soil resistivity profile. The highest RDC values are for configurations 1, 2, 3 and 4, and these configurations were found to have higher $R_{impulse}$ under negative impulses than under positive impulses. On the other hand, for lower RDC (configurations 5 and 6), the results were found to be inconsistent, where $R_{impulse}$ were found to be independent of impulse polarities for configuration 5, and $R_{impulse}$ were found higher under positive impulses than that negative impulses, as shown in Figures 26 and 27 for configurations 5 and 6, respectively.

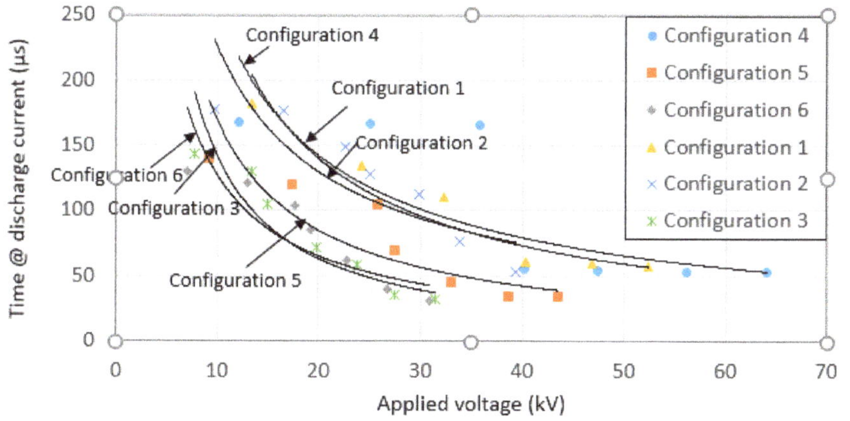

Figure 20. Time taken for current trace to discharge to zero vs. applied voltage for various grounding systems under negative impulse polarity.

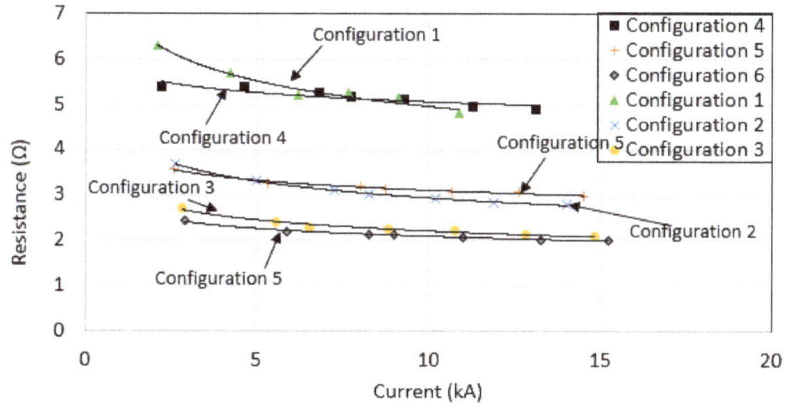

Figure 21. $R_{impulse}$ vs. peak current for various grounding systems under negative impulse polarity.

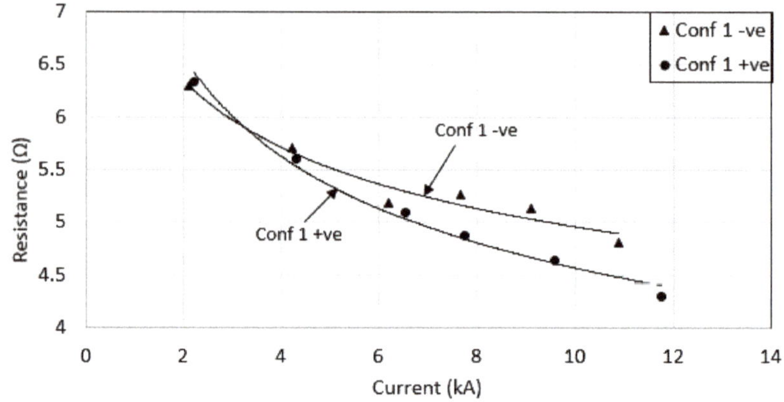

Figure 22. Impulse resistances versus current peak for configuration 1.

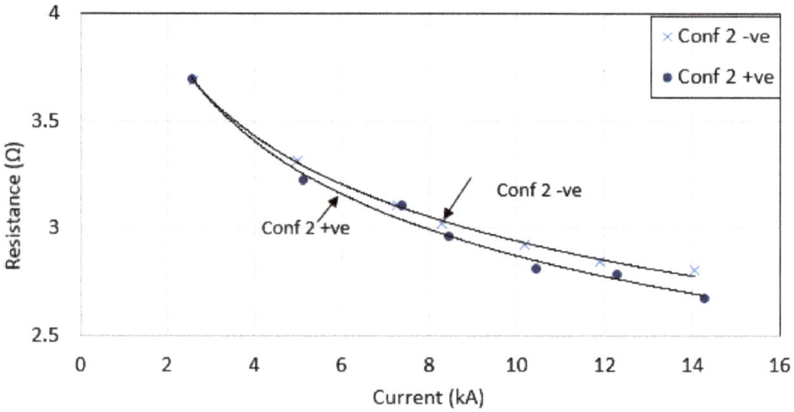

Figure 23. Impulse resistances versus current peak for configuration 2.

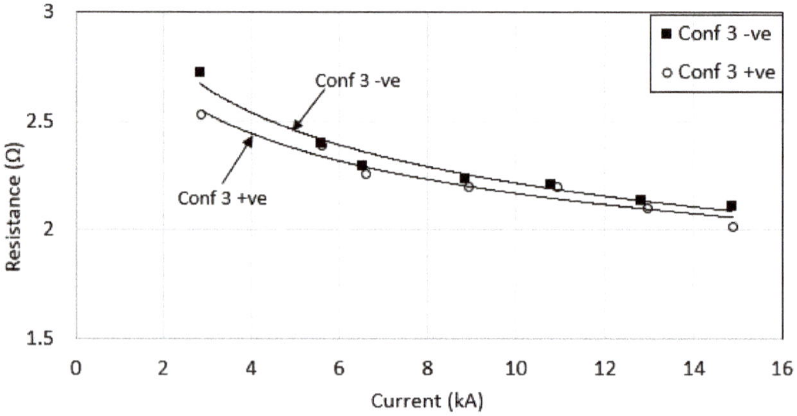

Figure 24. Impulse resistances versus current peak for configuration 3.

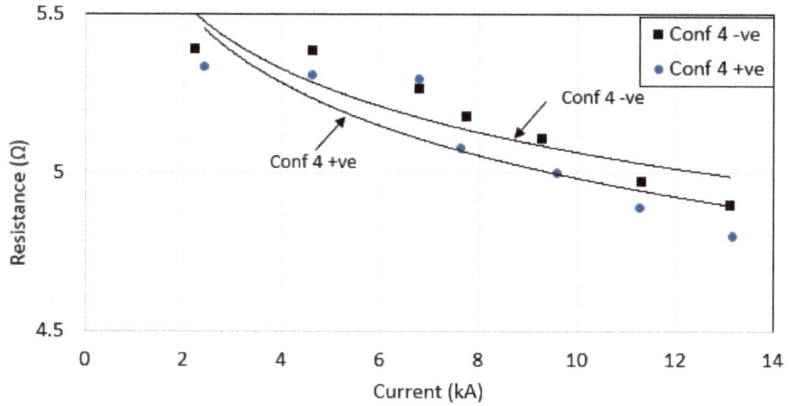

Figure 25. Impulse resistances versus current peak for configuration 4.

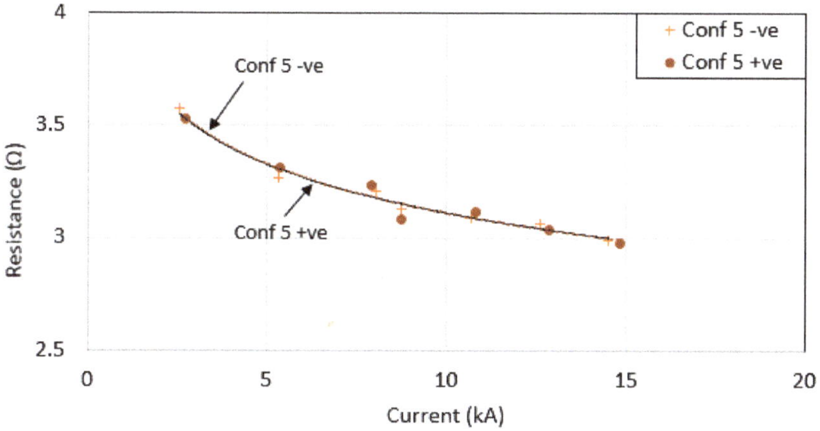

Figure 26. Impulse resistances versus current peak for configuration 5.

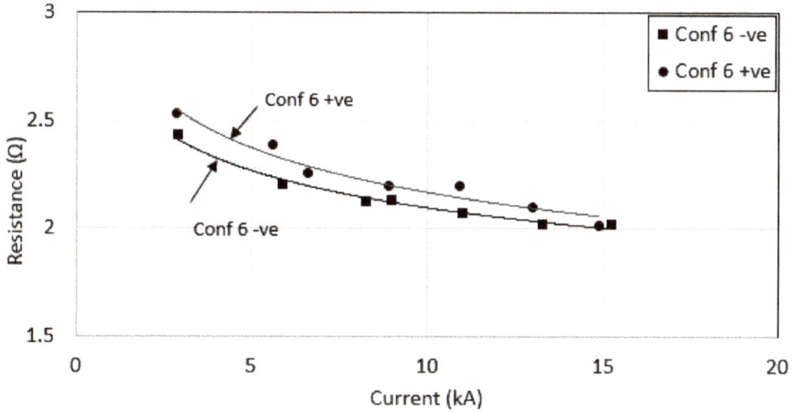

Figure 27. Impulse resistances versus current peak for configuration 6.

4. Conclusions

Experiments were performed on different grounding configurations using various configurations at installed in low resistivity profile soils at the same field site. It was found that under high impulse currents, more than 5 kA, less current dependence of $R_{impulse}$ was observed. The characteristics of various configurations were also investigated under lower current magnitudes, below 2 kA. High nonlinearity was observed for grounding systems with high RDC (above 313.2 Ω). The results revealed that the new GDSR earth electrode is more suitable to be used for low fault currents, such as 11 kV systems and below. When the grounding systems of various configurations were tested under negative impulse polarity, higher $R_{impulse}$ was seen in negative impulses than positive impulses for grounding systems with high RDC. This study also shows that under positive polarity conditions, only configuration 4 has a dependence on the time to discharge to zero for current trace, whereas a dependence of time to discharge to zero for current trace is not seen in the other configurations. On the other hand, for negative polarity, the measured time to discharge to zero for current traces decreases with applied voltage for all configurations, and the results are more consistent than for positive impulse polarities.

Author Contributions: A.W.A.A., M.S.R., N.M.N., and S.A.S.A. involved in setting up the experiment, A.W.A.A. and M.S.R. sorted out the data, A.W.A.A. prepared the draft and N.M.N. helped with validation of analysis. N.N.A. helped with the draft revision, N.N.A., N.M.N. secured the funding.

Funding: This research was funded by Telekom Malaysia Research and Development (TMR&D), grant number MMUE180008, MMUE170005 and MMUE180027.

Conflicts of Interest: The authors declare no conflict of interest.

References

1. Towne, H.M. Impulse characteristics of driven grounds. *Gen. Electr. Rev.* **1928**, *31*, 605–609.
2. Bellaschi, P.L. Impulse and 60-cycle characteristics of driven grounds. *Electr. Eng.* **1941**, *60*, 123–127. [CrossRef]
3. Bellaschi, P.L.; Armington, R.E.; Snowden, A.E. Impulse and 60-Cycle Characteristics of Driven Grounds-II. *Electr. Eng.* **1942**, *61*, 349–363. [CrossRef]
4. Sonoda, T.; Takesue, H.; Sekioka, S. Measurement on Surge Characteristics of Grounding Resistance of Counterpoises for Impulse Currents. In Proceedings of the 25th International Conference on Lightning Protection, Rhodes, Greece, 18–22 September 2000; pp. 411–415.
5. Sekioka, S.; Hayashida, H.; Hara, T.; Ametani, A. Measurement of Grounding Resistance for High Impulse Currents. *IEE Proc. Gener. Transm. Distrib.* **1998**, *145*, 693–699. [CrossRef]
6. Takeuchi, M.; Yasuda, Y.; Fukuzono, H. Impulse Characteristics of a 500 kV Transmission Tower Footing Base with Various Grounding Electrodes. In Proceedings of the 24th International Conference on Lightning Protection, Birmingham, UK, 14–18 September 1998; pp. 513–517.
7. Yunus, M.; Nor, N.M.; Agbor, N.; Abdullah, S.; Ramar, K. Performance of Earthing systems for Different Earth Electrode Configurations. *IEEE Trans. Ind. Appl.* **2015**, *51*, 5335–5342. [CrossRef]
8. Haddad, A.; Griffiths, H.; Ahmeda, M.; Harid, N. Experimental Investigation of the Impulse Characteristics of Practical Ground Electrode Systems. In Proceedings of the International Conference on High Voltage Engineering and Application, New Orleans, MS, USA, 11–14 October 2010.
9. Yang, S.; Zhou, W.; Huang, J.; Yu, J. Investigation on Impulse Characteristic of Full-Scale Grounding Grid in Substation. *IEEE Trans. Electromagn. Compat.* **2018**, *60*, 1993–2001. [CrossRef]
10. Harid, N.; Griffiths, H.; Haddad, A. Effect of Ground Return Path on Impulse Characteristics of Earth Electrodes. In Proceedings of the 7th Asia-Pacific International Conference on Lightning, Chengdu, China, 1–4 November 2011; pp. 686–689.
11. Guo, D.; Clark, D.; Lathi, D.; Harid, N.; Griffiths, H.; Ainsley, A.; Haddad, A. Controlled Large-Scale Tests of Practical Grounding Electrodes—Part I: Test Facility and Measurement of Site Parameters. *IEEE Trans. Power Deliv.* **2014**, *29*, 1231–1239. [CrossRef]

12. Mousa, S.; Harid, N.; Griffiths, H.; Haddad, A. Experimental Investigation on High-Frequency and Transient Performance of a Vertical Earth Electrode. In Proceedings of the Universities Power Engineering Conference (UPEC), Soest, Germany, 5–8 September 2011; pp. 1–4.
13. Clark, D.; Guo, D.; Lathi, D.; Harid, N.; Griffiths, H.; Ainsley, A.; Haddad, A. Controlled Large-Scale Tests of Practical Grounding Electrodes—Part II: Comparison of Analytical and Numerical Predictions with Experimental Results. *IEEE Trans. Power Deliv.* **2014**, *29*, 1240–1248. [CrossRef]
14. Guo, D.; Lathi, D.; Harid, N.; Griffiths, H.; Haddad, A.; Ainsley, A. Large-Scale Earthing Test Facilities at Dinorwig Power Station. In Proceedings of the International Conference on Conduction Monitoring and Diagnosis, Beijing, China, 21–24 April 2008; pp. 808–811.
15. Duan, L.; Zhang, B.; He, J.; Xiao, L.; Qian, L. Experimental Study on Transient Characteristics of Grounding Grid for Substation. In Proceedings of the 25th International Conference of Lightning Protection (ICLP), Estoril, Portugal, 25–30 September 2016.
16. Etobi, N.; Nor, N.M.; Abdullah, S.; Othman, M. Characterizations of a Single Rod Electrode under High Impulse Currents with Different Polarities. In Proceedings of the 1st IEEE International Conference on Electrical Materials and Power Equipment (ICEMPE), Xi'an, China, 14–17 May 2017.
17. Abdullah, S.; Nor, N.M.; Ramar, K. Field measurements on Earthing Systems of Different Soil Resistivity Values under High Impulse Conditions. *Electr. Eng.* **2017**, *99*, 1005–1011. [CrossRef]
18. Reffin, M.; Nor, N.; Ahmad, N.; Abdullah, S. Performance of Practical Grounding Systems under High Impulse Conditions. *Energies* **2018**, *11*, 3187. [CrossRef]
19. Nor, N.M.; Haddad, A.; Griffiths, H. Factors Affecting Soil Characteristics under Fast Transients. In Proceedings of the International Conference of Power Systems Transients (IPST), New Orleans, MS, USA, 14 August 2003.
20. Nor, N.M.; Haddad, A.; Griffiths, H. Characterization of Ionization Phenomena in Soils under fast impulses. *IEEE Trans. Power Deliv.* **2006**, *21*, 353–361. [CrossRef]
21. Petropoulos, G.M. The High-Voltage Characteristics of Earth Resistances. *J. Inst. Electr. Eng. Part II Power Eng.* **1948**, *95*, 59–70.
22. *IEEE Standard 142: IEEE Recommended Practice for Grounding of Industrial and Commercial Power Systems*; Institute of Electrical and Electronics Engineers: Piscataway, NJ, USA, 2007.
23. Simmons, P. *Electrical Grounding and Bonding, Based on 2017 National Electricity Code (NEC)*, 5th ed.; CENGAGE Learning: Boston, MA, USA, 2012.

energies

MDPI

Article

Seasonal Influences on the Impulse Characteristics of Grounding Systems for Tropical Countries

Muhd Shahirad Reffin [1], Abdul Wali Abdul Ali [1], Normiza Mohamad Nor [1,*], Nurul Nadia Ahmad [1], Syarifah Amanina Syed Abdullah [1], Azwan Mahmud [1] and Farhan Hanaffi [2]

1 Faculty of Engineering, Multimedia University, 63100 Cyberjaya, Malaysia; muhd_shahirad@yahoo.com (M.S.R.); walikdr17@gmail.com (A.W.A.A.); nurulnadia.ahmad@mmu.edu.my (N.N.A.); synina@gmail.com (S.A.S.A.); azwan.mahmud@mmu.edu.my (A.M.)

2 Faculty of Electrical Engineering, Universiti Teknikal Malaysia Melaka, 76100 Durian Tunggal, Malaysia; farhan@utem.edu.my

* Correspondence: normiza.nor@mmu.edu.my

Received: 12 March 2019; Accepted: 5 April 2019; Published: 8 April 2019

Abstract: One of the most important parameters of the performance of grounding systems is the soil resistivity. As generally known, the soil resistivity changes seasonally, hence the performance of grounding systems, at DC and under high impulse conditions. This paper presents the performance of grounding systems with two different configurations. Field experiments were set up to study the characteristics of the grounding systems seasonally at power frequency and under high impulse conditions. A review of field testing on practical grounding systems was also presented. It was found that the soil resistivity, RDC and impulse characteristics of grounding systems were improved over time, and the improvement was higher for electrodes that have more contact with the soils.

Keywords: grounding; grounding electrodes; high impulse conditions; seasonal; soil resistivity

1. Introduction

Grounding systems are necessary to discharge high fault currents to ground and ensure the safe operation of power systems at all time. It was reported in IEEE Standard 81 [1] that due to the soil settling process and compactness of the soil, the earth impedance of ground electrode decreases slightly over a year or more after installation, due to the soil settling process and improved compaction in the soil. A few research investigations have also been reported on the influence of soil resistivity on practical grounding systems at power frequency and high impulse current [2,3]. As generally known, soil composition, inhomogeneity, hydrological, geological process can vary seasonally, and some studies have analysed the seasonal influence on grounding systems in terms of resistance values at low voltage, and low currents by field experiments [4–7]. However, limited studies have been published on the seasonal influence on the performance of grounding systems under high impulse conditions. He et al. [8] observed the seasonal influence on grounding systems under high impulse conditions for two seasons: winter and summer. They observed that the soil resistivity at the top layer was stable, whereas a higher soil resistivity of the bottom layer was observed during winter than during summer [8]. This resulted in higher resistance at the power frequency in winter than during summer. When the impulse factor was measured as the ratio of $R_{impulse}$ to RDC, a close impulse factor was obtained for both seasons. However, the investigations on the impulse characteristics of grounding systems were completed for two seasons, and no continuous measurement was made in between the seasons.

Further, the time period required for the checking and maintenance of grounding systems for both at power frequency and transient conditions has not been intensively studied or suggested. IEEE Standard 80:2013 [9] suggests the ground resistance be checked periodically after completion of construction, however, no specific time period is mentioned. Similarly, IEEE Standard 142: [10] and IEEE Standard 81 [1] state that power frequency tests should be conducted periodically, however, again no specific time period is suggested. Due to the lack of study on the seasonal performance of grounding systems and suggested time periods that can be found in literature, this paper therefore aims to address this shortfall.

In this study, field tests were used to investigate the characteristics of two practical grounding systems under high magnitude current surges throughout the year. Seasonal influences on the steady-state and impulse resistances were investigated, which represents the condition when the grounding systems were left over a period of time in real practice. These measurements allow a better understanding of the seasonal performances of grounding systems and provide information on how frequently grounding systems need to be checked and maintained after installation.

2. Experimental Arrangement

2.1. Review of Field Testing and Measurements

Different test arrangements may result in different and unreliable results when measuring the impulse characteristics of practical grounding systems. A review of field testing and measurement of impulse tests on grounding systems was firstly performed, to present various test set-ups and arrangements adopted in previously published works [2,8,11–27]. The study of soil behavior under impulse condition by means of field testing showed that the resistance decreased with increasing current [2,12–27]. It was noted from these studies [2,11–27] that the main concerns in the field measurements are the costs, logistics and time challenges. The guidelines were also found to be limited, due to the limited standards for field testing and measurements of grounding systems under high impulse conditions.

There are a few well established methods suggested in the standards on the measurements of earth resistance of earthing systems at low voltage and low frequency currents, namely the two-point method, three point method, ratio method, staged fault tests and fall-of-potential method [1,9,10]. However, it is now well accepted that soil characteristics under high magnitude impulse currents would become 'non-linear' and different than measured at low voltage and low frequency currents [2,8,11–27]. Due to the different characteristics of earthing systems under high impulse currents than when under low voltage, low frequency currents, there is a need to assess the practical earthing systems under high impulse conditions.

Field measurements undoubtedly can provide important results concerning the impulse characteristics of earthing systems since they represent the closest scenario to when high currents are practically discharged to grounding. With the improvement of impulse voltage/current generators which can be mobilized to the sites, impulse characterisations of earthing systems under high impulse conditions by field measurements have now become popular. Since then, a lot more studies have been directed towards impulse tests on earthing systems using field measurements. However, the measurement and testing methods found in these papers [11,13–27] differ from one another, which could be due to the lack of standards emphasising the required guidelines, as well as the great dependence on the available configurations and test site.

Since it is now well accepted that the impulse characteristics of earthing systems are different from those at low voltage low frequency currents, it is equally important to assess the performance of earthing systems under high impulse conditions. IEEE Standard 81 [1] provides some guidance and recommendations for measurements of earthing systems under high impulse conditions by field measurements, however they do not address it quantitatively, probably due to the little research work that has been carried out on grounding assessment under high impulse currents by field measurements. Other than the costs, logistics and time challenges in the field measurements, technical challenges are the main issues faced by researchers in performing field measurements of the impulse characteristics of earthing systems. This is due to the limited testing standards for field testing and measurements of earthing systems under high impulse conditions. Another technical challenge is due to the limited number of impulse generator manufacturers who are willing to tailor their designs to allow them to be mobilized to the field sites. So far, to the authors' knowledge, standards on impulse tests on practical earthing systems available in the literature are limited.

2.1.1. Distance of Remote Earth from the Electrode under Tests

In IEEE Std 81 [1], brief guidelines are provided for measurements at field sites using a mobile impulse generator. The work presented in IEEE Std 81 [1] is based on the tests conducted at the Georgia Institute of Technology. It was suggested in IEEE Standard 81 [1] to have the same leads and reference ground arrangement as that used for low-frequency Fall-of-Potential (FOP) tests. Figure 1 shows the test arrangement for the experimental test set-up of earthing systems under high impulse conditions. The distance of the electrode under test to the earth probe is 62% of the distance between the electrode under test to the remote earth. As for the ground mat, it was recommended in IEEE Std. 80 [9] that the dimensions of the electrode under test to the remote earth be extended by 3 to 4 times the diagonal dimensions of the ground mat. However, the IEEE standard guidelines are brief, and many other authors [11,13–22] have adopted other test set-up arrangements in their measurements, where all these distances were not as specific as those highlighted in the IEEE Standard [1].

2.1.2. Impulse Generator

As for the mobile impulse generator, no generally accepted standard has been presented so far. Some studies [18] used laboratory facilities, without any special design changes to the impulse generator for the testing of earthing systems by field measurements. On the other hand, some studies [13,15,16,19] have used impulse generators purposely built for their tests of earthing systems under high impulse conditions by field measurements. Marimoto et al. [16] have exclusively developed a weatherproof mobile impulse voltage generator in an effort to test the grounding systems of power substations.

No detailed information has been published in the standards on the specific methods, procedures and precautions of the measurements for testing the earthing systems at field sites. It was briefly highlighted in [1] that the current and voltage leads should be isolated from earth to avoid any interference, which can be done by hanging the leads over polyvinyl chloride (PVC) conduits. Different methods of hanging these leads to isolate the earth have also been found in some other studies [16,19]. However, other studies have not mentioned any consideration of the isolation of the leads [17–21]. This shows that there is a need to clarify the proper measurement methods.

2.1.3. Remote Earth/Auxiliary Earth

Another important parameter that needs to be considered in the experimental arrangement forthe testing of earthing systems by field measurements is the auxiliary ground, which is needed to carry the return current to the impulse generator via the ground under tests. So far, to the authors' knowledge, no specific standard has discussed the design requirements of this auxiliary ground.

Some authors have constructed auxiliary grounds around the electrode under tests. In this arrangement, a bigger remote earth size is expected since it is installed around the electrode under test. Thus, the earth resistance values of the ring electrode may be expected to have lower earth resistance values than the electrode under test. References [20,21] have used a remote earth installed around the electrode under tests. A few clarifications are needed when having the remote earth around the electrode under tests such as the appropriate radius of remote earth that should be constructed around the electrode under tests.

On the other hand, some authors [12,13,15–19] used separate earthing systems, which are constructed away from the electrode under tests as the auxiliary ground. It is stated in IEEE Standard 80 [1] that the earth resistance of remote earth should be lower than that the electrode under tests.

However, many times the steady state earth resistance values of the remote earth have not been mentioned or addressed by some authors. Though the remote earth constructions look bigger in the published work [12,13,15–19], the earth resistance values of the remote earth were not specifically mentioned in the papers. Yunus et. al. [17] studied the effects of earth resistance values of remote earths on the electrodes under test for when the remote earth was placed away from the electrode under test. Some significant observations were that the results were found to be different than the findings in most literatures, with higher earth resistance values of the remote earth than that the electrode under tests [17], and where the impulse earth resistance value was found to be higher than that measured at low voltage and low frequency currents (DC earth resistance value) when higher remote earth resistance values were used. They also found that the earth resistance values of electrodes under test increased with increasing currents in the higher earth resistance values of the remote earth [17].

Further, Abdullah et al. [22] compared both arrangements (remote earth around the electrode under test and remote earth placed at some distance away from the electrode under tests). They [22] found close agreement between the results of these two arrangements. This shows that both test arrangements are acceptable, as long as the condition that the earth resistance value of the remote earth be smaller than that of the electrode under testing is met.

2.1.4. Placement of Current Transducer

Other concerns for the field measurements and testing of the earthing systems under high impulse conditions are the placement of current transducers during the field measurements. In some published papers [23–26], the experimental arrangement was not shown at all, though some parameters of the generators and test circuit are described. Due to the different possible arrangements of the remote earth, surrounding the electrode under tests and away from the electrode under test, as highlighted in Section 3, the placement of the current transducer (CT) would expectedly be different too.

Here, the placement of the CT for the remote earth placed at a distance away from the electrode under test is discussed first. In IEEE Standard 81 [1], the current transducer was placed at the electrode under test, at the same point where the cable of voltage divider was also connected. Some studies [15,21,24,25,28] also positioned the CT in series to the electrode under tests.

On the other hand, for the arrangement where the remote earth is placed at some distance away from the electrode under tests, Chen and Chowdhuri [27] measured the current at the remote earth, and not in series to the electrode under tests. So far, to the authors' knowledge, such an arrangement was only found in their paper [27]. As for other papers, i.e., [13,15,16], they did not clearly show the CT placement, though there were some current measurements. Due to a lack of test set-up arrangements, the authors feel that there is a need for a proper standard for impulse tests on earthing systems by field measurements. As for other kinds of arrangement where the remote earth is installed around the electrode under tests, the CT is more commonly placed at the remote earth [12,20,22].

2.2. *Test Set up for This Study*

In the investigation described in this paper, a mobile impulse generator which is capable of generating high voltages up to 300 kV and high impulse currents up to 10 kA was used (see Figure 1). The impulse generator was powered by a diesel generator. Current measurements were achieved with a current transformer of sensitivity of 0.01V/A and with an attenuation of the probe of $\times 10$. A resistive divider with a ratio of 3890:1 was used for voltage measurements. Two commercially available digital storage oscilloscopes (DSOs), powered by batteries, were used to capture the voltage and current signals separately. In order to avoid any interferences and flashover in the circuit, all equipment and leads were placed above ground. For the diesel generator and impulse generator, an epoxy was used as a frame to separate them from the ground. For the DSOs, the batteries are placed on an insulation table, made of epoxy. All the leads and mesh cables were isolated from the ground by hanging the leads on epoxy insulation rods. Figure 2 shows the experimental arrangement of the impulse tests of grounding systems under high impulse conditions.

Figure 1. Equipment used for field measurements.

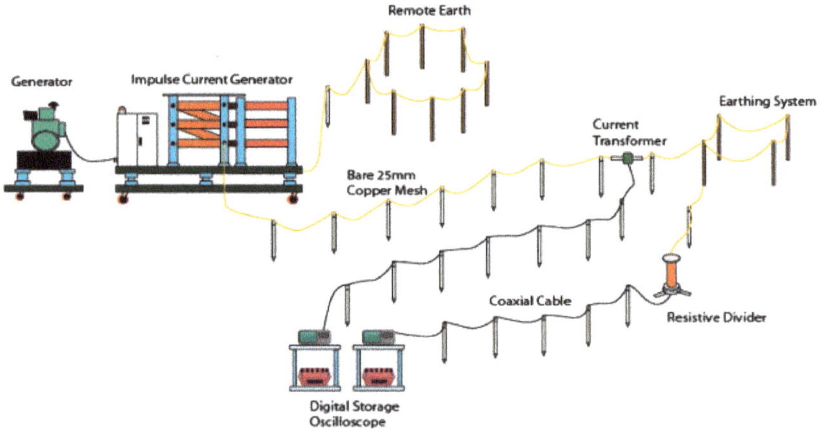

Figure 2. Test set-up for field measurements.

2.3. Electrodes under Tests

In this study, two configurations were adopted: a grounding device with spike rods (configuration 1) and one with four rod electrodes (configuration 2), shown in Figures 3 and 4, respectively. Configuration 2 represents a conventional electrode, where the design does not emphasize any enhancement of the ionization process. Configuration 1 was developed based on the evidence from the work by Petropolous [28] that when spikes or needles were attached to the spherical electrode, lower impulse resistance values were noted for the same voltage due to the higher field intensities at the spikes. This indicates the effectiveness of the earth electrodes with spikes as compared to electrodes without spikes. In this paper, a new grounding electrode with spike rods is introduced. By having spike rods at the electrodes, a high electric field intensity will be concentrated at the spikes, which encourages soil ionisation and breakdown processes to take place, hence allowing more current to be discharged to the ground. This will result in a more effective grounding system. For this reason, and to provide more contact between the electrodes and surrounding soil, a grounding device with spike rods (consisting of two rods (inner shaft (120) and outer shaft electrodes (110)) was adopted. The grounding device with spike rods is 1.5 m in length. The diameter of the inner rod is 3 cm, and outer rod is 5 cm, with a gap between inner and outer rods of 1 cm. There are five spikes (123), each one of 20 cm length. During the installation, the spike rods (123) are kept closed from the surface. For the installation, due to its large diameter, a pre-bore with a diameter of 4 cm was firstly performed using an auger. Upon completing a pre-bore hole with a depth of 1.5 m, the grounding device with spike rods (100) is positioned into the hole, in a generally vertical configuration, with all the spike rods (123) concealed and protected from damage within the shaft. The outer shaft (110) is subjected to impact force while it is driven through hammering into the ground. During driving, the top end is protected by a dolly or capping to protect the top end of the shaft against any damage from driving or hammering the electrodes. Once the grounding device with spike rods (100) reached the required distance, 1.5 m, the inner shaft (120) was turned using the provided winch (121) in such a way that the grounding spike rods (123) protruded out and pierced into the soil mass. An indication that all the spike rods (123) protrude out is when the winch stops, and is not able to turn the inner shaft (120) anymore. Another type of ground electrode used in this study consists of four ground rod electrodes, with each rod of 20 mm diameter, and a depth of 1.5 m. Copper mesh of 25 mm width and thickness of 2 mm was used to connect the rod electrode at all four points, as shown in Figure 4.

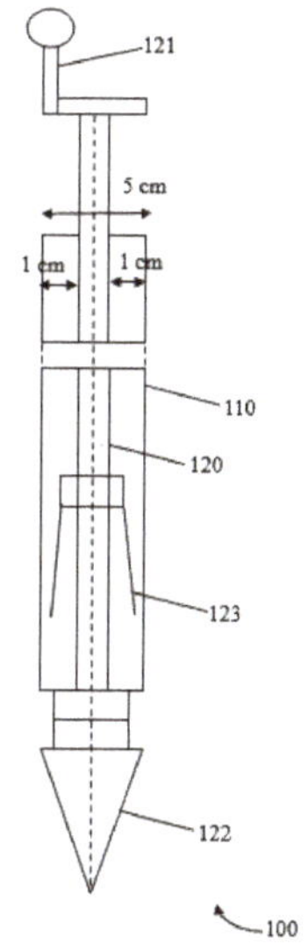

Figure 3. Configuration 1, consists of a grounding device with spike rods.

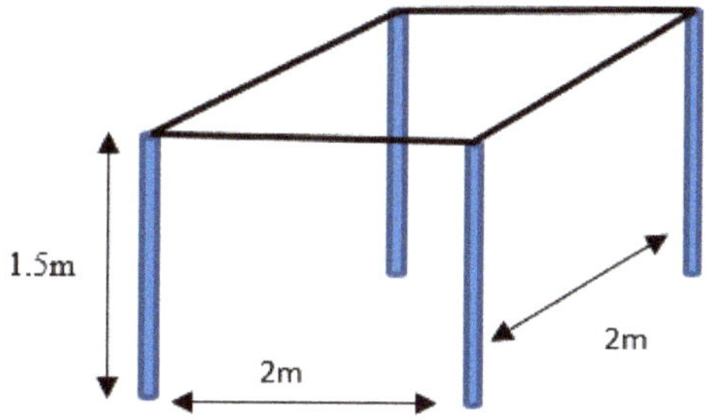

Figure 4. Configuration 2 for the tests, consisting of a 2 m × 2 m grounding grid.

In this study, the Finite Element Method (FEM) was utilized to obtain the voltage profiles for both configurations. Figures 5 and 6 show the point of electric profile is taken for configurations 1 and 2, respectively. The soil resistivity was taken at the beginning of the tests using a Wenner Method, which was computed as a two-layer medium model, using a Current Distribution, Electromagnetic Fields, Grounding and Soil Structure Analysis (CDEGS) software. It was found that the resistivity at the top layer is 119.5 Ωm with a depth of 7.18 m, and the bottom of a two-layered soil is 391 Ωm, with an infinite depth. Both configurations were injected at 5 kA. The corresponding trends of the electric field for configurations 1 and 2 are shown in Figures 7 and 8, respectively. As can be noted, the shapes of the electric fields for both configurations are different when a similar soil resistivity layer was used. A higher and more non-uniform electric field was noted for configuration 1, where the maximum surface potential is 820 MV (see Figure 7). For the case of configuration 2, the potential is rather uniform, with 650 MV at the rod, and 180 MV at the copper strips (see Figure 8). The FEM simulation indicates that configuration 1 is preferred, since it can increase the electric field significantly, and has non-uniform electric field, thus reducing the earth resistance value more significantly under transient conditions, compared to configuration 2.

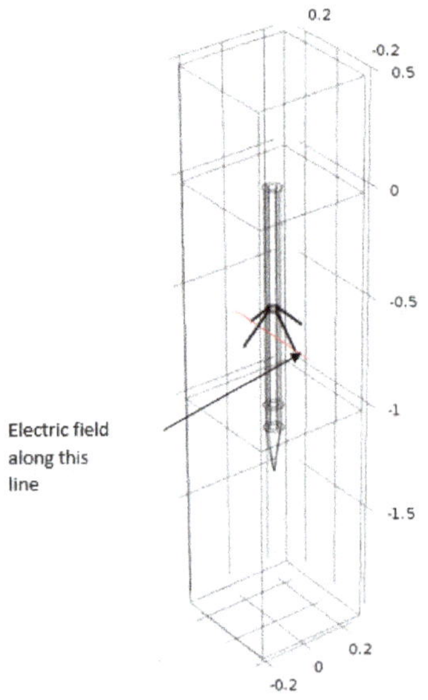

Figure 5. Electric profile is taken along the spike rods.

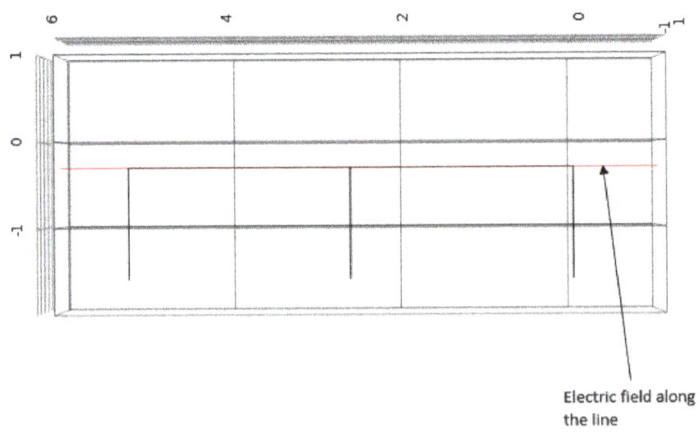

Figure 6. Electric profile taken along the conventional rods.

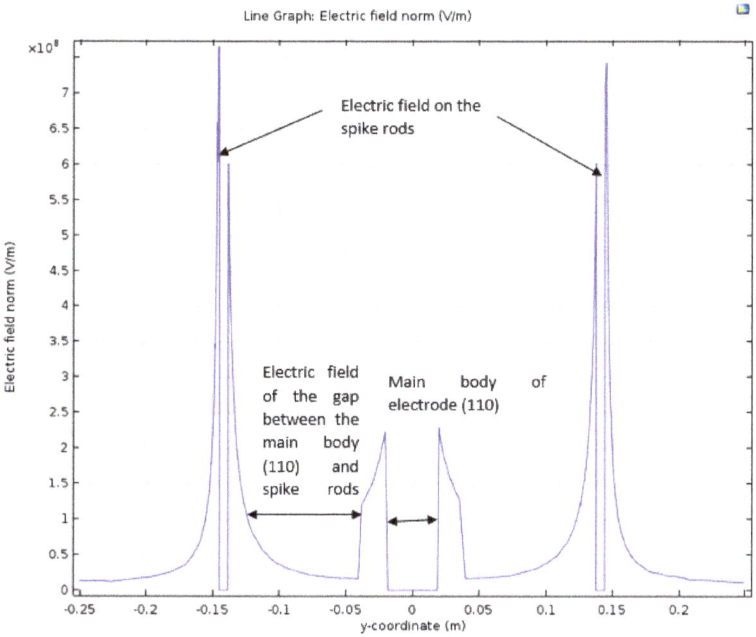

Figure 7. Electric field along the spike rods.

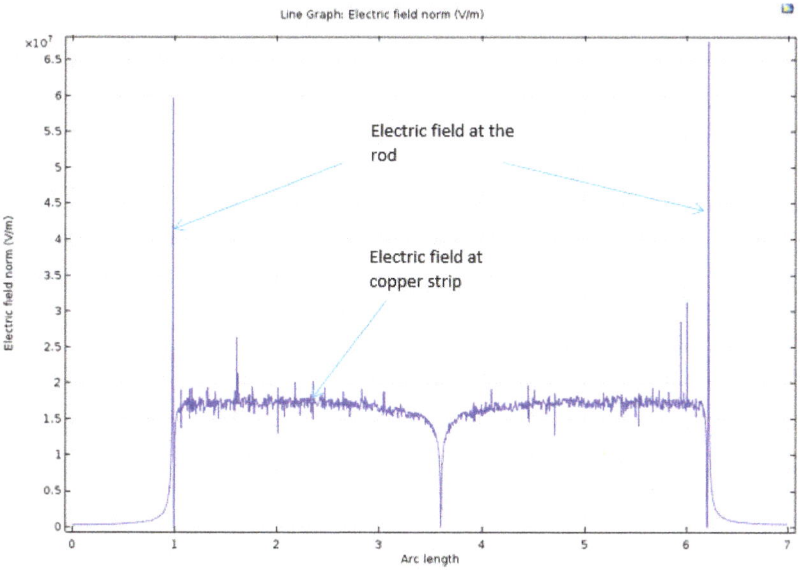

Figure 8. Electric field along the conventional rods.

2.4. Auxillary or Remote Electrodes

A larger size of auxillary or remote earth grounding system was used, so that a lower earth resistance value was achieved, in comparison to the test electrodes, presented earlier in Section 2.2. In this study, the remote earth was installed 50 m away from the electrode under tests. Figure 9 shows the remote earth used in this study, which consists of a 20 m × 30 m grounding grid. The spacing between copper strips is 5 m apart for the 20 m side and 15 m apart for the 30 m side. Hard copper strips of 30 mm width and 2 mm thick were used, which were buried 300 mm below the earth's surface, and welded exothermically to 12-rod electrodes, where each one is 20 mm in diameter, and 1.8 m long. Using a Fall-of-Potential (FOP) method, the earth resistance values of the remote earth were measured, and found to range between 8.5 Ω to 10 Ω throughout the year, which is always lower than the electrodes under tests.

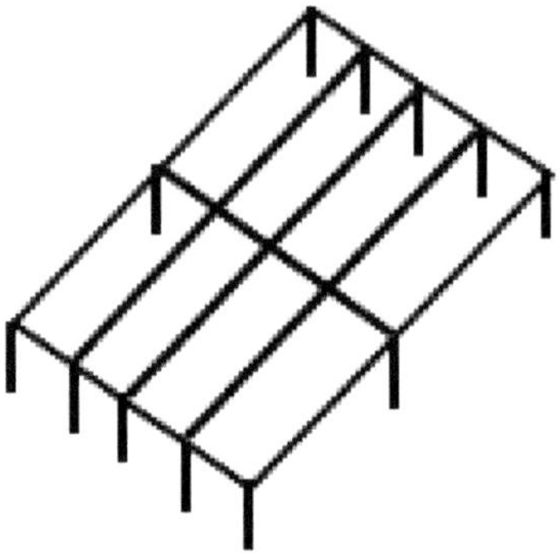

Figure 9. A 20 m × 30 m remote earth mesh.

3. Results and Analysis

Using the Wenner method, the soil resistivity was measured in March 2018 and March 2019 to see variations in soil profile over the year. Configurations 1 and 2 are 30 m apart. The soil resistivity was measured at the site over 150 m-long lines within the test site, and these configurations were installed within this 150 m-long line. The results were then modeled into 2-layer soil model using Current Distribution, Electromagnetic Interference, Grounding and Soil Structure Analysis (CDEGS), which results are summarised in Table 1. As can be seen, the soil resistivity slightly varied over the year, with the height of the first layer also being reduced after a year. FOP, as outlined in IEEE Standard 81 [1] was applied to measure the DC and low-current resistances of configurations 1 and 2 throughout the year, which are shown in Table 2. The resistances of both configurations were found to reduce after the installation, where a higher reduction was seen for configuration 1, with a decrease by more than 20% throughout the year. This could be due the presence of five spike rods, which provide more contact within the soil, as compared to configuration 2. It was also noted that for the measurement in September 2018, high RDC values were noted for both configurations, with respect to any other months. It was experienced by the authors that during the measurement, the weather was hotter than the rest of the months.

Table 1. Soil resistivity profile.

Month of Measurement	Soil Resistivity of Layer 1, ϱ_1 (Ωm)	Soil Resistivity of Layer 2, ϱ_2 (Ωm)	Height of Layer 1 (m)
March 2018	119.5	391.0	7.2
March 2019	111.4	454.2	5.2

Right after RDC measurements, impulse tests were performed on both test electrode configurations on the same day. Impulse currents and the corresponding voltages were measured and shown in Figures 10 and 11 for configuration 1, at charging voltage of 30 kV and 210 kV, respectively, for test no. 1. As can be seen in the figures, faster times to discharge to zero for both voltage and current traces was seen at higher voltage magnitudes. Similar voltage and current traces were seen for configuration 2, and for other test nos., where faster times to discharge to zero for both voltage and current traces at higher currents were observed. The time to discharge to zero for voltage and current was plotted against the applied voltage for all the tests, the time to discharge to zero decreased with applied voltage, as shown in Figures 12 and 13 for configuration 1 and 2, respectively. Both configurations were found to have the slowest time to discharge to zero for test no.4, which has high RDC. Both configurations with the lowest RDC (in Dec. 2018) were found to have the fastest discharge time to zero, indicating a good conductivity of the grounding systems. Time to discharge to zero was also found to be higher for configuration 1, in comparison to configuration 2 (see Figures 14 and 15) for test no. 1 to 3, and no. 3–7 respectively). The results also indicated that the lower the RDC, the faster the time for current and voltage to discharge to zero.

Table 2. Measured resistance of electrodes at power frequency, low-current tests.

Test No.	Date of Measurement	DC Resistance, RDC (Ω)			
		Conf. 1	Percentage Difference from the First Reading (%)	Conf. 2	Percentage Difference from the First Reading (%)
1	21/03/2018	91	0	60.9	0
2	07/05/2018	69.1	24.1	55.2	9.36
3	02/08/2018	69.7	23.4	57.5	5.58
4	03/09/2018	84.2	7.25	80.3	−31.9
5	17/10/2018	71.9	20.99	57.1	6.24
6	21/11/2018	70.7	22.3	55.9	8.2
7	17/12/2018	70.1	23	57.1	6.24
8	25/2/2019	69	24.2	55.5	8.9

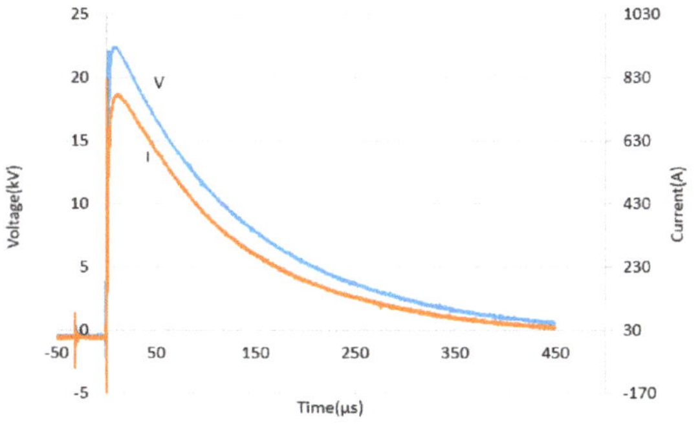

Figure 10. Voltage and current traces for configuration 1 at a charging voltage of 30 kV.

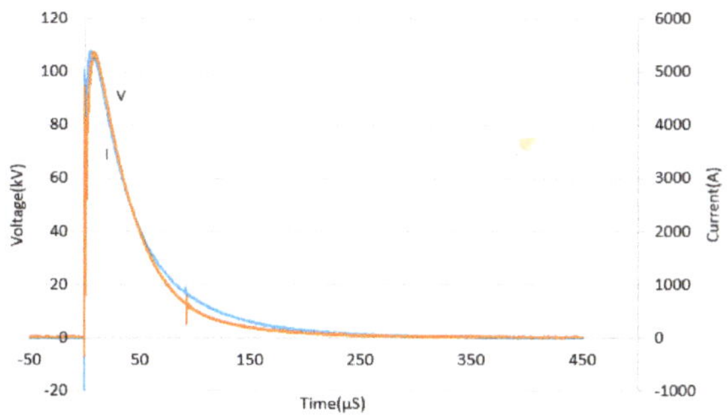

Figure 11. Voltage and current traces for configuration 1 at a charging voltage of 210 kV.

Impulse resistance values were measured as the ratio of the voltage at current peak to the current peak, and plotted versus the peak current, as shown in Figures 16 and 17 for configurations 1 and 2, respectively. The resistance values obtained from these tests were found to decrease with current magnitude, indicating the effect of impulse currents on the characteristics of test electrodes for configurations 1 and 2. It was noted that the impulse resistance are lower a few months after installation, showing improvement in the grounding systems for both configurations. In order to determine the effectiveness of the test electrodes in comparison to its RDC, the percentage of reduction of impulse resistance from its corresponding RDC was measured as (1), where the $R_{impulse}$ was taken as the average impulse resistance measured from varying the charging voltage of 30 kV until 210 kV:

$$\text{Percentage of resistance reduction} = \left(\frac{R_{DC} - R_{impulse}}{R_{DC}}\right) \times 100\% \quad (1)$$

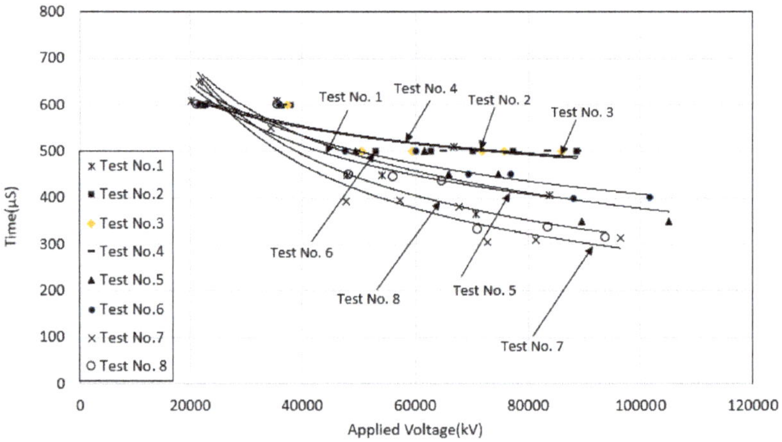

Figure 12. Time to discharge to zero for voltage and current traces for configuration 1 throughout the year.

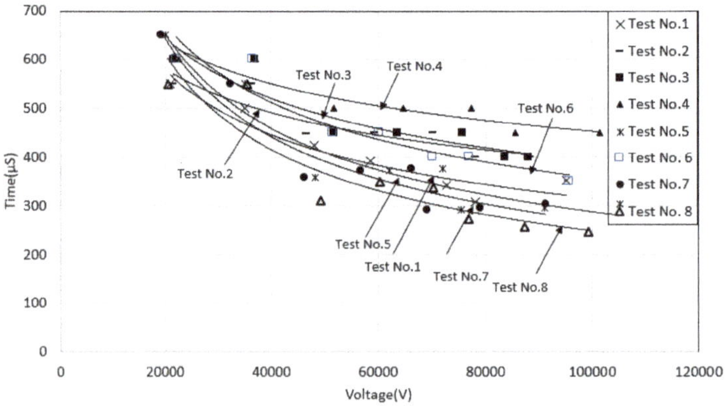

Figure 13. Time to discharge to zero for voltage and current traces for configuration 2 throughout the year.

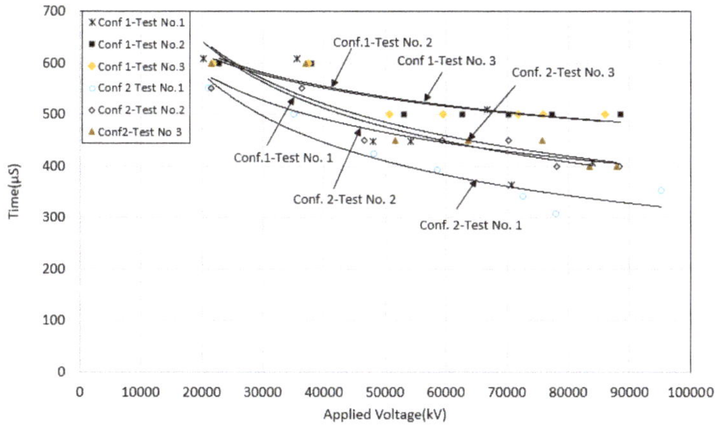

Figure 14. Time to discharge to zero for voltage and current traces for configuration 1 and 2 for tests no. 1 to 3.

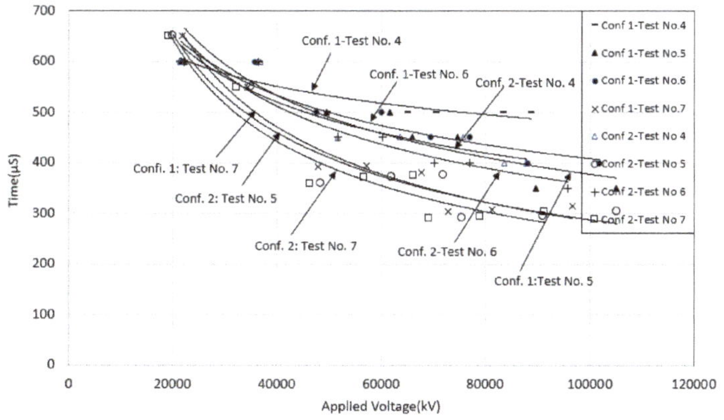

Figure 15. Time to discharge to zero for voltage and current traces for configuration 1 and 2 for tests no. 4 to 7.

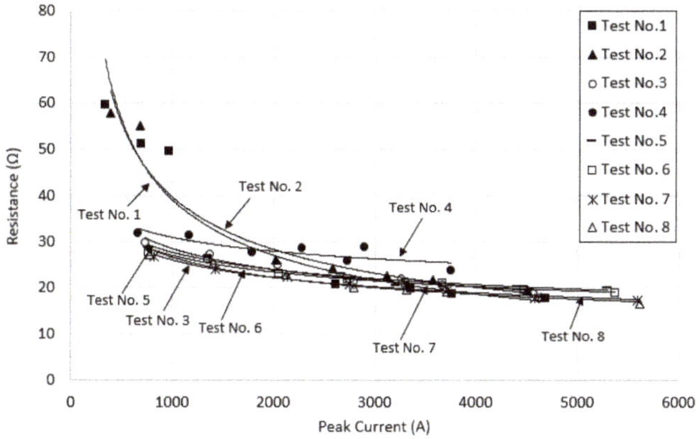

Figure 16. Impulse resistance vs. peak current for configuration 1 throughout the year.

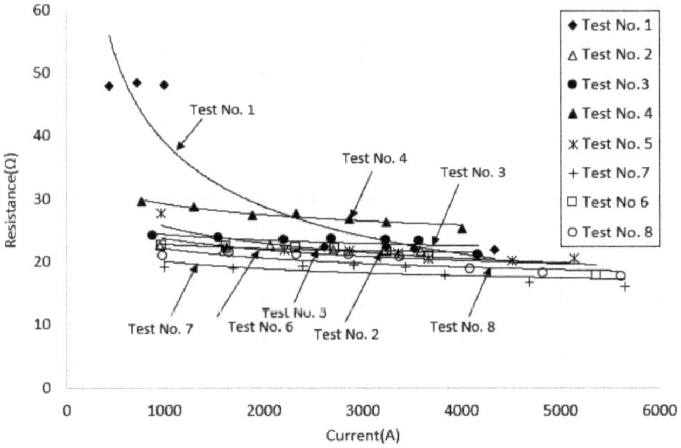

Figure 17. Impulse resistance vs. peak current for configuration 2 throughout the year.

Table 3 summarises the impulse factor for both configurations, measured throughout the year. As can be seen from the table, percentage of resistance reduction is higher for configuration 1, in comparison to configuration 2. A higher percentage of resistance reduction for configuration 1 after a months of installation was also noted, showing the effectiveness of the earth electrode design with spike rods.

Table 3. Percentage of impulse resistance reduction from RDC throughout the year.

Test No.	Configuration 1			Configuration 2		
	R_{DC}	Average $R_{impulse}$ (Ω)	Resistance Reduction from R_{DC} (%)	R_{DC}	Average $R_{impulse}$ (Ω)	Resistance Reduction from R_{DC} (%)
1	91	33.94	62.7	60.9	33.4	45.1
2	69.1	32.4	53.1	55.2	22.0	60.1
3	69.7	23.5	66.3	57.5	23.3	59.5
4	84.2	28.2	66.5	80.3	27.6	65.6
5	71.9	23	68	57.1	22.2	61.1
6	70.7	22.5	68.2	55.9	21.4	61.7
7	70.1	21.1	69.9	57.1	18.4	67.8

4. Conclusions

Power frequency and impulse tests by field measurements on a grounding device with spike rods and a small grid were performed throughout the year. It was found that under high impulse conditions, the RDC of both configurations became lower after months of installation. The percentage of reduction of RDC from the first installation was found to be more pronounced for configuration 1 than configuration 2, suggesting the effectiveness of the new grounding electrode with configuration 1. This could be due to the spike rods in the grounding electrode, providing more and better contact of the electrodes with the soil, hence reducing the RDC. The characteristics of both configurations were also investigated under high impulse currents. The percentage of earth resistance reduction of $R_{impulse}$ from its RDC was found to be higher for configuration 1. In addition, a higher percentage of earth resistance reduction was exhibited for configuration 1 after months of installation, indicating an improvement in the grounding systems for electrodes with spikes. The results from the study provide an important guideline for power engineers concerning the time period that needs to be considered for maintenance of grounding systems, where it is desirable to perform the maintenance of grounding systems after one year of installation. The results indicate that it may be desirable to consider using electrode rods with spike rods, so that earth resistance values can be reduced under transient conditions. This scenario gives a more effective grounding system, which enhances the ionisation process and increases the dissipating current.

Author Contributions: M.S.R., A.W.A.A., N.M.N., S.A.S.A. and N.N.A. involved in setting up the experiment, M.S.R. and A.W.A.A. sorted out the data, M.S.R. and A.W.A.A. prepared the draft and N.M.N. helped with validation of analysis. A.M., N.M.N. and N.N.A. helped in securing funding. F.H. helped with the FEM simulation.

Funding: This research was funded by TELEKOM MALAYSIA RESEARCH AND DEVELOPMENT (TMR&D), grant number MMUE170005, MMUE180008 and MMUE180027.

Conflicts of Interest: The authors declare no conflict of interest.

Subscripts and Abbreviations

DC	Direct current
$R_{impulse}$	Resistance values measured under impulse condition
RDC	Resistance values measured at low voltage, low frequency currents
DSO	Digital Storage Oscilloscope
FEM	Finite Element Method
CDEGS	Current Distribution, Electromagnetic Fields, Grounding and Soil Structure Analysis
FOP	Fall-of-Potential

References

1. ANSI/IEEE Std 81-2012. *IEEE Guide for Measuring Earth Resistivity, Ground Impedance, and Earth Surface Potentials of a Ground System*; IEEE: Piscataway, NJ, USA, 2012.
2. Reffin, M.S.; Nor, N.M.; Ahmad, N.N.; Abdullah, S.A. Performance of Practical Grounding Systems under High Impulse Conditions. *Energies* **2018**, *11*, 3187. [CrossRef]
3. Tu, Y.; He, J.; Zeng, R. Lightning Impulse Performances of Grounding Devices Covered with Low-Resistivity Materials. *IEEE Trans. Power Deliv.* **2006**, *21*, 1701–1706. [CrossRef]
4. Gustafson, R.J.; Pursley, R.; Albertson, V.D. Seasonal Grounding Resistance Variation on Distribution Systems. *IEEE Trans. Power Deliv.* **1990**, *5*, 1013–1018. [CrossRef]
5. Abdullah, N.; Marican, A.; Osman, M.; Abdul, N. Rahman Case Study on Impact of Seasonal Variations of Soil Resistivities on Substation Grounding Systems Safety in Tropical Country. In Proceedings of the 7th Asia-Pacific International Conference on Lightning, Chengdu, China, 1–4 November 2011; pp. 150–154.
6. Gonos, I.; Gonos, I.F.; Moronis, A.X.; Stathopulos, I.A. Variation of Soil Resistivity and Ground Resistance during the Year. In Proceedings of the 28th International Conference on Lightning Protection (ICLP), Kanazawa, Japan, 17–21 September 2006; pp. 740–744.
7. Androvitsaneas, V.P.; Gonos, I.; Stathopulos, I.A. Performance of Ground Enhancing Compounds during the Year. In Proceedings of the 34th International Conference on Lightning Protection (ICLP), Rzeszow, Poland, 2–7 September 2018; pp. 1–5.
8. He, J.; Wu, J.; Zhang, B.; Yu, S. Field Testing for Observation of Seasonal Influence on Grounding Device at Impulse Condition. In Proceedings of the Asia-Pacific Symposium on Electromagnetic Compatibility, Singapore, 21–24 May 2012; pp. 445–448.
9. *ANSI/IEEE Std 80-2013: IEEE Guide for Safety in AC Substation Grounding*; IEEE: Piscataway, NJ, USA, 2013.
10. *IEEE 142-2007: IEEE Recommended Practice for Grounding of Industrial and Commercial Power Systems*; IEEE: Piscataway, NJ, USA, 2007.
11. Bellaschi, P.L. Impulse and 60-Cycle Characteristics of Driven Grounds. *IEE Trans. Power Appar. Syst.* **1941**, *60*, 123–128.
12. Kosztaluk, R.; Loboda, M.; Mukhedkar, D. Experimental Study of Transient Ground Impedances. *IEEE Trans. Power Appar. Syst.* **1981**, *100*, 4653–4660. [CrossRef]
13. Sekioka, S.; Hara, T.; Ametani, A. Development of a Nonlinear Model of a Concrete Pole Grounding Resistance. In Proceedings of the International Conference on Power Systems Transients, Lisbon, Portugal, 3–7 September 1995; pp. 463–468.
14. Dick, W.K.; Holliday, H.R. Impulse and Alternating Current Tests on Grounding Electrodes in Soil Environment. *IEEE Trans. Power Appar. Syst.* **1978**, *PAS-97*, 102–108. [CrossRef]

15. Sekioka, S.; Sonoda, T.; Ametani, A. Experimental Study of Current Dependent Grounding Resistance of Rod Electrode. *IEEE Trans. Power Deliv.* **2005**, *20*, 1569–1576. [CrossRef]
16. Morimoto, A.; Hayashida, H.; Sekioka, S.; Isokawa, M.; Hiyama, T.; Mori, H. Development of Weatherproof Mobile Impulse Voltage Generator and Its Application to Experiments on Nonlinearity of Grounding Resistance. *Trans. Inst. Electr. Eng. Jpn.* **1997**, *117*, 22–33. (In English) [CrossRef]
17. Yunus, S.; Nor, N.M.; Agbor, N.; Abdullah, S.; Ramar, K. Performance of Earthing Systems for Different Earth Electrode Configurations. *IEEE Trans. Ind. Appl.* **2015**, *51*, 5335–5342. [CrossRef]
18. Ramamoorty, M.; Narayanan, M.M.B.; Parameswaran, S.; Mukhedkar, D. Transient Performance of Grounding Grids. *IEEE Trans. Power Deliv.* **1989**, *4*, 2053–2059. [CrossRef]
19. Yang, S.; Zhou, W.; Huang, J.; Yu, J. Investigation on Impulse Characteristic of Full-Scale Grounding Grid in Substation. *IEEE Trans. Electromagn. Compat.* **2017**, *60*, 1993–2001. [CrossRef]
20. Clark, D.; Guo, D.; Lathi, D.; Harid, N.; Griffiths, H.; Ainsley, A.; Haddad, A. Controlled Large-Scale Tests of Practical Grounding Electrodes- Part II: Comparison of Analytical and Numerical Predictions with Experimental Results. *IEEE Trans. Power Deliv.* **2014**, *29*, 1240–1248. [CrossRef]
21. Harid, N.; Griffiths, H.; Haddad, A. Effect of Ground Return Path on Impulse Characteristics of Earth Electrodes. In Proceedings of the 7th Asia-Pacific International Conference on Lightning, Chengdu, China, 1–4 November 2011; pp. 686–689.
22. Abdullah, S.; Nor, N.M.; Etopi, N.; Reffin, M.; Othman, M. Influence of Remote Earth and Impulse Polarity on Earthing Systems by Field Measurements. *IET Sci. Meas. Technol.* **2017**, *12*, 308–313. [CrossRef]
23. Towne, H.M. Impulse Characteristics of Driven Grounds. *Gen. Electr. Rev.* **1928**, *31*, 605–609.
24. Vainer, A.L. Impulse Characteristics of Complex Earth Grids. *Elektrichestvo* **1965**, *3*, 107–117.
25. Vainer, A.L.; Floru, V.N. Experimental Study and Method of Calculating of the Impulse Characteristics of Deep Earthing. *Electical Technol. Ussr (Gb)* **1971**, *2*, 18–22.
26. Liew, A.C.; Darveniza, M. Dynamic Model of Impulse Characteristics of Concentrated Earths. *IEE Proc.* **1974**, *121*, 123–135. [CrossRef]
27. Chen, Y.; Chowdhuri, P. Correlation between Laboratory and Field Tests on the Impulse Impedance of Rod-type Ground Electrodes. *IEE Proc. Gener. Transm. Distrib.* **2003**, *150*, 420–426. [CrossRef]
28. Petropoulos, G.M. The High-Voltage Characteristics of Earth Resistances. *J. IEE* **1948**, *95*, 172–174.

MDPI
St. Alban-Anlage 66
4052 Basel
Switzerland
Tel. +41 61 683 77 34
Fax +41 61 302 89 18
www.mdpi.com

Energies Editorial Office
E-mail: energies@mdpi.com
www.mdpi.com/journal/energies

www.ingramcontent.com/pod-product-compliance
Lightning Source LLC
LaVergne TN
LVHW071938080526
838202LV00064B/6632
* 9 7 8 3 0 3 9 2 8 7 1 6 1 *